Pipeline Leak Detection Handbook

Pipeline Leak Detection Handbook

Morgan Henrie PhD, PMP, PEM
CEO/President, MH Consulting, Inc

Philip Carpenter PE
President, Serrano Services and Systems

R. Edward Nicholas
President, Nicholas Simulation Services LLC

ELSEVIER

AMSTERDAM • BOSTON • HEIDELBERG • LONDON
NEW YORK • OXFORD • PARIS • SAN DIEGO
SAN FRANCISCO • SINGAPORE • SYDNEY • TOKYO

Gulf Professional Publishing is an imprint of Elsevier

G|P
P|

Gulf Professional Publishing is an imprint of Elsevier
50 Hampshire Street, 5th Floor, Cambridge, MA 02139, United States
The Boulevard, Langford Lane, Kidlington, Oxford, OX5 1GB, United Kingdom

British Library Cataloguing-in-Publication Data
A catalogue record for this book is available from the British Library.

Library of Congress Cataloging-in-Publication Data
A catalog record for this book is available from the Library of Congress.

ISBN: 978-0-12-802240-5

For Information on all Gulf Professional Publishing visit our website at https://www.elsevier.com/

Publisher: Joe Hayton
Senior Acquisition Editor: Katie Hammon
Senior Editorial Project Manager: Kattie Washington
Production Project Manager: Kiruthika Govindaraju
Cover Designer: Maria Inês Cruz

Typeset by MPS Limited, Chennai, India

Contents

Chapter 1

Introduction

1.1 INTRODUCTION

This book is an introduction to the problem of quickly detecting leaks, ruptures, and spills from commodities such as natural gas, liquefied natural gas, liquefied petroleum gas, refined petroleum products, and crude oil pipeline transportation systems. Pipelines as a whole, given the tremendous quantity of transported products, are perhaps the safest mode of commodity transport. However, unplanned commodity loss, due to breaches in pipeline integrity, does occur and is a very undesirable side effect of transporting fluids by pipeline. Many of these negative impacts may be severe in nature, ranging from unexpected system downtime to environmental damage, property damage, loss of company good will, loss of investor confidence, government fines, injury, and loss of life. Significant financial costs have occurred due to pipeline integrity breaches and resulting spills. As an example, the 2010 gas line incident in San Bruno, CA resulted in a $1.4 billion fine, loss of life for 8 people, and total destruction of 38 homes.

Pipelines are virtually everywhere. Although most are buried, there are also many miles of pipeline that are constructed above the ground and under water (in rivers, lakes, seas, and oceans). We discuss how pipelines develop leaks, ruptures, and resulting spills. Commodity releases are often detected by people, but they are also detected by pipeline leak detection technology. Leak detection technology, the primary focus of this book, has been designed, implemented, operated, and maintained in an effort to detect when these events occur so the operator can respond in a timely manner. This book focuses on pipelines used by the petroleum industry, yet many of the various aspects of this book are applicable to other pipeline infrastructures.

Fortunately, the unintended escapement of commodity from pipelines due to a pipeline system integrity breach is a relatively rare problem. Pipelines have a long history of providing safe and economical commodity transportation. As shown here, the total existing worldwide length of cross-country pipelines is truly phenomenal. On a worldwide basis, existing pipelines for all commodities run approximately 2 million miles (3.2 million km).

Pipeline Leak Detection Handbook. DOI: http://dx.doi.org/10.1016/B978-0-12-802240-5.00001-7

The worldwide oil demand in 2013 was approximately 90 million barrels (1.4309×10^7 m^3) of combined crude oil and refined products, and approximately 3500 billion m^3 of gas are consumed throughout the world daily. All of this must be moved through a series of pipelines 24 hours per day, 7 days per week, safely and efficiently.

In summary, most of this transportation activity occurs safely, efficiently, quietly, and with little fanfare. But accidents do happen. Leaks, ruptures, and spills do occur. Detecting these events as quickly as possible provides a means to minimize the negative consequences of these occurrences.

1.2 WHY ARE PIPELINES IMPORTANT?

When one considers modern society's industrial lifestyle and today's standard of living throughout the developed world, one must understand that this would not be possible today without the petroleum or water industry pipeline infrastructure. Virtually anywhere you travel in the industrialized or developed world, the very fabric of what keeps things moving is grounded in the commodities transported through these pipelines.

Pipelines provide water to virtually every building and natural gas to many homes as well as offices, commercial buildings, electric utilities, hospitals, and so forth. Pipelines also transport crude oil and natural gas from wells and production locations through distribution systems to processing systems such as refineries, and then gas and refined products to end users such as homes, factories, and power plants.

Natural gas pipelines provide approximately 25% of the energy consumed in the United States, and an even greater extent in other countries. Refined products are also raw inputs for a vast array of commercial products such as clothing, cosmetics, and pharmaceuticals. Modern plastics made from oil are used extensively in numerous products affecting all facets of our lives.

Pipelines are essential lifelines for almost every activity of modern life. Despite these benefits, however, pipelines also bring a common threat: the potential to leak.

1.3 PIPELINE BASICS

Development of oil, natural gas, and other petroleum commodities has a long history going back to AD 347, with China drilling the first oil wells. When the first petroleum-based commodity pipeline was built is undetermined at this time. Some sources place the first petroleum pipeline construction in the 11th century, whereas others identify the first petroleum commodity pipeline construction as occurring in 1860s. Although the origins of the first pipeline are in question, it is clear that from the humble beginnings of the first pipeline until today, the construction of pipeline systems has spread throughout the world. Every nation in the world relies on pipelines.

Modern history clearly shows an ever-increasing reliance on and demand for this transportation mode. As an example, the first oil pipeline in the United States was built in 1865, following the 1859 discovery of oil in Pennsylvania. This 2-in. pipeline stretched a remarkable 5 miles and moved approximately 2000 barrels per day. In 1879, great strides occurred in oil industry pipeline construction and operation when a 6-in. pipeline, 109 miles in length, began operation.

The first large-diameter (24 in. or more), long-distance petroleum commodity–based pipeline in the United States was constructed in 1925. During the Second World War, many large-diameter, long-distance pipelines were constructed and placed into service. The size and length of pipelines continued to increase as time progressed.

Gas pipelines have an equally long history. In the 19th century, early gas pipelines were constructed in major cities to provide gas for lighting purposes. From this start, the construction and use of gas pipelines proliferated across the country over the ensuing decades [1].

One indication of how dependent the developed world is on pipelines is the total pipeline length that is reported as either being in service today or being constructed. In 2015, it was estimated that there were 118,623 miles (190,905 km) of oil and gas industry pipelines planned or under construction. Within the 120 countries surveyed by The World Factbook [2], an estimated 2,140,931 miles (3,445,495 km) of these pipelines are in service.

As an example of the breadth, depth, and diversity of pipelines within a country, in the United States there are 192,388 miles (309,618 km) of hazardous liquid or carbon dioxide pipelines, 2,149,230 miles (3,458,850 km) of gas distribution pipelines, 299,000 miles (368,540 km) of onshore natural gas pipelines, and 171,000 miles (275,198 km) of offshore natural gas transmission and gathering system pipelines in production.

From a production standpoint, in 2012, crude oil production was 89.08 million barrels per day. The petroleum industry refineries produced 82.24 million barrels of refined products. It is also estimated that 3.471 trillion standard cubic meters of natural gas were produced. Table 1.1 provides a listing of total production volume for the top 10 petroleum-producing countries in the world in this time frame.

As Table 1.1 shows, the combined production of these countries is 53,620,000 barrels per day. Each of these barrels must be transported from the well to a refinery and, ultimately, to the consumer in the form of a refined or other commercially produced product. Pipelines are an essential component of the delivery mechanism.

1.4 PIPELINE DESIGN ESSENTIALS

In this book, we generally assume that most readers are familiar with pipeline design and operating principles. However, it is useful to review some of

TABLE 1.1 Top Petroleum Producers [3]

Rank	Country	Production Volume (Barrels per Day)
1	Saudi Arabia	11,150,000
2	Russian Federation	10,210,000
3	United States of America	9,023,000
4	Iran	4,231,000
5	China	4,073,000
6	Canada	3,592,000
7	United Arab Emirates	3,087,000
8	Mexico	2,934,000
9	Kuwait	2,682,000
10	Iraq	2,638,000
	Total	53,620,000

the key points before diving into the details of pipeline leak detection. To that end, we provide a brief discussion of physical pipeline components, pipeline data acquisition and control, and pipeline hydraulics.

1.4.1 Physical Components

Pipelines are fixed systems that transport fluid commodities from one location to another. In our definition, pipeline systems include all physical devices, components, computer systems, telecommunication systems, and the pipe itself, which is required to move the petroleum product between various locations.

In principle, the fundamental architecture is simple: a pipe connects a commodity source at high pressure to another at a lower pressure. However, this fundamental architecture allows for a lot of additional complexity. Pumps or compressors may be required to provide additional motive potential in the form of a pressure increase. Tanks may be included to provide temporary storage at system boundaries. Valves of various types may be used to divert flow, prevent backflow, or confine commodity in the pipeline. And more complex topologies are certainly not uncommon. For example, multiple pump or compressor stations may be required and pipelines may branch, tee, or even be networked. Therefore, pipelines can span the range from being very simple, short in length, and with minimal physical components to very complex and integrated systems that span mountains, seas, and very long distances.

Fig. 1.1 shows a simple pipeline example of a liquid (top) pipeline and a gas (bottom) pipeline.

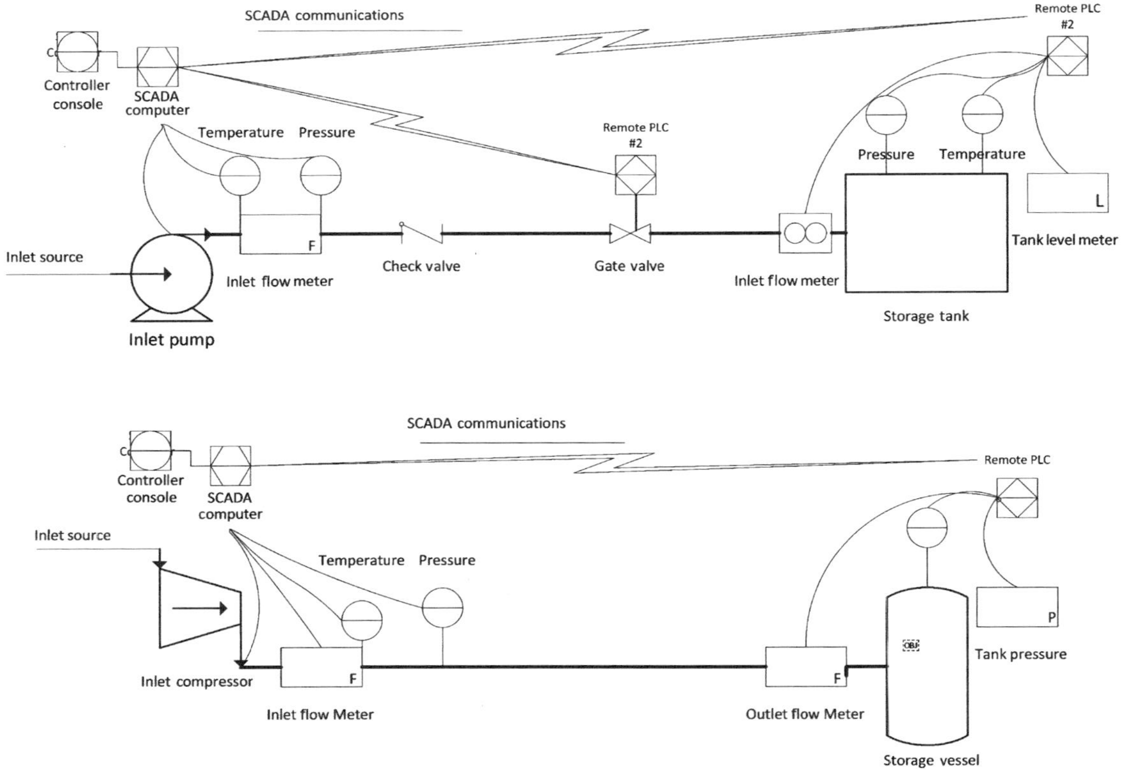

FIGURE 1.1 Examples of liquid and gas pipelines.

All pipelines have at least one inlet where the commodity enters the pipeline system and at least one outlet where the commodity leaves the system. In Fig. 1.1 the inlet is on the left and the outlet is on the right of the figure. Each pipeline system also requires some motive force that provides the energy to move the commodity from the inlet to the outlet. For liquids, the motive source can be provided by many possible types of pumps. Gas pipelines rely on compressors that pack and pressurize the gas to provide the same function. Pumps and compressors may be controlled locally or remotely from the pipeline control center.

Referring to Fig. 1.1, the pipeline system may include physical devices such as check valves, which prevent the fluid from flowing backwards, and isolation valves to segregate portions of the pipeline for various reasons. Isolation valves may be operated by closing them either locally using a manual hand crank or pushbutton motor operator, or remotely via commands from the pipeline control center.

Pipelines also may include tanks for storage, emergency relief, holding, transfer, or other purposes. All are designed to temporarily hold the commodity and may provide an inlet source to the pipeline as well. Tanks add complexity to leak detection systems, the topic of this book, which must account for the commodity entering and leaving as well as commodity changes (such as evaporation and mixing), which may occur within the tanks.

1.4.2 Data Acquisition and Control

A set of physical pipeline components is of little use if it cannot be monitored, operated, and controlled. To achieve this requires field monitoring instrumentation. Field monitoring instrumentation, such as pressure instruments, temperature instruments, and flow meters, provide the pipeline controller information regarding what the pipeline operating characteristics are. They are also very important data inputs for internal pipeline leak detection systems, as described later in the book.

Given their geographic size, nearly all pipelines are monitored and controlled from a single site. Operating and monitoring the pipeline operation is accomplished through the control environment. The control environment includes the human (the controller) who is monitoring and managing the pipeline system. It also includes the Supervisory Control and Data Acquisition (SCADA) system, which is discussed in more detail later. The control environment also includes one or more remote site data concentrators such as Remote Terminal Units (RTU) or Programmable Logic Controllers (PLC) and a telecommunication infrastructure. An example of the control environment is shown in Fig. 1.1.

To understand how the control environment works, we start by looking at what is occurring along the pipeline and work our way to the SCADA computer and, ultimately, the controller.

As a note, the following discussion applies equally well to liquid and gas commodity pipelines.

Pipeline design requires monitoring and control at locations such as the pump or compressor stations, remote gate valves, and system inlets and outlets. At these locations, essential operating physical states, such as pressures, temperatures, valve positions, pump or compressor operating status, and flow rates are continuously measured. Depending on the location that is being monitored and controlled as well as the system complexity, the number of field points may range from only a few to several hundred or even thousands.

Although having these data locally is important, achievement of the most effective and efficient pipeline operation requires the pipeline to be controlled from a single location. Achieving this central monitoring and control capability requires the transfer of remote data to the controller location, and commands enacted at the control location to the remote locations. This is accomplished through a combination of data concentrators, the communication infrastructure, the SCADA computer, and the human machine interface (HMI).

Data concentrators are located at the remote sites and can be a range of devices such as computers, PLCs, or other devices that connect to the field instruments. The data concentrator continuously monitors and gathers the field information and transfers all current data to the SCADA computer over the SCADA communication infrastructure. This can occur on a predetermined schedule, when a significant field data change has occurred, or upon request from the SCADA computer.

Pipeline SCADA communication infrastructures can include virtually every type of telecommunication system such as fiber optics, microwave radios, very-high-frequency radios, ultrahigh-frequency radios, phone lines, dedicated wide area networks, local area networks, various satellite telecommunication infrastructures, or combinations of any subset of these. Regardless of the technology used, the communication infrastructure connects the central control SCADA computer to the remote site data concentrators. For later reference, it is worth noting that the quality of the field data and the communications infrastructure can impact the quality of the leak detection system.

The heart of the SCADA system is how the pipeline controller interacts with the pipeline system through the central SCADA computer. The pipeline controller has the responsibility of continuously monitoring the entire system, responding to system alarms, and introducing system changes by sending commands to the field devices. Note that HMI stands for Human Machine Interface and refers to the user interface by which one interacts with a software system.

The SCADA computer system can take many forms, such as a client/server, a distributed computer network, or a dedicated master–slave system. Regardless of the actual SCADA computer form, this part of the system has the tri-purpose of: (1) obtaining all monitored and measured pipeline field physical data; (2) implementing an HMI for the controller to monitor and control the pipeline; and (3) providing a means to send commands from the control center to remote devices and equipment. These commands take the form of starting/stopping the pumps/compressors, opening/closing valves, and changing pressure set points.

In summary, pipelines include the physical system that receives the commodity, such as from another pipeline, well, or tank. The received commodity is then transported from the inlet to an outlet location, which may be another pipeline or some form of tankage. Along the pipeline infrastructure are devices that measure various physical states such as pressure, temperature, flow rates, valve positions, and pump/compressor status. The pipeline physical state information is gathered in a local data concentrator that ultimately sends the local data to the SCADA computer at the central location. As a side note, SCADA computer systems and the SCADA telecommunication infrastructure are frequently redundant. This provides a higher level of availability in case there is a failure of one SCADA computer or a telecommunication circuit. In addition, the judgment of the operator is a critical aspect of the pipeline control process and, as discussed later, is also a critical component in the operation of the pipeline leak detection system.

1.5 PIPELINE LEAKS, RUPTURES, SPILLS, AND THEFT

Across the world, nations can generally be classified as either developed or developing. Important to this differentiation is that developed nations have a relatively high level of economic growth, security, longer life spans, and better health care. Although many factors contribute to the differentiation between developed and developing nations, all developed nations enjoy the utilization of an extensive pipeline infrastructure.

Pipeline infrastructures support the gathering, transportation, and distribution of many essential commodities that support the developed nation social fabric. Such transported items include potable water, waste water, crude oil, refined petroleum products, natural gas, propane, carbon dioxide, and anhydrous ammonia for use in fertilizers, to name just a few. Pipelines bring drinking water to our homes and businesses and carry waste away to where it can be safely handled. Petroleum pipelines are especially important because the commodities they transport provide fuel sources to electric generation plants and raw material inputs to many manufacturing and production processes, and they are sources of heat for our homes and businesses.

TABLE 1.2 PHMSA Significant Incidents: Hazardous Liquid Pipelines [4]

Category	Count
Total significant incidents 1989–2008	2180
3-Year average (2006–08)	109
5-Year average (2004–08)	118
10-Year average (1999–2008)	123
20-Year average (1989–2008)	141

1.5.1 Breach of Integrity Incident Rates

Leak events are generally quite rare, but they do occur. It is a given that at some point nearly all pipeline systems will experience an unforeseen release of commodity: a leak. Whether the leak is small and gradual or large and sudden, the consequences can be dire or minor depending on the fluid characteristics, location, and circumstances surrounding the leak event.

Although the pipeline industry provides a very safe transportation method, significant events continue to happen, as reported to the US Department of Transportation (DOT) Pipeline and Hazardous Material Safety Administration (PHMSA), as shown in Table 1.2.

Research has identified that leak events are primarily caused by:

- External interference or third-party activity
- Corrosion
- Construction defect and mechanical or material failure
- Ground movement or natural hazards in general
- Operational error or hot-tap by error[1]
- Other or unknown causes [5]

The objective of transporting commodity by pipeline is that everything that enters the system stays within the system until it reaches its destination point, that is, the prevention of a leak and subsequent spill, if it is a liquid pipeline. Note that the difference between a liquid pipeline leak and a spill is that a leak is the liquid escaping the pressure boundary. A spill is the accumulation of commodity in the surrounding environment that has escaped the pressure boundary through the leak.

Part of the negative impact of a leak is the value of the lost commodity. However, this cost is often dwarfed by other costs of the spill, which could

1. "Hot-tap by error" refers to maintenance penetration into a pressurize pipelined when the pipeline was assumed to be unpressurized.

include, but are not limited to, pipeline downtime, third-party or employee injury and death, environmental or property damage, and loss of corporate good will. Given these impacts, various means of detecting the occurrence of a leak and/or resulting spill have been implemented over time.

The earliest form of leak detection, and the one that continues to provide the highest degree of accuracy and reliability, is direct observation [6]. If someone sees liquid, for example, crude oil, escaping from the pipeline system and reports it to the proper authorities, then this is a highly accurate indication that a leak has occurred. Although this was the first leak detection system, it is still a very effective and common means of detecting leaks today. From 2010 through 2015, 58.23% of all PHMSA-reported incidents were detected by visual observation.

Because pipeline operators are unable to have someone continuously watching every foot of pipeline, various other techniques have been implemented to identify when a leak has occurred. These methods are referred to as leak detection systems or leak detection technology. Leak detection technology research and development continue today in an effort to develop systems that can detect leaks or spills faster, with lower leak rates and smaller spill sizes, with more precise location capabilities, and with fewer false alarms.

To reiterate, within the hazardous liquid pipeline industry, owners and operators acknowledge that a major risk is the occurrence of commodity leaks and resulting spills. Although always at risk, the industry continues to strive for zero events. As the industry has noted:

> *Liquid pipeline spills along rights-of-ways have fallen over this decade, in terms of both the number of spills and the barrels of product spilled per 1,000 miles travelled. The frequency of releases decreased from 2 incidents per thousand miles in 1999–2001 to 0.7 incidents per thousand miles in 2006–2008, a decline of 63 percent. Similarly, the amount of barrels released per 1,000 miles decreased from 629 in 1999–2001 to 330 in 2006–2008. [7]*

The Department of Transportation Office of Pipeline Safety provides further evidence that the number of hazardous liquid pipeline incidents have declined, as shown in Fig. 1.2.

Note that the rate of spill reduction appears to have changed since 2007. As Fig. 1.2 shows, the total number of annually reported spills was in a steady decline between 2002 and 2007. Since 2007, however, the overall average number of spills appears to have leveled off at a rate of approximately 100 incidents per year. Although the year-to-year reported number varies above and below the 2007 value, it appears, on average, that the curve has flattened out. No specific reason has been identified that clarifies why the 2002 through 2007 decline trend did not continue.

Although the number of spills has declined since 2002, and as the flattening of the decline curve since 2007 helps demonstrate, leaks and ruptures

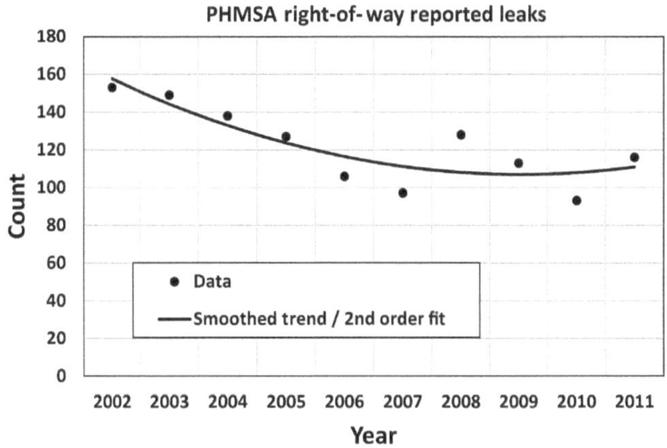

FIGURE 1.2 Number of hazardous liquid pipeline spills per year [4].

may continue to occur and operators must continue to provide leak detection systems in an effort to mitigate the consequences. This ultimately reduces the pipeline operator's risk.

1.5.2 Commodity Theft

Historically, leak detection systems have been focused on detecting commodity releases that occur unexpectedly. A recent phenomena is the loss of petroleum product from pipelines due to theft. A characteristic of theft is that it is performed by an intelligent agent who wishes to remain undetected and who expects to contain the extracted product. Consequently, external leak detection systems will not detect a theft.

1.6 LEAK DETECTION APPROACHES

Leak detection is accomplished by a wide range of approaches that have various strengths, weaknesses, and costs. These systems include direct observation approaches of various kinds and technology-based systems that are generally classified as either internal or external technology.

Direct observation accounts for the identification of the majority of commodity releases. This involves someone detecting the commodity release and reporting it to the pipeline control center. The person observing the release may be an employee, third-party, or someone from the general population of people living, working, or traveling in the area where the commodity release has occurred.

The technology-based internal leak detection system utilizes pipeline physical measurements to infer that a commodity release has occurred. The physical measurements may include flow rates, pressures, and temperatures. One example of this is a basic flow balance system that simply subtracts what leaves the pipeline from what enters the pipeline. If more commodity enters than leaves the pipeline, then a commodity release is inferred. Other internal leak detection systems range from comparatively simple pressure and flow deviation–based systems to much more complex real-time transient model (RTTM) leak detection applications.

An RTTM develops a model of what should be occurring, assuming no leak, within the pipeline and compares the modeled pipeline to measurements obtained from the actual pipeline. If there is a difference between the modeled and measured pipeline, then the RTTM may infer that a commodity release is occurring.

External leak detection systems differ from internal leak detections systems because they use a variety of means to detect the presence of the pipeline commodity outside of the pipeline or a change in the surrounding environment resulting from the commodity leaving the pressure boundary. Consequently, the systems are not reliant on measured pipeline parameters. Some of these internal systems include cables that sense rapid change in temperatures within a very small area, hydrocarbons that are absorbing of diffusing various light sources, and sounds that a leak or rupture would induce into the surrounding area. Although a variety of external leak detection methods have been developed, they all share a commonality in that a breach in pipeline integrity has occurred and the transported commodity is detected external to the pipeline integrity shell.

1.7 THE BOOK STRUCTURE

This book is structured to provide the reader with basic and advanced information and tools related to a range of leak detection systems.

- We start in Chapter 2, Pipeline Leak Detection Basics, by describing leak detection basics. This provides the reader a grounding in the terms, technology, and approaches to leak detection.
- In Chapter 3, Mass Balance Leak Detection, we introduce and categorize mass balance–based leak detection systems.
- Chapter 4, Real-Time Transient Model–Based Leak Detection, addresses the special category of RTTM mass balance leak detection systems.
- Chapter 5, Statistical Processing and Leak Detection, discusses the challenge of extracting a leak signature from a noisy signal.
- Chapter 6, Rarefaction Wave and Deviation Alarm Systems, describes the detection and processing of the negative pressure wave associated with the onset of a commodity release.
- Chapter 7, External and Intermittent Leak Detection System Types, discusses external leak detection systems.
- Chapter 8, Leak Detection System Infrastructure, discusses the system infrastructure that is required to support a leak detection system.

- Chapter 9, Leak Detection Performance, Testing, and Tuning, addresses the topic of evaluating and quantifying leak detection system performance.
- Chapter 10, Human Factor Considerations in Leak Detection, describes the human factors related to leak detection, including interaction with leak detection technology systems and the direct observation of leaks.
- Chapter 11, Implementation and Installation of Pipeline Leak Detection Systems, presents topics related to the implementation and installation of a leak detection system.
- Chapter 12, Regulatory Requirements, reviews regulatory requirements related to leak detection systems.
- Chapter 13, Leak Detection and Risk-Based Integrity Management, addresses leak detection and risk-based integrity management.

1.8 TERMINOLOGY

In closing this chapter, we need to ensure that the reader understands some of the terminology we commonly use. As with most books and new study areas, understanding the meaning of various terms, words, and phrases is essential to obtaining the most from reading this book. As such, throughout this book we rely on common terminology to transfer our meaning in a consistent manner and format. Although new terms will be introduced as they occur within specific contexts, the following provides a listing of the most common terms we refer to and use on a constant basis.

Controller: The individual in the pipeline control room responsible for performing day-to-day pipeline control actions and for responding appropriately to leak detection system alarms.

Commodity: A general term we use to refer to any fluid moving through the pipeline.

Direct Observation: A human sensing the presence of the leak by any method of observation whether by smell, sight, sound, or any other way.

External Leak Detection (ELDS) Systems: Leak detection systems that monitor the commodity once it is external to the pipeline.

Internal Leak Detection (ILDS) Systems: Leak detection systems that monitor measurements of the pipeline state and flows to deduce that a leak may be occurring.

Leak: An unintended breach in the pipeline pressure boundary that allows the contained commodity to escape from the pipeline. The key attribute is that a leak is where the commodity is leaving the pipe pressure boundary due to a breach such as a flange failure, puncture, corrosion erosion, and so forth.

Leak Detection System (LDS): A system designed to detect any breach of integrity and alert the operator to the event.

Mass Balance Section (MBS): A section of the pipeline that is monitored independently of other sections for a leak by a mass balance approach.

Operator: The legal entity responsible for maintaining and performing day-to-day operations of a pipeline system.

Rupture: The sudden and catastrophic failure of the pipe pressure boundary. Rupture size is significant in relationship to the pipe cross-sectional area.

SCADA: SCADA systems are computer applications that provide remote monitoring and control of the pipeline system.

Spill: The accumulation of the liquid commodity after it has left the pipeline pressure boundary.

Other specific terms are highlighted as they occur in the document.

1.9 NOMENCLATURE

Throughout this book is an extensive set of equations. Table 1.3 lists the most commonly used nomenclature found in these equations.

TABLE 1.3 Symbols and Nomenclature

Symbol	Description
a	Speed of sound
A	Inside cross-sectional area of the pipeline
A	Availability
ASV	Adjusted spill volume
c	Speed of light
c_V	Specific volumetric heat capacity at constant volume
c_P	Specific volumetric heat capacity at constant pressure
CRT	Cable response time
D	Inside diameter of the pipeline
E, E_{Pipe}	Young's Modulus of the pipe
E_0, E_1	Estimated stopping times (in terms of counts) for null and alternate hypotheses
f	Friction factor
f_0, f_1	Probability densities for null and alternate hypotheses
f_{SCAN}	Scan frequency
f_L	Probability density function

(*Continued*)

TABLE 1.3 (Continued)

Symbol	Description
g	Acceleration of gravity
G	Maximum normed residual
H_0, H_1	Null and alternate hypotheses
I	Cable current
k	Thermal conductivity
K_c	Commodity bulk modulus
L_{TOT}	Total pipeline length
MAD	Data set median absolute deviation
M_z	Modified Z-score
$\mathbf{M_C}$	Covariance matrix
$MTBF$	Mean time between failures
$MTTR$	Mean time to repair
N	Newton
N, n	Integer count
P	Probability
P_C	Conditional probability
$P(y\|x)$	Probability that x is true given that y is true
p	Pressure
PT	Spill propagation time
q	Heat flux per unit area
Q_{Leak}, q_{Leak}	Leak rate, leak volumetric flow
q_S	Heat flux at the inside pipe surface
r	Radial distance from the pipeline center
R	Cable resistance
r_{FA}	False alarm rate
R_{Leak}	Leak incident or event rate
S	As a subscript, refers to inside pipe surface
SRC	Spill remediation cost
std	As a subscript, refers to value at standard conditions (STP)
SV	Spill volume

(*Continued*)

TABLE 1.3 (Continued)

Symbol	Description
t	Time
t	Time to detect
t_P	Time between runs, periodicity time
t_T	Pig transit time
t_A	Analysis time
T	Temperature
$t_{\alpha/(2N),N-2}$	Critical value of the t-distribution with $N-2$ degrees of freedom and a one-sided significance level of α/N
x	Distance
U	Uncertainty or noise
u	Velocity
V	Cable voltage or potential
v	Velocity
V	Volume
VB	Volume balance
V_a	Detection tube air velocity
V_ρ	Cable velocity of propagation
$\mathbf{W}$	Watts$\left(\frac{J}{\text{sec}}\right)$
$\mathbf{W_D}$	Decorrelation matrix
WT_{PIPE}	Wall thickness of the pipe
$\mathbf{X_o}$	Input array with colored noise
$\mathbf{Y_o}$	Whitened output array
z	Elevation
Z_t	Partial derivative of variable Z with respect to time at distance x, $\left.\frac{\partial Z}{\partial t}\right)_x$
Z_x	Partial derivative of variable Z with respect to distance at time t, $\left.\frac{\partial Z}{\partial x}\right)_t$
Z_α	Number of Gaussian standard deviations required to achieve a one-tailed confidence of α
Λ	Likelihood ratio
α	Thermal diffusivity $\alpha = \frac{k}{\rho c_P}$
α	Coefficient of thermal expansion. $\alpha = \left.\frac{1}{V}\frac{\partial V}{\partial T}\right)_p$
α	Type I error probability

(Continued)

TABLE 1.3 (Continued)

Symbol	Description
β	Type II error probability
ε	Uncertainty or noise
ε_i	Random noise in autoregressive Markov time series
ν	Poisson ratio (negative ratio of transverse to axial strain); value is 0.27 to 0.3 for steel
μ	Mean or average of some data set
σ	Standard deviation of some data set
ρ	Density
κ_{Leak}	Leak rate per unit distance
φ	Autoregressive Markov series factor

REFERENCES

[1] Furchtgott-Roth D. [Issue Brief] Pipelines are safest for transportation of oil and gas. Manhattan Institute for Policy Research; 2013.

[2] Factbook. The World Factbook, https://www.cia.gov/library/publications/the-world-factbook/fields/2117.html#180 [accessed 26.02.15].

[3] Zaidi S. Top 10 oil producing countries, http://toponthelist.info/2013/01/16/top-10-oil-producing-countries/ [retrieved 03.02.15].

[4] PHMSA. http://www.phmsa.dot.gov/pipeline/library/data-stats [accessed 26.02.15].

[5] Papadakis GA. Major hazard pipelines: a comparative study of onshore transmission accidents. J Loss Prevent Proc 1999;12 91–107, p. 92.

[6] DOT. Leak detection technology study, For PIPES Act, H.R. 5782. The U.S. Department of Transportation; December 31, 2007.

[7] AOPL. Pipelines and safety. Association of Oil Pipelines, http://www.aopl.org/files/Pipelines_and_Safety.pdf; [accessed 25.01.14].

Chapter 2

Pipeline Leak Detection Basics

Pipelines are in the business of transporting commodities that are dangerous, or valuable, or both. Unfortunately, accidents and incidents that involve breaches of integrity (ie, leaks and spills) do happen, and in certain locales, outright theft of the commodity itself can be a problem. It is in the interest of operators, regulators, and third parties to ensure that when such incidents occur, a rapid response on the part of the operator limits the damage by shutting down the pipeline operation, isolating the leak site (typically by closing pipeline valves) and dispatching response crews to the site to contain the damage and clean the site. Leak detection systems (LDSs) are a key component of this response.

2.1 THE CHALLENGES OF DETECTING PIPELINE LEAKS

Pipelines present a combination of circumstances that make detection of leaks, spills, and ruptures challenging. The first of these is the sheer size of these systems. The typical pipeline segment is roughly 50 to 100 miles (approximately 80–160 km) long. Some are much larger, extending over many hundreds or even thousands of miles. On the basis of absolute linear footage, pipelines are some of the largest artificial constructs that have ever been built by humans. They share a commonality with other dedicated transportation facilities such as canals, roads, and railroads in that they are physically dispersed over very long distances. Because of these size and dispersion issues, it is very difficult to efficiently monitor every mile in an attempt to detect and locate leaks or spills.

Another problem is that they are often hidden. The typical pipeline is buried beneath the surface of the ground. This is done for a number of very good reasons, including cost. Although it may seem counterintuitive, in the long-term it is often cheaper to go to the expense of burying pipelines. Unless properly confined, an above-ground line will tend to expand and contract with changes in temperature; as these changes occur, the pipe will tend to move back and forth along its length like the writhing of a snake, sweeping away the vegetation as it moves. In addition, an exposed line is subject to damage through sabotage or other insult, such as collision from an

Pipeline Leak Detection Handbook. DOI: http://dx.doi.org/10.1016/B978-0-12-802240-5.00002-9

automobile. Furthermore, most people do not actually want to look at industrial facilities such as pipelines. Many jurisdictions require that pipelines must be buried when possible.

However, a leak from a buried line, especially a slow leak, can continue for a long time without being detected. Because the line is buried, and depending on the local terrain and fluid properties, the flow of a liquid commodity from the leak site can diffuse slowly to the surface, drain downward into the soil, move along the top of the water table, or run along the buried pipe for a very long distance. Commodity escaping the buried line is also hidden from view, and it can stay hidden for a very long time. All of these effects can result in a large cumulative volume spilled before it is detected.

Another issue is that once they start, pipeline leaks tend to *continue* until they are discovered. It is as if a truck runs off the road and every truck and car behind it mindlessly follows it into the weeds, where they all tend to accumulate in a large, growing pile of wrecked vehicles. This goes on until some responsible individual notices the steadily increasing, enormous pileup and takes action to do something about it. (Note that this is not necessarily true in the case of *theft*. We return to this later.)

There are also transient effects, which can be very important, especially in systems with highly compressible commodities such as gas and multiphase pipelines. If we increase the pressure at the point of the pipeline system where fluid is entering the line, then more commodity will be entering the pipeline than will be leaving it. This is referred to as *pack*. Let us say we have a system that looks for leaks by watching for a mismatch between the flows entering and leaving the system. If we have a transient of some sort, then packing may look like a leak. This effect will potentially continue for some time. At some point, of course, the rates into and out of the system will equalize again, but during this period of transient operation it is more difficult to find a small leak unless we have some knowledge of how rapidly the system is packing up or down. The sheer size of the pipeline system works against us because the pack will be proportional in size and duration to the volume of the pipeline system.

This brings us to the issue of uncertainty in the leak signal. Much of this uncertainty is driven by instrument measurement and calculation errors. In general, any LDS is going to be highly dependent on the quality of its input data. One kind of LDS might use some type of external sensor that can directly detect the presence of spilled or escaped hydrocarbon. However, it is not unusual for such detectors to continuously output a signal that only becomes significant if the signal exceeds a threshold. If the threshold is set too low, then the system will be subject to a large number of false alarms (false positives). Alternately, if set too high, then the system will fail to detect a leak when it should.

In the next section, we look at the application of external and internal leak detection approaches to a nonpipeline metaphor that nearly anyone can relate to: the toll road. Although our example may seem to be removed from the topic of pipeline leak detection, it provides an easy introduction to many of the concepts that we discuss in more detail later in this book.

2.2 THE TOLL ROAD AND THE FREE-RIDER PROBLEM

Imagine that we have a 100-km-long cross-country toll road (shown in Fig. 2.1) where all vehicles are required to utilize smart radio frequency identification tags (RFID tags). These tags can be detected by smart tag counters at points of tollway entry and egress. Individual drivers pay according to the distance that they travel by accounting for the locations where they enter the toll road system and the locations where they leave. They are not billed if they are not detected as they leave the system, and the records of their entry are wiped at the end of the day if they are not detected leaving by that time. The system is closed to unmetered traffic. However, it is bounded by untolled frontage roads, and there are several locations where the unfenced grassy strips between the toll road and the frontage road might allow unprincipled drivers to evade paying the toll.

One might be concerned that the system was losing revenue if drivers were somehow leaving it without their comings and goings being properly measured. Several errors could contribute to this, such as the possibility that the tag detectors are not working (a measurement problem) or, possibly, drivers are exiting the system at some un-tolled location (a vehicle leak!). People who enter the system and evade paying are referred to here as free riders. We want to measure how well our system is accounting for vehicles and, in particular, we want to determine whether or not we have a free-rider problem.

2.2.1 Directly Detecting Free Riders

We could tackle the free-rider problem by stationing police vehicles at various locations along the frontage road. This is an application of a *direct* (because our escaped driver detectors—the police—catch the runaways as they exit the system) or *external* (because the police are actually outside the toll road when they catch the vehicle leakage from the system) leak detection method.

This works well if the police happen to be at the location where a driver illegally exits the toll road, but that is precisely the problem. What if the police are not at the right location? We have 100 km to cover and only a limited number of police cars that we can devote to our detection problem. If we only have five police cars, and if each vehicle has a visibility range of 2 km, then this means that we have $2 \times 5/100 = 0.1$, or only 10%, coverage for our system. The problem here is that our external system uses discrete detectors with limited detection coverage per detector, which means that full

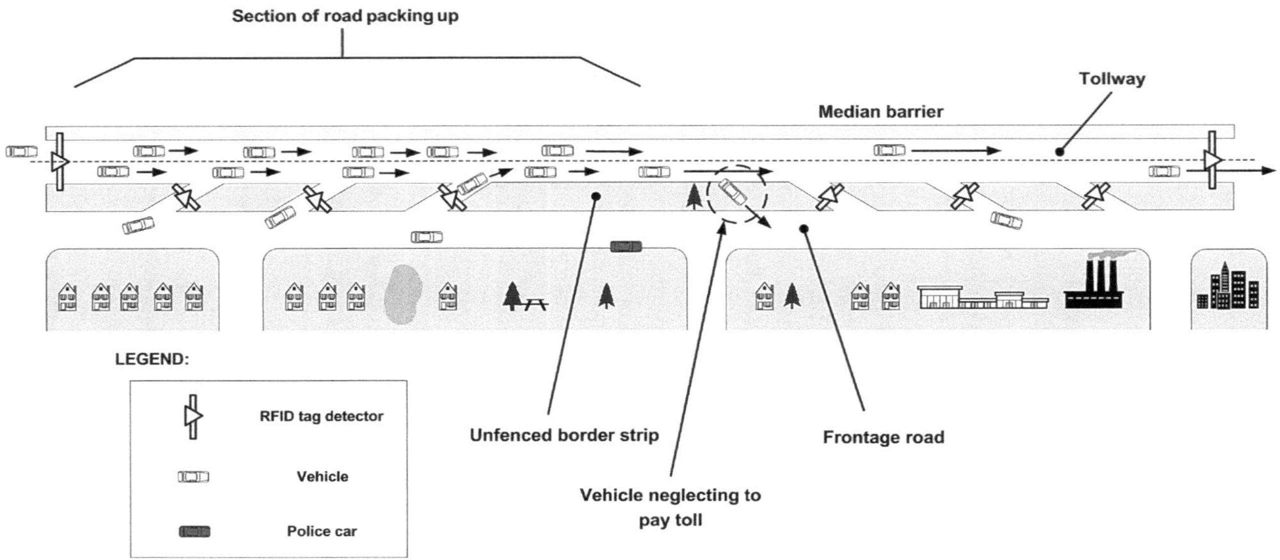

FIGURE 2.1 The toll road.

coverage becomes expensive in terms of total detectors. If illegally departing drivers leave the tollway at independent and randomly chosen locations, then our detection system can catch only 10% of them. The probability that our system will actually detect or warn of (alarm) a leakage of vehicles is expressed mathematically as *P(Alarm | Leak)* and is generally referred to in this book as the *leak detection probability*. The terminology $P(x \mid y)$ is referred to as a conditional probability, and is generally taken to mean the probability of x given the fact that y is true.

Let us assume that our tollway leakage is like a pipeline leak in that it is localized and persists in time. This could be a result of something simple, such as the possibility that locations where it is easy to drive across the grass without bottoming out are rare, so that when one car finds a good location, others will tend to copy the behavior and try to exit the road at that same location. If the police park at the same place every day, and if those places are not the location where cars are leaving the highway, then the probability that any continuing leakage will never be detected is 90%.

What can we do to make this work better? We need to boost the leak detection probability so that it is much higher than 10%. We could simply increase the number of detectors (ie, increase the number of police parked at the side of the frontage road). This would definitely work because it increases the coverage on the tollway and therefore boosts the efficiency of our system. It is also an expensive way to go because we would have to increase the number of police 10-fold. Is there a better way?

As noted, the police are stationed in cars, which can move. What happens if they do not just park on the frontage road and are instead allowed to drive along it? How does this help us? A lot, actually. Now, every time a car tries to make its escape, there is a 10% chance that the police will be driving by. This applies to every car that uses the same escape route. This is a repeated Bernoulli trial problem: every pass by a police car is the equivalent of a new, fixed probability trial. It is not difficult to show that if 10 cars attempt to cross the grassy border at the same place every day, then the odds that the police will identify the vehicle leak as well as the leakage location are 65% by the end of the first day. At the end of the second day, this grows to 88%; on the third day, it is 96%. This little tweak has made the leak detection probability a function of time and has also significantly improved the odds of detecting the vehicle loss. The nice thing about this approach is that the probability of success for a set of repeated Bernoulli trials asymptotically approaches 100%, which ensures that the leak will ultimately be caught if we perform enough trials.

We have illustrated several important principles. One is that the value of the leak detection system is in proportion to the coverage. In the case of external detectors, more detectors generally equal better detection. A second critical point is that the most sensitive LDS in the world provides little value if it can only measure leaks over a small fraction of the system. An important qualifier here is that the effectiveness may also be in proportion to the local consequences of a spill. If the damage to people, structures, or the environment is greater in some places than others, then it makes sense to expend more effort to detect and respond to leaks in those areas. An analogous location for a pipeline is the high consequence area. In the United States, this is a regulatory term for a place where the impact of a leak has significant potential for unacceptable damage.

Another important principle is that the probability of detecting the leak generally increases as the time you spend looking for it increases. Patience is often a virtue in the field of leak detection.

Finally, an essential principal is that every LDS has some envelope in which it detects leaks well. For conditions or situations that are outside of that envelope, leak detection sensitivity will be much poorer or nonexistent. In short, leak detection systems have performance maps.

2.2.2 Detecting Free Riders by Counting Cars

Let us try a different approach. Another way to measure the free rider or unaccountable vehicle loss from the system would be to sum the rates at which vehicles come into the tollway at the entrance ramps and subtract the sum of the rates at which they leave the system at all of the exit ramps. We call this an internal detection system because there is no external observer: the measured flows are automatically collected from inside the system by the RFID detectors themselves. If we think of the vehicle entrance and exit rates as a set of flows, then the sum of the entrance and exit flows is the vehicle flow balance (VFB):

$$VFB = VF_1 + VF_2 + VF_3 + \cdots + VF_N = \sum_{i=1}^{N} VF_i$$

EQUATION 2.1

where VF_i is the vehicle flow expressed in terms of vehicles per unit of time. The vehicle flow is positive for vehicles entering the tollway (at on-ramps) and negative for vehicles leaving the system at off-ramps. In principle, a nonzero and positive VFB might indicate whether we have a problem with vehicles leaving the system and not paying the tolls.

This is not the full story. There are legitimate reasons why the VFB might not always sum to zero. At certain times of the day, such as the morning or evening rush hour, more vehicles will be entering the toll system at one end than leaving it at the other. Vehicles will enter at a number of locations at one end (from the suburbs in the morning or the center of the city in the afternoon or evening) and then will need time to travel some distance before they leave it again (to various downtown locations in the morning or to the suburbs again much later in the day).

In addition, drivers tend to want to leave more distance between each vehicle the faster they drive. Conversely, this means that the more vehicles there are on any stretch of road, the slower everyone will drive. The combination of these factors will cause the entire roadway to have many more vehicles traveling slower and in a much denser condition during rush hour periods than during other times of the day.

Thus, anyone looking at the instantaneous VFB of the system will observe that at the beginning of rush hour, there are many more vehicles entering the toll system than are leaving it. The flow balance is a nonzero positive value. The difference occurs because the vehicles are *packing up* the system. As such, if we do not account for the packing, then to the simple vehicle leakage detection system described, it looks like we have a vehicle leak. In fact, if we refer back to Fig. 2.1, we can see that the tollway is experiencing exactly this situation: more vehicles are entering at the suburban end of the tollway than are leaving the city, and the number of cars inside the tollway is increasing as a result.

Toward the end of rush hour, the pack will start to be relieved because, at this point, the rate of vehicles leaving the system is now greater than the rate of vehicles entering the system. Now, the VFB is negative, and it looks like we have an unaccountable generation of cars within the system (obviously the opposite of a leak).

Clearly, this simple approach does not work very well. At the very least, it is limiting. It is worth noting that we can still use the simple flow balance to detect free riders in the system if we set some minimum imbalance rate that is comfortably greater than the largest flow balances that we normally see in the system during periods of high packing. This minimum rate acts as a *threshold*. Of course, the free-rider loss rate that we can be expected to detect by this method is likely to be quite high, because it is exactly equal to the minimum imbalance rate we have set. This illustrates another important principle of leak detection, which is that the smallest detectable leakage rates are normally limited by estimation, measurement, communication, and data processing errors. Because the flow balance is a poor estimator of the true transient balance of vehicles and has an error equal in magnitude to the rate of vehicle pack, the minimum detectable leak size in a system that uses the flow balance is determined by the normal instantaneous packing rate.

What if we want to measure a rate of paying ridership loss that is smaller than this? An alternative is to recognize that the packing effects are transient and tend to balance or cancel out over time. Thus, we could simply average the flow balance over some period of time that is long compared to the period over which the packing fluctuations typically occur. However, the shortest period of time that will be able to make this work is likely to be one or two rush hour cycles, because that is approximately how long it would take for the transient effects to balance out. This could be a period of hours or even days. Standard statistical principles would tend to suggest that the longer we average our flow balance on the system, the smaller the resulting error.

This brings us to our next important principle of leak detection: the smaller the leak rate that you wish to detect, the longer you must wait.

These performance limitations are irritating, and some try to surmount them by simply ignoring them and lowering the threshold below the implied packing bounds. This will indeed increase the sensitivity of our free-ridership detection system, but at an important cost. If we lower the effective flow balance needed to trigger an alarm below the largest normally encountered value of the flow balance, then we will increase the odds that we will trigger an inappropriate alarm on our system. Because these alarms do not correlate to an actual vehicle leakage, we refer to them in this book as either false alarms or false positives. There is another important principle here: inappropriately lowering the detection threshold will increase the number of false positives. The false-positive rate is another performance metric that we return to in Chapter 9, Leak Detection Performance, Testing, and Tuning.

We implemented our free-rider detection system because it was assumed to add value. However, it should be inherently obvious that every false positive reduces the value of the alarms issued by the system. If we lack confidence that the alarms produced by the system actually correspond to a stream of free riders some place on the tollway, then it will be necessary to validate each alarm. This will require some external agent to analyze each alarm (at some cost in response time and effort) to determine whether or not it actually corresponds to a loss of paying drivers. In a probabilistic sense, the probability that any one alarm created by the system actually corresponds to a stream of free riders exiting the highway at some location is equal to the number of true positives (ie, alarms that actually correspond to a loss of vehicles) divided by the total number of alarms (the sum of the true positives and the false positives), as shown in Eq. (2.2):

$$P(\text{Leak}|\text{Alarm}) = \frac{N_{\text{Alarms}}(\text{Leak})}{N_{\text{Alarms}}(\text{Leak}) + N_{\text{Alarms}}(\text{No leak})}$$

EQUATION 2.2

where $N_{\text{Alarms}}(\text{Leak})$ is the number of true positives and $N_{\text{Alarms}}(\text{No leak})$ is the number of false positives. $P(\text{Leak}|\text{Alarm})$ is the probability that a leak actually exists given that an alarm is present, and it is referred to in this book as the *alarm efficiency*. The alarm efficiency is not the same as the leak detection probability, which is the probability that a leak will be detected and alarmed, $P(\text{Alarm}|\text{Leak})$. Consequently, $P(\text{Leak}|\text{Alarm}) \neq P(\text{Alarm}|\text{Leak})$.

In Chapter 13, Leak Detection and Risk-Based Integrity Management, we see that in developed countries the leak incident rate for a typical pipeline segment is generally very low, well under one real leak event per year. If we presume that the false alarm rate is significantly higher than this, then, by inspection, we can see that the ratio in Eq. (2.2) will be much less than 1. This illustrates yet another important principle: the value of alarms generated by the detection system tends to be inversely proportional to the number or rate of false alarms created by the system.

Let us get back to the packing error. It would be nice if we could simply eliminate this problem completely. We can perform a more principled balance on the tollway by including the packing effect. In fact, if we assume that vehicles follow a conservation law, which we call the law of Conservation of Highway Vehicles, then we can show that in the absence of a free-rider problem the VFB minus the rate at which vehicles are accumulating in the system (we call this the *vehicle packing rate*, or VPR) must be zero. If it is not zero, then this means that the RFID tag detectors are not measuring properly, the accounting algorithm has an error in it, or that devious drivers are somehow leaving the system at some unknown location. We call this new packing rate adjusted parameter the *tollway vehicle balance* (TVB) and define it as:

$$TVB = VFB - VPR$$

EQUATION 2.3

All of the parameters in this equation are measured in vehicles per unit of time.

Of course, this still leaves us with the problem of determining one of the critical parameters of the vehicle balance calculation: the VPR. How might we do this? Let us consider a more mathematical definition: the VPR is the rate of change in the total number of vehicles on the tollway or:

$$VPR = \frac{dn_{\text{Veh}}}{dt} = \frac{d}{dt}\int_0^L \rho_{\text{Veh}}\, dx$$

EQUATION 2.4

where t is time, n_{Veh} is the number of vehicles, and ρ_{Veh} is the local vehicle density expressed in units of 1/length. Parameter x is length along the toll road, and we simplistically assume a linear roadway that is L units long.

Therefore, one way to evaluate the packing rate would be to station a set of packing rate monitors at intervals along the roadway. (Note that the monitors are on the toll road itself, making them components of our internal system.) To simplify, each packing rate monitor is nothing more than an observer standing at the side of the road with a calculator. Once per minute, our packing rate monitor calculates the vehicle density by counting the number of vehicles between a point halfway up the road to the last packing detector and halfway down the road to the next packing detector. The packing rate monitors are stationed so that they can clearly see up and down the road to the points where the neighboring monitors' measurement territories begin. At any time, the local packing rate is equal to the number of vehicles counted in this iteration within the evaluator's stretch of roadway minus the number of vehicles counted during the last time period. If that last time period was 1 minute ago, then this gives us the local packing rate in vehicles per minute. The packing rate for the system is the sum of the packing rates for every packing monitor over the entire roadway. A positive rate indicates the system is packing or gaining vehicles, whereas a negative rate indicates it is unpacking or losing vehicles.

This works! Of course, you may object because this is an expensive way to calculate the toll road packing rate! If we assume that each observer is responsible for approximately 2 to 4 km of road, then we need at least 25 monitors to cover the entire highway. It also requires a lot of counting on the part of the monitors, but we will optimistically assume that this can be automated. So, let us look at the alternatives. One possibility is to limit the number of packing monitors to a reasonable number. Of course, this means that because no one monitor can count more than what he or she can see, we will never get a true count. Instead, we can extrapolate to get the rate in the unmonitored section by taking the average packing rate that we get with this more limited pack monitoring set. We then multiply that rate by the ratio of that portion of the system that has no monitors divided by the portion of the system with monitors.

We have effectively performed a statistical sampling of the vehicles on the tollway. This is actually not a bad way to go, but note that it is only a sample. The portions of the system that have no packing rate monitors may be higher or lower in vehicles than the portions with monitors, and thus there will be some resultant measurement error in the calculation of the average packing rate. This error will increase as we continue to decrease the number of monitoring personnel. In line with our earlier observations, this means that if we wish to avoid an excessive number of false alarms, then we will either have to raise the minimum threshold or wait longer to catch our free-rider leak events.

There is another possible way to determine the packing rate of the system, and that is to use a limited number of internal pack measurements combined with an accurate and principled simulation or modeling process to calculate the packing rate for the entire tollway. Although this seems like a stretch, this is actually a feasible and commonly used approach in pipeline leak detection systems, where the physical conservation and transport equations that address the commodity transport problem are fairly well understood. (It may also be possible to do this for our vehicle transport/free-rider detection system, because specialists working in the traffic engineering field have actually developed rigorous equations and models to describe the dynamics of automobile transport on road systems.) This modeling approach allows us to minimize the additional measurements needed to determine the packing rate. This provides us another important principle: proper application of governing physical principles can allow us to perform better leak detection.

The use of conservation, measurement, and statistical principles may seem indirect and, in some ways, complex when compared to the more direct approach taken in the last section. However, as we shall see later, the pipeline analogy of this approach—mass balance leak detection—presents a significant set of advantages that make it one of the most widely utilized approaches for leak detection.

2.3 LEAK LOCATION AND OTHER ISSUES

Although it is important to detect the leak, this does us little good if we are unable to locate and stop it. Referring back to our two tollway free-rider detection system approaches, it should be clear that although the external detection approach may require a lot of external monitors to function, the leak location is positively identified once the leak is detected. On the other hand, the version of the internal free-rider detection system described here can potentially detect the vehicle leak using only the RFID readers needed to perform billing in the system. However, it has no real ability to locate the leak. We can provide an internal vehicle balance system with the ability to locate the leak (with some error) by providing additional measurements. The simplest way to do this would be to implement additional RFID detectors between the start and end of the toll road. Then, we will be able to isolate the leak by determining which set of detectors contains the imbalance. The more detectors that we implement, the smaller the location error. This shows that the ability to locate the leak improves as we add instrumentation.

We can also make our two systems work together. It may be expensive to hire police to drive up and down the frontage road all day long. However, we need the RFID detectors to operate the tollway because they are part of our billing system. Therefore, it may make sense to use the internal vehicle balance system to initially detect the loss and alert a dispatcher or other responsible party who can then call the police (our external system) to confirm, locate, and contain it. This illustrates another pair of important principles. The first is that multiple LDSs may provide superior performance compared to a single system alone by allowing the advantages of the disparate system types to complement each other. The second is that full implementation of the leak detection capability is likely to require additional communication, analysis, and decision-making functions that go beyond simply detecting the leak and providing an alarm.

We should also keep in mind that all leak detection is inference. In all cases, we are trying to infer the existence of a leak by sifting through and analyzing the data and information that provides evidence regarding whether a leak is probably present. Because the evidence may be only partially supportive, or because of the random nature of the leak incident itself, this effectively means that all LDSs are probabilistic in nature. This implies that statistical and pattern recognition techniques can prove to be very useful in these kinds of systems.

This is even true of our external vehicle detection system. Although it is tempting to think of the detections coming from our moving patrol cars as being strictly deterministic (ie, the patrol officer either sees the escaping vehicle or does not), the binomial nature of all the patrol cars working in unison in combination with the random nature of the leak location ensures that the chances of detection will be probabilistic in time.

Even the implied deterministic/binary nature of our prime detector—the officer in the police car—should be called into question. At any time, the officer may be looking ahead or into the rear view mirror, or elsewhere. If a vehicle crosses the grassy border onto the frontage road while the patrol officer is looking the other way, then how is the officer to determine whether a new vehicle way down the frontage road entered from the toll road or from a side street?

This problem also occurs in pipeline "direct detection" or external systems that identify the presence of components such as hydrocarbon or other commodity detectors. Because most of these sensors produce analog outputs, it may be difficult to determine whether or not a particular output value is significant when compared to the fluctuating output of the sensor when operating under normal conditions.

It is important to remember that LDSs are *categorization* systems. At some fundamental level, they will always effectively reduce to thresholding systems that rely on the ability to issue alarms when some combination of analog or real numbered inputs exceeds one or more critical thresholds. Note that multiple thresholds may be embedded at various levels in some data processing hierarchy. For example, thresholds may be set at a low level in various detectors in the LDS so that each of the individual detectors emit an alarm or some binary on/off signal. In turn, the LDS may issue a leak alarm only after the total number of low-level binary activations exceeds a higher-level threshold.

Given the fact that pipeline systems have a wide range of topological and design heterogeneity, it is implied that a significant level of sophistication and effort may be required to configure, set thresholds, and otherwise *tune* all of the expected detection, location, display, and other functions for any new LDS. These issues can raise challenges for implementation, operation, and maintenance of leak detection systems.

2.4 LEAK DETECTION AND THEFT

As noted previously, an important principle that has proven to be very useful when tuning or setting thresholds or other key LDS parameters is that normal leaks are random in time and space, are relatively rare, and persist.

With only a few exceptions, they last until they are detected, at which point the line is shut down and either drained or repaired. This persistence can be a very useful parameter in terms of determining LDS alarm thresholds.

However, in recent years, pipeline LDSs have been increasingly used to detect commodity theft. Theft of commodity is not normally much of an issue in the developed world. It is, however, a significant challenge in large areas of the developing world. An LDS can be a useful adjunct in terms of detecting and minimizing the negative impact of pipeline commodity theft. It is important to recognize that there is an important difference between a normal leak and theft of commodity during normal operation. In the first case, the loss of commodity is a rare incident relatively random in time and space, which starts and persists.

However, in the second case, the loss is aided and abetted by one or more intelligent agents who do not want to get caught. In a sense, it is as if the commodity wants to escape! Consequently, the pattern of loss may be much different for theft-related commodity loss events. These losses may span the spectrum of being relatively uncoordinated and easily detected to very sophisticated and difficult to detect. In the first case, we might expect frequent attacks at multiple locations grouped at inconvenient but predictable times of the day. More sophisticated attacks might draw a specific amount of commodity for a predetermined time that is designed to be less than a conventional LDS's threshold and persistence requirements. Consequently, the LDS pattern recognition approach, or tuning, for a theft detection system may need to be modified from the methodology chosen for conventional leak detection. This brings us to our final principle: pipeline LDSs are highly situational and must be appropriately tailored to the pipeline configuration, operational parameters, and problem at hand.

2.5 FUNCTIONAL REQUIREMENTS

So, what is the LDS expected to do? The previous sections should have suggested that there may be a fairly high level of complexity for the demands that may be placed on the systems. Therefore, we suggest a set of user-oriented functions that must be provided by the installed LDS. The most complex of these demands, is, of course, the following:

- The LDS should detect valid leaks and not create false positives.

Additional *secondary* user-oriented requirements may be imposed by the pipeline operator to make the LDS more useful. These *may* include (and are not limited to) the ability to:

- estimate the leak rate;
- estimate the location of the leak;

- provide additional data in the form the tables, trends, or other charts to assist the leak detection analysts in diagnosing alternate causes of the leak alarm;
- offer alternate causal hypotheses that would apply if the alarm was not caused by a leak to focus the leak detection analysts in terms of diagnosing the leak;
- calculate a rupture or leak probability or some other urgency-oriented metric;
- present the user with a set of leak alarm response alternatives.

Finally, modeling or other internal components of the LDS may provide a set of tertiary functions that can prove useful in the normal operation of the pipeline. These may include:

- Calculation of pressures, temperatures, and flows at locations in the pipeline where no measurements are present
- Presentation of such calculated variables in the form of tabular, trend chart, profile chart, or other useful formats
- Tracking of pipeline pigs and commodity batches
- Self-testing or performance testing of the LDS based on recorded supervisory control and data acquisition (SCADA) data sets
- Incident analysis or other modeling performed offline based on recorded data or user-created data sets
- Many other functions not listed here

In general, the system must rest on a set of *LDS functional foundations*. In some way, every LDS must:

- obtain field data via one or more real-time critical field measurements;
- communicate the data to one or more locations where the data can be appropriately processed;
- process the data in some way to produce pipeline rupture or leak alerts or alarms to downstream systems.

It should be clear that this automatically implies the existence of responding systems or personnel taking the appropriate steps to control, contain, and clean the breach of integrity conditions. Without such responders, the LDS provides minimal benefits to the operators.

2.6 THE FUNDAMENTAL PRINCIPLES SUMMARIZED

In Sections 2.2, 2.3, 2.4 and 2.5, we identified a number of important pipeline leak detection principles. Table 2.1 provides a summary of the principles we have discussed so far in a somewhat reorganized form, along with corresponding corollaries, implications, and issues that may go along with those principles.

TABLE 2.1 Fundamental Pipeline Leak Detection Principles

Principle	Corollaries and Issues
1. The value of a leak detection system is generally proportional to the amount of the pipeline that is being monitored by the LDS.	• More instruments at more locations generally means better leak detection • Special consideration or additional weighting should be given to locations along the pipeline that have higher consequences in the event of a leak
2. Leak detection systems are multifunctional in nature	• At a high level, the systems must detect leaks • Secondary functions may include locating the leak; estimating the leak size; estimating the leak start time; providing supporting or diagnostic data in the form of trends, tables, and profiles, including an indication of the reliability or importance of the alarm; and providing alternate causal hypotheses for alarms • Tertiary functions that are useful in normal pipeline operation may be enabled by the modeling or other subsystems of the LDS • To achieve these functions, the LDS must obtain field data, communicate the data to a place where it can be processed, process the data in some way to detect the leak, and issue alerts or alarms to other systems that or human agents who can take action to control and contain the leak
3. Leak detection systems tend to be highly situational	• Selection of the leak detection system approach should take into consideration the commodity, pipeline design, pipeline operating characteristics, local damage issues, regulatory requirements, and other factors
4. All leak detection is inference	• All inference requires evidence • The incoming evidence signals are always noisy or unreliable to some degree, and the leak detection inference process is therefore inherently probabilistic in nature • The detectable leak size is limited by the contribution of all errors • Errors can result from measurement, communication, data processing, physical modeling, and other sources • The detection probability generally increases over time • Large leaks are detected faster than small leaks • Statistical and pattern recognition techniques are critical to development of appropriate thresholds for a pipeline LDS

(*Continued*)

TABLE 2.1 (Continued)

Principle	Corollaries and Issues
5. Commodity loss signals can vary in nature depending on the cause of the leak	• Normal leak signals tend to persist until the leak is detected and brought under control • However, they can temporarily disappear if the line is shut down or becomes slack • Signals associated with commodity theft may not persist or may exhibit the other characteristics not associated with normal leaks
6. There are many different methods and approaches that can be used to detect pipeline leaks and ruptures.	• Different approaches have different strengths and weaknesses • Depending on the fundamental LDS approach, it may be possible to use simulation methods to improve the leak detection performance • Using more than one independent LDS approach can increase costs but can also provide benefits by exploiting the differing strengths of those approaches; note that in some jurisdictions, a multipronged approach is mandatory
7. It is difficult to efficiently and effectively deploy and tune a pipeline leak detection system if you do not have some understanding or measurement of its performance	• LDS performance is often expressed using a multi-dimensional set of parameters, including probability of detection, time to detect, etc. • Critical metrics that should be considered include maps that address detectable leak size, probability of detection over time as well as false-positive rates, as appropriate to the system; other metrics may also be important • Excessive false-positive rates either reduce the value of the LDS or add unnecessary support costs for the system, or both • Excessively high thresholds imply that you will miss leaks that you really should be able to detect

2.7 ARCHITECTURAL FOUNDATIONS

All LDS design and architecture inevitably rests on the functional foundations discussed above. The system must gather evidence from one or more pipeline sites that will be affected by the leak. It must bring all of the required evidence to some location where it can be appropriately processed. That processing will include an analysis of the data to see if there is any reason to infer the presence of a leak. If there is, then it must notify some individual or automation component capable of taking action to isolate the leak and initiate some level of spill response.

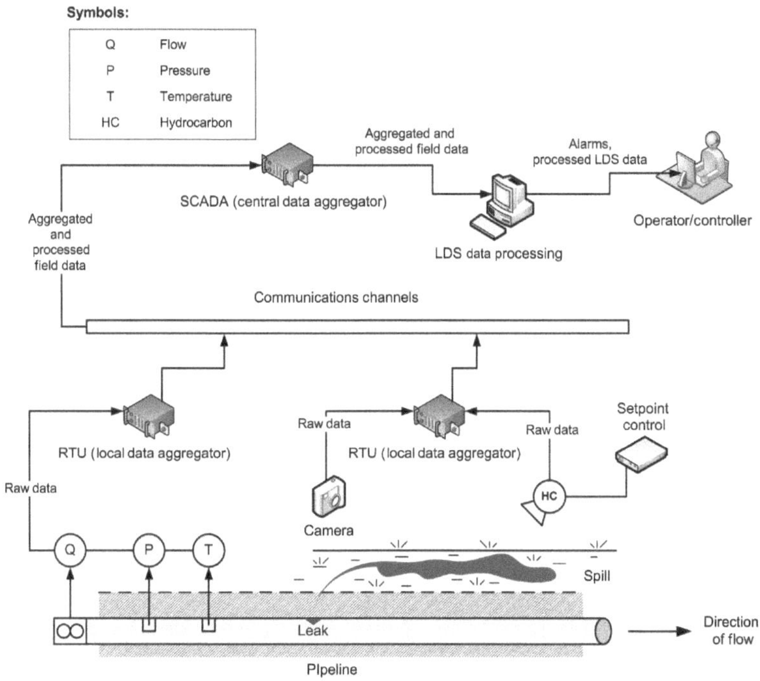

FIGURE 2.2 Leak detection system architecture.

Refer to Fig. 2.2. From an architectural view point, we can think of the system as consisting of the following essential components:

- *Instruments*: Measurements providing evidence of the rupture, leak, or spill. As we have previously inferred, there really is no such a thing as a fundamental "leak," "rupture," or "spill" detector. LDSs infer the presence of leaks based on more fundamental measurements of local variables that are obtained by field or site instrumentation such as "pressure" or "temperature" or "hydrocarbon gas IR signal" or "flow rate."
- *Communications Channels*: It is highly unlikely that the leak will occur or be detected at the point of pipeline control. Consequently, some means must be provided to move the collected field evidence to other locations where processing of the data can occur or control actions can be implemented, or both. Communications channels can be fast or slow, can be electronic in nature, or consist of methods as primitive as the movement of documents from one desk to another.
- *Data Processing Elements*: Field measurements that correspond to a leak must be turned into a rupture, leak, or theft alarm. This processing may occur in a centralized location, but it may also occur in a distributed

fashion, with some processing being done at field locations (turning an analog signal to a local status or alarm through setpoint control is an example of this), and other more sophisticated processing, such as data conditioning, modeling, statistical analysis, and rupture, leak, or theft alarm generation being done at the pipeline control center. This is often the portion of the system that most people think of when they think of pipeline LDSs. In fact, commercial LDSs are often marketed this way. This approach is fundamentally flawed. The LDS, along with any planning, design, and analysis revolving around it correctly, must include the other critical elements discussed here.

- *Human Responders, Control Automation, Policies, and Procedures*: It is common to neglect the elements of control automation and human control response when implementing a pipeline LDS. In fact, very few pipeline LDSs implement automated control (ie, automatic isolation and pipeline shutdown) in response to a leak alarm. This is primarily because the operators do not trust the LDS. Many leak detection installations are unreliable, principally because they issue too many false alarms. The reasons for this are varied, but the consequence is that the failures in the data processing portion of the LDS are compensated for by pipeline controllers or other pipeline operations support staff. Again, the prudent pipeline operator will take steps to properly integrate the actions and responses of the controllers and operations support staff when responding to leak detection technology alarms.

As noted, many commercially provided pipeline LDSs provide only one or a portion of the components discussed so far. Vendor-supplied systems may provide only the LDS data processing element (especially true when we are thinking in terms of mass balance systems), but this does not necessarily need to be the case. Instrumentation vendors who focus on the instrumentation function may emphasize only a unique leak detection–oriented data collection component while still emphasizing the leak detection role of their products. This might, for instance, be the case for a vendor that provides flow measurements or internal mobile leak detection devices, such as leak detection pigs or smart balls.

There may be other auxiliary components of the LDS that may not apply to all LDS approaches. The principle component of this type is the data aggregator/concentrator.

- *Data Aggregator/Concentrators*: These are typically multifunction components that are used to acquire data from multiple sensors and control the pipeline system. These typically include remote terminal units and the SCADA system. Data aggregators/concentrators are typically required by large, stationary LDSs, such as internal mass balance systems and external distributed cable-based or discrete commodity detector systems.

It is not uncommon for many of the functions of the LDS to effectively piggyback on other multi-use (and expensive!) functions or components of the pipeline system, such as the backbone communication system, or the SCADA system. As a result, virtually no commercial vendors supply all of the required components for a pipeline LDS in the sense that we use the term in this book. The approach in this book is to emphasize the functionality of the LDS at a system level. This is done for a very good reason: because the typical vendor provides only a portion of the LDS, the performance of the system is impossible to understand in isolation. Virtually all regulation recognizes that it is the pipeline operator who is responsible for the successful implementation of the LDS. Especially given the potentially piggybacked nature of an LDS system installation, only the operator can know and understand the full functionality of the system and its components. In particular, it is up to the operator to recognize that changes to multiuse subsystems such as the communications backbones, field sensors, and SCADA system have serious potential to change and degrade the performance of the LDS.

This is a good place to emphasize that although Fig. 2.2 has a typical SCADA/electronic data acquisition look and feel to it, the reader should focus at this stage on the implied functionality of each LDS component and not necessarily on how it is implemented. For example, every LDS requires sensors. Those sensors could be in the form of remotely monitored pressure, temperature, and flow measurements. Alternately, the human eyeball is also a time-proven method of detecting spilled commodity. All leak detection systems require a communications component, but this communications component does not need to be implemented in the form of a backbone fiber optic or electronic communications system utilizing a standard data acquisition method such as Modbus. It could be in the form of email or even the physical transport (in someone's shirt pocket, for example) of an electronic flash memory thumb drive of key data files. Similarly, every pipeline LDS must have a leak detection data-processing component to it. That component might be in the form of automation process hosted on a PC, but it could also consist of a data analysis process performed by an operating company employee.

All of these methods have their advantages and disadvantages. All, under the right circumstances, can function very well. In the final analysis, it is up to the operator to take a principled systems approach and ensure that all of the required LDS functions are addressed in an effective manner.

2.8 A TAXONOMY OF PIPELINE LEAK DETECTION SYSTEMS

There are a large number of approaches that have been used by pipeline operators to detect ruptures, leaks, and spills. In any complex field, whether it is biology, music, or leak detection systems, a classification approach can be useful as a start in understanding the organization and relations between the types of objects addressed by the field. It is important to realize that there are a number of approaches that can be taken to classify leak detection approaches. In addition, other parties working in the pipeline leak detection field may have taken a somewhat different path than the one taken here.

Such a taxonomic approach is shown in Fig. 2.3. We begin by noting that leaks can be detected in two possible ways. One possibility is that the integrity breach might be detected incidentally, such as by a field operative going about his or her regular duties, by a member of the public, or by an pipeline controller. The pipeline operator should not underestimate the importance of incidental leak detection because a large number of leaks and resulting spills are detected in exactly this way.

The other possibility is that they can be detected by design, such as by pipeline LDS technology or procedural approaches. In the context of this book, these are all leak detection systems. Therefore, we are not limited to dedicated technology solutions; we also include purposeful and procedural detection by operating company personnel or its agents. In a manner similar to the way we approached our tollway free-rider detection system, we further subdivide leak detection system categories into external and internal sensor-based systems.

It is important for the reader to understand the difference between external and internal systems. The external route to detecting a leak is one method that most people immediately think of because it feels like a very direct approach: catch the leaked commodity that has escaped from the pipe using some observer or instrument conveniently positioned along the pipeline. An externally based system utilizes field instrumentation or components that are physically located outside of the pipe and that are designed to measure variations in conditions at locations that are external to the pipe. We might be measuring air or soil temperatures, looking for the presence of hydrocarbons, water, natural gas, or some other trace of the commodity, or watching for an infrared radiation signal. We could be looking for acoustic signatures or some very complex signal, such as a camera or video image. The important issue is that the origin of the signal is outside of the pipe, which we can also take to mean that we are looking for commodity that has left the pipeline environment.

Internal systems, however, measure process values for pressures, temperatures, flows, or other variables that correspond to the state of the pipeline commodity inside the pipe.

We divide external systems into two sub-categories: leak detection systems based on stationary sensors and those that are based on mobile sensors. External systems are discussed in detail in Chapter 7, External and Intermittent Leak Detection System Types.

As noted previously, internal systems rely on sensors designed to measure pipeline process variables. Internal sensor-based technology is used widely throughout the pipeline industry. Several chapters of this book are dedicated to these systems, including Chapter 3, Mass Balance Leak Detection, Chapter 4, Real-Time Transient Model–Based Leak Detection, Chapter 5, Statistical

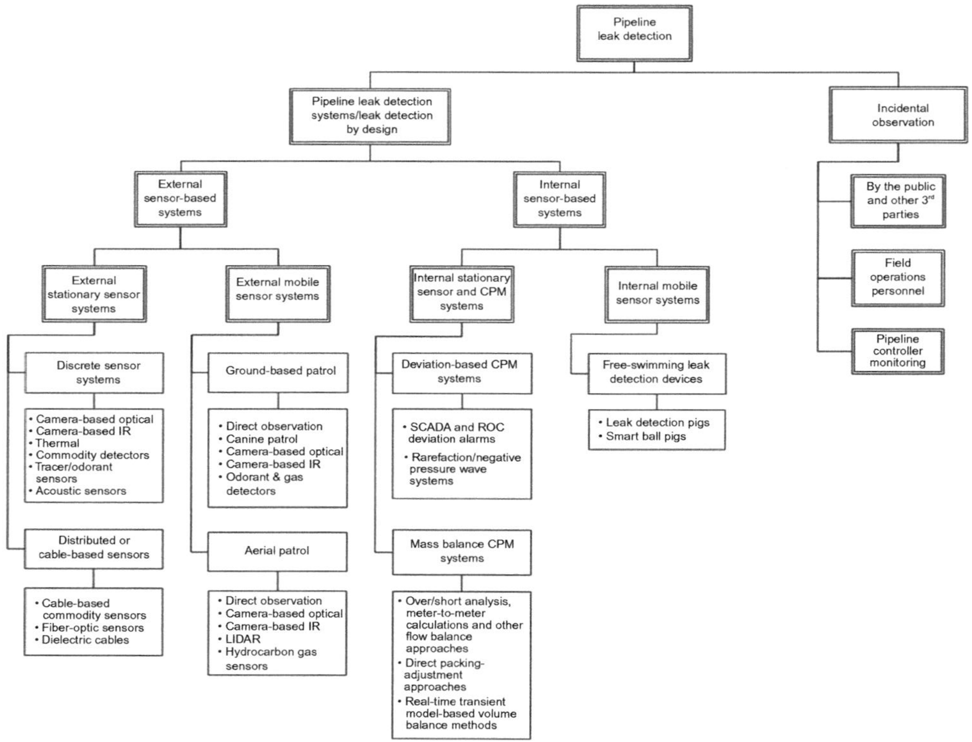

FIGURE 2.3 Pipeline leak detection system taxonomy.

Processing and Leak Detection, and Chapter 6, Rarefaction Wave and Deviation Alarm Systems.

In summary, pipeline leak detection systems come in many different shapes, sizes, and complexities. Each system type has associated positive and negative attributes. The participation of operations personnel and other people in observation and data or alarm processing tasks may represent critical LDS functions. And finally, the determination of an LDS approach that is best for any particular pipeline is situational and must be addressed via a thorough analysis of the full pipeline system, company internal requirements, and any applicable regulatory requirements.

Chapter 3

Mass Balance Leak Detection

This chapter discusses leak detection by mass balance. We lay the formal foundation for this leak detection approach, discuss the validity of using conservation of standard volume as a proxy for conservation of mass, describe the impact of measurement and other uncertainties, and describe various types of mass balance leak detection systems that are differentiated by approximations and approaches used to compute the changes in mass of the pipeline system. This chapter also lays the groundwork for the examination of real-time transient model (RTTM)-based systems in Chapter 4, Real-Time Transient Model–Based Leak Detection.

3.1 LEAKS AND CONSERVATION OF MASS

The American Petroleum Institute developed the term *Computational Pipeline Monitoring* (CPM) to refer to software-based algorithmic modeling tools that enhance the ability of a pipeline controller to recognize anomalies such as leaks on a pipeline [1]. The Canadian petroleum industry labels a CPM system a *Computerized Leak Detection System* (CLDS). Mass balance leak detection systems are perhaps the most common CPM/CLDS system, having been used in one form or another since the advent of the use of computers for the monitoring and control of pipelines in the 1970s.

Conservation of mass is a fundamental physical law and forms the basis for many internal leak detection systems. Simply put, one expects that if one can measure all of the fluid entering the pipeline and subtract from that all of the fluid leaving the pipeline, then the difference between the two will be exactly equal to the change in the amount of fluid in the pipeline. In the absence of a leak, we expect that the pipeline system mass will be perfectly conserved.

A leak is flow out of the pipeline system that is neither measured nor expected. From the perspective of our perfectly constructed system, a leak causes an apparent violation of conservation of mass.

The consequence is that, in our ideally constructed system, the conservation of mass provides a perfect leak detector. If we can perfectly calculate the mass balance of the system assuming no leak flow, it

Pipeline Leak Detection Handbook. DOI: http://dx.doi.org/10.1016/B978-0-12-802240-5.00003-0

would be exactly zero as long as there is actually no leak present. However, if a leak is present, then the calculated mass (im)balance will be precisely equal to the size of the leak.

The conservation of mass principle is applicable to the full pipeline or any portion of the pipeline system. The only requirement is that we must be able to perfectly determine the amount of mass entering and leaving that portion of the pipeline through the expected supplies and deliveries and perfectly calculate the rate of change of mass within that portion of the pipeline system.

So, given that physics has provided us with an apparently perfect means of detecting a leak, the only challenge becomes calculating the mass balance of the pipeline system. To quote Shakespeare's Hamlet, "Aye, there's the rub." This chapter is therefore devoted to elucidating the practical issues associated with computing the balance of mass in a real pipeline system (or portion of the system) and the various techniques that have been formulated to systematize the process.

Before delving into the complexities associated with computing the mass balance of the pipeline, we should first recognize that on many pipelines, mass flow is neither measured nor calculated. Instead, standard volumes (volume at specified conditions such as atmospheric pressure and 60°F) are measured and calculated. In oil pipelines, flow is usually measured in barrels per hour or day at standard conditions and, similarly, the amount of fluid in the pipeline is expressed in standard barrels. Gas pipeline flow is often expressed in volumetric units such as standard cubic feet per hour and the amount of fluid in the pipeline is expressed as standard cubic feet of gas.

It is common to substitute, without formal justification, a principle of conservation of standard volume as a proxy for conservation of mass and assume that they are the same. We will examine the validity of this substitution. It turns out that for most practical purposes, we can make this substitution but it is useful to understand the extent to which this common assumption is valid and the basis of its validity. We examine first the leak detection problem in terms of mass conservation, and then return to the topic of utilizing standard volumes for leak detection.

3.2 PIPELINE MASS BALANCE SECTION

We start by demonstrating that to perform leak detection using mass balance, we need some way to compute the rate of change of mass over a portion of the pipeline and also a way to compute all of the flow rates into and out of that pipeline section.

To facilitate this discussion, we define a mass balance section (MBS) as any portion of the pipeline for which we will monitor for leaks using a mass balance system. Ideally, an MBS will be bounded by flow meters and/or flow meter proxies (such as tank volumes) that provide a means of calculating all mass flow rates into and out of the MBS. The entire pipeline may be a single MBS or, if there are sufficient flow rate measurements, it may be subdivided into a number of MBSs.

Depending on the boundaries of the MBS, other methods might be used to obtain the mass flow rates into or out of the control volume. For example, in an RTTM-based system (see Chapter 4: Real-Time Transient Model–Based Leak Detection), the RTTM computed mass flow rates at pipe ends bounding the MDS might be used as proxies for flow meters. As might be the case in some natural gas pipelines, some delivery flow rates might be estimated from time of day, solar insolation, and temperature.

A good flow meter is far better than any flow meter proxy, but we are getting ahead of ourselves.

3.3 LEAK DETECTION BY MASS BALANCE: FOUNDATIONAL PRINCIPLES

We turn now to formalizing our statement of leak detection by mass balance and define some key terms for our discussion. Consider some section of the pipeline as a control volume as illustrated in Fig. 3.1.

For convenience, we call this control volume a pipeline MBS. This control volume may span many miles, but mass flows into the pipeline at only N discrete locations and out of the pipeline at M locations. The mass flow rates into the control volume are $\dot{m}_{\text{in},1}\ldots\dot{m}_{\text{in},N}$ and the mass flow rates out of the control volume are $\dot{m}_{\text{out},1}\ldots\dot{m}_{\text{out},M}$. The rate of change of mass inside the control volume is $\dot{m}_{\text{MBS}}$. Conservation of mass requires that the rate of change of mass of the control volume must be equal to the sum of the mass flow rates in minus the mass flow rates out (Eq. 3.1).

$$\sum_{i=1}^{N} \dot{m}_{\text{in},i} - \sum_{j=1}^{M} \dot{m}_{\text{out},j} = \dot{m}_{\text{MBS}}$$

EQUATION 3.1 Conservation of Mass of Mass Balance Section

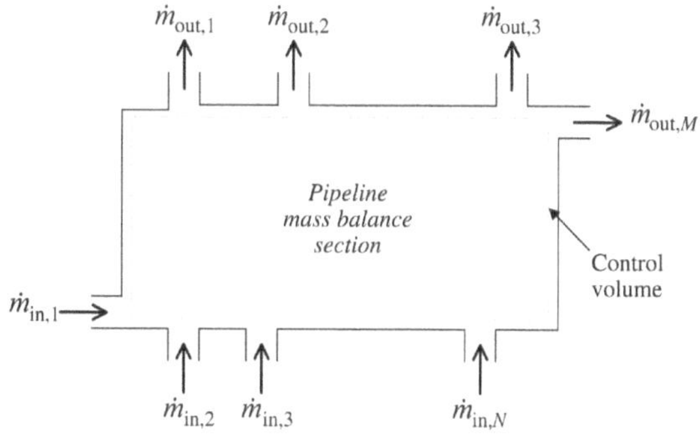

FIGURE 3.1 Control volume representation of a section of pipeline.

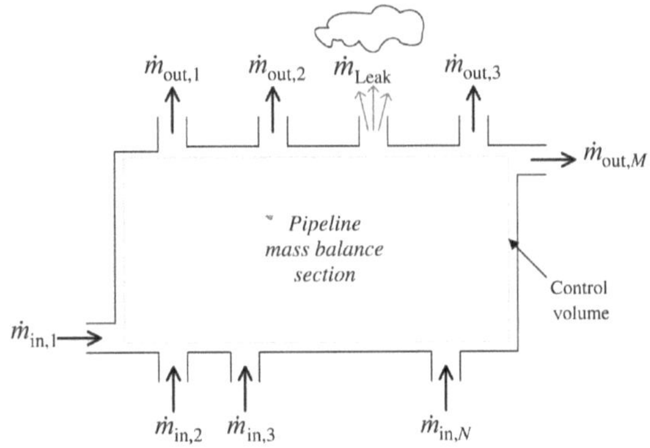

FIGURE 3.2 Control volume representation of a leaking pipeline.

Now, consider the same control volume with a single leak of mass flow rate $\dot{m}_{\text{Leak}}$ illustrated in Fig. 3.2. From the perspective of conservation of mass, the leak is simply another mass flow rate out of the pipeline, except for the following:

1. The timing of the leak is unknown
2. The location of the leak is unknown (unless seen and reported)
3. The mass flow rate of the leak is unknown

Now, conservation of mass requires the following:

$$\sum_{i=1}^{N} \dot{m}_{\text{in},i} - \sum_{j=1}^{M} \dot{m}_{\text{out},j} - \dot{m}_{\text{leak}} = \dot{m}_{\text{MBS}}$$

EQUATION 3.2 Conservation of Mass of Mass Balance Section with Leak

Rearranging terms, we state our fundamental equation of leak detection by mass balance.

$$\dot{m}_{\text{Leak}} = \sum_{i=1}^{N} \dot{m}_{\text{in},i} - \sum_{j=1}^{M} \dot{m}_{\text{out},j} - \dot{m}_{\text{MBS}}$$

EQUATION 3.3 Fundamental Equation of Leak Detection by Mass Balance

It is common to refer to the last term of Eq. (3.3) as the mass packing rate, MPR_{MBS}. The mass packing rate is a critical term and, for now, we move forward with the assumption that we have some means to calculate it. We return to the packing rate later.

It is convenient to define a mass flow balance, MFB_{MBS}, as the difference between the known inlet and outlet mass flow rates as shown in Eq. (3.4):

$$MFB_{\mathrm{MBS}} = \sum_{i=1}^{N} \dot{m}_{\mathrm{in},i} - \sum_{j=1}^{M} \dot{m}_{\mathrm{out},j}$$

EQUATION 3.4 Mass Flow Balance Definition

Finally, we define the observable mass balance as:

$$MB_{\mathrm{MBS,Observable}} = MFB_{\mathrm{MBS}} - MPR_{\mathrm{MBS}}$$

EQUATION 3.5 Observable Mass Balance Definition

The *observable mass balance* is usually referred to simply as *mass balance*. We have chosen to call it *observable* because it is defined in terms of the known (observable) mass flow rates into and out of the pipeline and ignores the leak flow rate, which is unknown in timing, location, and size. In contrast, the *true mass balance* would include the leak flow rate.

Depending on the presence or absence of a leak, the observable mass balance (under ideal conditions) is given by Eq. (3.6):

$$MB_{\mathrm{MBS,Observable}} = \begin{cases} 0 & \text{No leak} \\ \dot{m}_{\mathrm{Leak}} & \text{Leak} \end{cases}$$

EQUATION 3.6 Observable Mass Balance: No Leak and Leak Cases

When a leak occurs, the pipeline will typically start unpacking ($MPR_{\mathrm{MBS}} < 0$) and the flows will rebalance so that more measured fluid is entering the pipeline than leaving it ($MFB_{\mathrm{MBS}} > 0$). The combination of these will be positive and, assuming that we can measure/compute both mass packing rate and mass flow balance precisely, the positive imbalance will exactly equal the leak.

Of course, in the real world, every term that goes into the computation of the observable mass balance has uncertainty associated with it. We defer discussion of these issues to Section 3.5.

3.4 VOLUME BALANCE AT STANDARD CONDITIONS AS A PROXY FOR MASS BALANCE

In Section 3.3, we developed the foundational principles for leak detection by mass balance. However, in most US pipelines, and in many throughout the world, flows are measured in volumetric units per unit of time at "standard" conditions and the amount of fluid in the pipeline is expressed in volumetric units at standard conditions. Standard temperature and pressure (STP) are often defined as follows:

In the United States:	60°F and 14.696 psia (1 atmosphere)
In Europe:	15°C and 100 kPa (0.987 atmosphere)

Some European countries use "normal" conditions, which are typically defined as 1 atmosphere (101.325 kPa) and 0°C.

In the United States, a common unit for oil flow rate is net barrels per hour, whereas in Europe it may be measured in metric tons per hour. Common gas flow units are standard cubic feet per hour (SCFH) in the United States or standard cubic meters per hour (Sm^3/h) in Europe.

When pipeline flow and inventory are measured and computed in volumetric rather than mass units, we need an alternative formulation for leak detection by mass balance. This alternative formulation, leak detection by standard volume balance, is presented in Section 3.5.

We must first examine to what extent standard volume balance is a useful proxy for mass balance.

3.4.1 Conservation of Standard Volume Is Not a Physical Principle

Unlike conservation of mass, conservation of standard volume is *not* a physical principle. Conservation of mass does *not* imply conservation of standard volume. This statement may come as a surprise to many readers because it has been common practice to substitute one for the other. Certainly, any mass flow rate can be converted to a volumetric flow rate at STP by Eq. (3.7).

$$F_i = \frac{\dot{m}_i}{\rho_{\mathrm{STD},i}}$$

EQUATION 3.7 Conversion From Mass Flow Rate to Volumetric Flow Rate at STP

where $\rho_{\mathrm{STD},i}$ is the mass density at standard conditions. However, unless the fluid in the pipeline is uniform throughout (in which case $\rho_{\mathrm{STD},i} = \rho_{\mathrm{STD}}$), there is no way to convert Eq. (3.3) to volumetric units at standard conditions.

On the surface, it seems intuitive that conservation of mass implies conservation of standard volume. Certainly, if we measure an amount of fluid entering a pipeline at STP, then we expect that when that fluid leaves the pipeline, it will still have the same volume at STP as it had when it entered the pipeline. If there is no mixing of fluids within a pipeline, then this is indeed true because one can readily convert from standard volume to mass

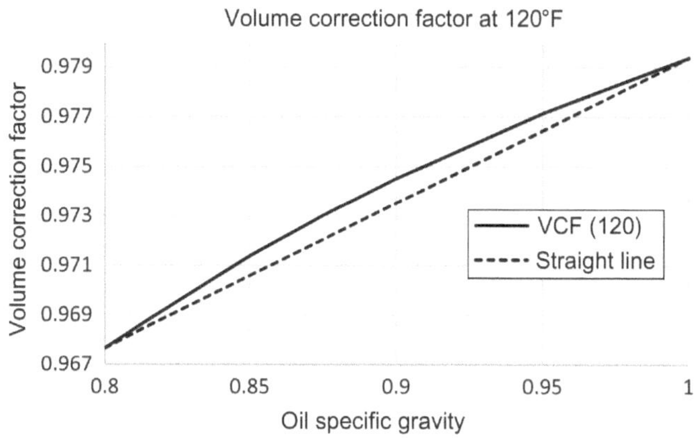

FIGURE 3.3 Volume correction factor of oil at 120°F as a function of specific gravity.

using Eq. (3.7). However, when mixing fluids with differing properties, as is often the case in pipelines, there is no longer a precise equivalence between standard volume and mass. Examples of mixing include:

- Injection of gas with a different composition into a stream of gas flowing through the pipeline
- Intermingling of oil in tanks at intermediate locations along a pipeline
- Mixing of inlet streams from different oil fields into a common pipeline stream
- Refineries along a pipeline extracting lighter end components and reinjecting heavier oil into the pipeline stream

Using oil pipelines as an example, the assumption generally made is that if we add one net (ie, measured at STP) barrel of oil to another net barrel of oil, then we get two net barrels of oil. While close to being correct, this is not true if one barrel of oil has different properties than the other. We illustrate this by examining how the volume correction factor (VCF) changes with specific gravity of the oil.

When working with oil, the VCF expresses the conversion of oil volume from the oil temperature to standard temperature (60°F in the United States). The American Petroleum Institute and others have developed an accurate calculation procedure for VCF as a function of oil specific gravity. Fig. 3.3 shows the VCF for oil at 120° Fahrenheit as a function of specific gravity from 0.8 to 1.0 (47.375 to 10° API). The dotted line is a straight-line interpolation between the endpoints. Note that VCF is not a linear function of specific gravity.

Using the two extremes of this curve, if we were to take one barrel of 0.8 specific gravity oil at 60°F and another barrel of oil of 1.00 specific gravity at 60°F and heat them both to 120°F, then we would have 2.0545 barrels of oil at 120°F ($1/0.96766 + 1/0.97937$). If we mix these same two barrels of oil, then the resulting specific gravity would be 0.875. At 120°F, the mixed barrels would have a volume of 2.0523 (2/0.97302). The mixed barrels occupy 0.1% less volume than the unmixed barrels. This is an extreme case of mixing of a very light oil with a very heavy oil, and it is reassuring that the shrinkage on mixing is so small.

We use this example to demonstrate that fluids do not necessarily conserve volume when mixed. When we inject two standard (net) barrels of oil, we would prefer that these barrels, on exiting the pipeline, be two standard barrels of oil, whether or not mixing has occurred within the pipeline. To the extent that this is violated conservation of standard volume is also violated.

Natural gas, fortunately, is nearly ideal at standard conditions (the compressibility factor is typically between 0.997 and 0.998 for natural gas at standard conditions). Therefore, any errors inherent in restating conservation of mass as conservation of standard volume are quite small, certainly less than 0.1% of the flow rate.

The issue of expansion or shrinkage of two volumes has been expanded upon in the literature. "In the blending of petroleum components with different physical properties, excess volumes occur because the components do not form ideal solutions. In ideal solutions, the total volume is equal to the sum of the volumes of the components. For a solution to approach ideality, the molecules of the materials blended together must be similar in size, shape, and properties. If the nature of the components differs appreciably, then deviation from ideal behavior may be expected. This deviation may be either positive or negative; that is, the total volume may increase or decrease when the components are blended [1]".

In conclusion, it is not correct to assume that standard volume is conserved in the same way that mass is conserved. However, we expect that the errors in this assumption are generally small, typically less than 0.1% of the volumes. Therefore, conservation of standard volume is a close approximation to the principle of conservation of mass, at least for most pipelines.

Based on several case studies (not presented here) of fluids ranging from crude oils to natural gas to mixtures of propane and butane, we conclude that the conservation of standard volume is a useful approximation to the law of conservation of mass for the purposes of pipeline leak detection. However, this approximation will result in mass balance uncertainties that will influence the detection of small leaks. It appears that these uncertainties are generally somewhat less than 0.1% of flow rate but consideration of the impacts should be evaluated on a case-by-case basis.

3.4.2 Formulation of Mass Balance Leak Detection in Terms of Volume at STP

With the cautions of Section 3.4.1 in mind, we proceed now to formulate leak detection by conservation of standard volume as a proxy for leak detection by conservation of mass. We use the terms *standard* volume and *net* volume interchangeably to refer to the volume of the pipeline fluid

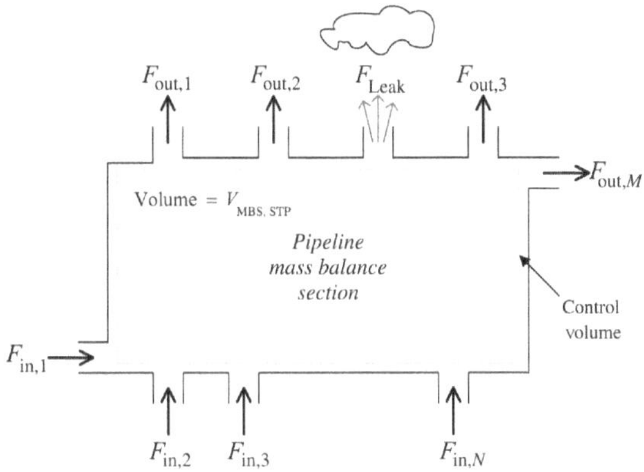

FIGURE 3.4 Representation of a leaking section of pipeline: volumetric units.

(gas or liquid) at STP (standard temperature and pressure). Fig. 3.4 is a representation of Fig. 3.2 with all mass flow rates changed to volumetric flow rates F_i at STP. Instead of the mass of the LDS, we work with volume of the MBS at STP, $V_{MBS,STP}$.

When working in standard volumetric units instead of mass units, we substitute Eq. (3.8) for Eq. (3.3).

$$F_{Leak} = \sum_{i=1}^{N} F_{in,i} - \sum_{j=1}^{M} F_{out,j} - \dot{V}_{MBS,STP}$$

EQUATION 3.8 Fundamental Equation of Leak Detection by Standard Volume Balance

where each of the terms is a rate of change of standard volume per unit of time (eg, SCFH or BPH).

When working with standard volumes instead of mass, we substitute the following terms:

- Packing rate for mass packing rate
- Flow balance for mass flow balance
- Observable volume balance for observable mass balance

Rather than continuing to specify "at standard conditions," we assume that flow rates, flow balances, packing rates, and volume balances in the remainder of this chapter and Chapter 4, Real-Time Transient Model–Based Leak Detection are expressed as rate of change of volumes at standard conditions. We state our definitions of packing rate, flow balance, and observable volume balance here:

$$PR_{MBS} = \dot{V}_{MBS,STP}$$

EQUATION 3.9 Packing Rate Definition

$$FB_{\mathrm{MBS}} = \sum_{i=1}^{N} F_{\mathrm{in},i} - \sum_{j=1}^{M} F_{\mathrm{out},j}$$

EQUATION 3.10 Flow Balance Definition

$$VB_{\mathrm{MBS}} = FB_{\mathrm{MBS}} - PR_{\mathrm{MBS}}$$

EQUATION 3.11 Observable Volume Balance Definition

In a manner analogous to the formulation in mass units, in the absence of a leak, the expected value of the observable volume balance is zero. And, in the event of a leak, we expect the imbalance to be equal to the leak size (see Equation 3.12).

$$VB_{\mathrm{MBS,Observable}} = \begin{cases} 0 & \text{No leak} \\ F_{\mathrm{Leak}} & \text{Leak} \end{cases}$$

EQUATION 3.12 Observable Volume Balance: No Leak and Leak Cases

where F_{Leak} is the leak flow rate.

3.5 IMPACT OF UNCERTAINTIES IN MASS/VOLUME BALANCES ON LEAK DETECTION

In Section 3.3, we demonstrated that, in an ideal world, the observable mass balance of an MBS would be exactly zero in the absence of a leak and exactly the leak size in the presence of a leak. Similarly, when working with volume balance units, we expect that the observable volume balance of an MBS would be zero in the absence of a leak and the leak size when a leak exists (see Section 3.4.2). The observable mass/volume balance is our leak signal.

We have demonstrated that leak detection by standard volume balance is usually a reasonable proxy for leak detection by mass balance. The choice between the two is often determined by the available measurements. When measured flow rates are in standard volumetric units, it is typical to substitute the volume balance formulation for the mass balance formulation. This section uses mass balance terminology rather than volume balance. However, the discussion applies to either formulation.

The observable mass balance is an imperfect leak signal for a number of reasons, including:

1. Each flow rate, $\dot{m}_{\mathrm{in},i}$ or $\dot{m}_{\mathrm{out},j}$, has an uncertainty that may be variable in time.
2. The mass packing rate calculation is necessarily imperfect because it relies on measurements such as pressures and temperatures along the pipeline, which also have uncertainties. It also relies on the accuracy of many physical properties of the pipeline, the equation of the state of the pipeline fluid, distribution of fluid composition within the pipeline,

temperature of the pipeline surroundings, heat transfer between the pipeline and surroundings, and other factors.

3. Pressure, temperature, and other measurements are only available at discrete points along the pipeline. Therefore, to compute the packing rate, some *model* of the pipeline state between measurement points is required. The model can be as simple as assuming that the pipeline volume is constant or as complex as that provided by an RTTM. Although the choice of the model will affect the accuracy of the line pack calculations, all models will inevitably have some error.
4. The leak may be some distance from the nearest pressure, temperature, and/or flow measurements. Until the effect of the leak propagates to the nearest measurement, the leak signal will not show any evidence of the leak.
5. The leak will distort the state of the pipeline in ways that cannot be easily determined from discretely spaced measurements.

As a result of these errors, we have uncertainty in the observable mass balance which we refer to as $U(MB_{\mathrm{MBS,Observable}})$. This sometimes is generically referred to as *noise*. Noise conveys the expression of randomness and statistical independence of the various error sources. However, there is no guarantee that all sources of uncertainty are either random or statistically independent of each other.

To accommodate this uncertainty (and to limit false alarms from the leak detection system), one typically imposes a leak detection threshold that is dependent, perhaps in a complicated fashion, on the uncertainty in the observable mass balance, which we represent as $Threshold(U(MB_{\mathrm{MBS,Observable}}))$. A practical statement of mass balance leak detection is Eq. (3.13).

$$\text{If } MB_{\mathrm{MBS,Observable}} > Threshold(U(MB_{\mathrm{MBS,Observable}}))$$
$$\text{then}$$
$$MB_{\mathrm{MBS,Observable}} = \text{Leak Size Estimate}$$

EQUATION 3.13 Practical Statement of Mass Balance Leak Detection

To be a bit more precise, we recognize that the observable mass balance has two fundamental components so we can instead state:

$$U(MB_{\mathrm{MBS,Observable}}) = U(MFB_{\mathrm{MBS}} - MPR_{\mathrm{MBS}})$$

EQUATION 3.14 Observable Mass Balance Uncertainty

Note that we resist the temptation here to assume that the uncertainties in the two terms are independent, which is an assumption that allows us to deal with them separately. Assuming independence is useful and, in fact, foundational to many efforts of quantifying leak detection uncertainties.

However, the problem of time-based correlation can invalidate many of the outcomes that would apply if independence were assumed. We discuss this issue in more detail in Chapter 5, Statistical Processing and Leak Detection.

We have noted that some sources of uncertainty are not randomly distributed in time. However, a portion of the observable mass balance uncertainty can usually be reduced by averaging over time. Therefore, it is common, if not the norm, for the leak signal to be averaged over a period of time to reduce the observed fluctuations and for more sensitive leak detection thresholds to be used for longer averaging time periods.

Processing the leak signal, in this case the observable mass balance, and differentiating a true leak from uncertainty in the signal is a topic addressed in Chapter 5, Statistical Processing and Leak Detection.

3.5.1 Determining the Flow Balance

Fundamental to the concept of mass balance leak detection is the ability to account for all fluid entering or leaving the pipeline system. Ideally, each flow entering or leaving the pipeline would be metered accurately. Unfortunately, in aging natural gas pipeline infrastructures, many of the flows entering or leaving the pipeline are not metered at all or, if they are, the data acquisition is not in real time. In liquid pipelines, flow measurements can also have high levels of uncertainty due to imprecise flow measurement, utilization of tank volume changes as proxies for flow meters, and missing flow measurement data. In such cases, one is faced with the choice of living with the associated uncertainties and the corresponding impact on achievable leak detection sensitivity or the expensive task of adding accurate and reliable flow measurements to the system.

These issues are measurement issues and not amenable to proper resolution by improvements within mass balance leak detection monitoring software. Of course, intelligent software mitigation approaches can be applied, such as:

- Monitoring for a rapid, positive, and persistent excursion in the observable mass balance that is different from expected pipeline mass balance variations as a possible leak
- Estimating flow rates from historical and/or weather data

These software-based mitigations are only partially effective. There is no substitute for accurate and complete metering.

From one perspective, a leak is simply an unmetered (and unexpected) flow leaving the pipeline. If one cannot account properly for the rest of the flows through real-time metering, then the resultant uncertainties in the mass balance will mask a leak, unless the leak is substantially larger than that uncertainty.

3.5.2 Determining the Packing Rate

Leak detection by mass balance requires a calculation of the pipeline packing rate. There is no physical measurement of pipeline line pack. Instead, one must make the assumption that it is constant or that one is able to calculate it using other field instrument measurements along the pipeline. Typically, the line pack is calculated using all available pressure and temperature measurements in combination with an equation of state for the fluid.

One means of classifying mass balance leak detection systems is the fidelity with which they attempt to compensate for the changing mass of the pipeline. On the simplest end of the spectrum, one assumes that the packing rate is zero. Then there are various steady-state approaches that assume the pressure and temperature in the pipeline can be computed by some sort of linear interpolation between measurements. At the more sophisticated extreme, an RTTM may be used to compensate for both slowly and rapidly propagating transients throughout the pipeline system.

The performance of a volume/mass balance leak detection system is directly impacted by the accuracy with which the pipeline packing rate is computed. For example, a system that assumes that line pack is constant (ie, packing rate is zero) will be limited in sensitivity to the extent that this fundamental assumption is not true.

The most accurate calculation of line pack is provided by an RTTM approach. Even in that case, the leak detection sensitivity is limited by the ability of the RTTM to accurately compute the packing rate, which, in turn, is limited by the following:

- The accuracy of the measurements used as inputs (boundary conditions) to the RTTM
- The choice of measurements used as inputs to the RTTM
- The sensitivity of the pipeline to effects of unmeasured variables and fluid properties (such as viscosity)
- The validity of the physical assumptions inherent in the equations used to represent the pipeline state
- The fidelity of the numerical solution approach and parameters used to solve the partial differential equations representing the pipeline state
- Fundamental uncertainties associated with pipeline configuration and its surrounding environment, such as the ground thermal properties

3.6 API 1130 APPLICABLE CLASSIFICATION OF MASS BALANCE SYSTEMS

For liquid pipelines, API 1130 [2] defines a set of non-RTTM-based leak detection approaches that vary in the complexity with which one attempts to compensate for the amount of fluid in the pipeline. As discussed in

Sections 3.6.1–3.6.4, API 1130 briefly describes these in increasing levels of sophistication.

3.6.1 Line Balance CPM

A system that monitors actual flow in minus actual flow out of a pipeline system and uses a positive imbalance as an indicator of a leak. Changes in pipeline line pack are ignored and no attempt is made to convert flows to standard conditions.

3.6.2 Volume Balance CPM

Volume balance CPM is the same as a line balance CPM, except that flow rates are converted to standard conditions. Using the terminology of Section 3.4.2 this is equivalent to monitoring the flow balance, FB_{MBS}, with a positive value being evidence of a leak.

3.6.3 Modified Volume Balance CPM

Modified volume balance CPM is similar to volume balance CPM, except that an effort is made to compensate for changes in the inventory of the pipeline. The methodology by which this may be done is not specified by the standard.

3.6.4 Compensated Mass Balance

Modified volume balance is enhanced by keeping track of batches in the pipeline and estimating inventory of each batch based on batch bulk modulus and temperature modulus.

3.6.5 Real-Time Model Based Systems

API 1130 [1] distinguishes the RTTM-based systems from line balance, volume balance, and similar mass balance approaches. The standard indicates that RTTM-based systems look for discrepancies between modeled and measured data values (particularly pressures and flows) and therefore are an enhancement over volume balance systems.

Instead, from the authors' perspectives, we consider RTTM systems as applications that provide a fundamentally better approach to computing line pack and packing rate; therefore, they are a fundamental enhancement to other volume balance approaches that calculate line pack less precisely.

It is true that in the early years of computational pipeline monitoring (1970s–1990s), some RTTM-based systems were designed primarily around examining individual discrepancies between measured and modeled pressures and/or flows. However, most modern RTTM-based systems are fundamentally mass/volume balance systems that may be assisted by monitoring discrepancies between modeled (computed) and measured flows, pressures, and or temperatures or between two independent computations of the same modeled value.

3.7 OUR CLASSIFICATION OF MASS BALANCE–BASED LEAK DETECTION SYSTEMS

As an alternative to the API 1130 classifications, we classify mass/volume balance leak detection systems as follows:

1. Flow balance leak detection systems that ignore pipeline inventory. These make the implicit assumption that inventory of the pipeline is either constant or a variable uncertainty that limits the sensitivity of the leak detection system.
2. Volume/mass balance systems that account for pipeline inventory using steady-state assumptions. These assume that the packing rate can be computed using a succession of steady-state calculations. These systems may or may not attempt to track batches or composition changes through the pipeline.
3. RTTM mass balance systems that compensate for pipeline inventory changes by recognizing that the pipeline is in a dynamic state with pressure, flow, and temperature transients dynamically propagating through the pipeline system. This dynamic behavior is calculated using an RTTM. We delve into RTTM-based leak detection systems in Chapter 4, Real-Time Transient Model–Based Leak Detection.

REFERENCES

[1] Shanshool J, Habobi N, Kareem S. Volumetric behavior of mixtures of different oilstocks. Petroleum & Coal 2011;53(3):2238. Available online at http://www.vurup.sk/petroleumcoal.

[2] Computational Pipeline Monitoring for Liquids, API Recommended Practice 1130, First Edition, September 2007, Reaffirmed, April 2012.

Chapter 4

Real-Time Transient Model–Based Leak Detection

Real-time transient model–based leak detection developed during the 1970s micro-/mini-computer revolution, when real computing power became available for real-time pipeline monitoring systems. Even with 16-bit computers, modeling major pipeline systems was possible despite limitations of 16-bit hardware, 64-KB memory, and the consequent numerical and software challenges. Since that time, computing hardware has developed at a very rapid pace; however, leak detection software has developed at a much slower pace, with prominent systems on the market today bearing the design concepts of their 1980s predecessors.

Real-time transient modeling provides a quantum leap forward in computing real-time pipeline packing rates and thereby fundamentally improving the potential of mass balance leak detection over systems based on steady-state analyses. However, there remains a great deal of art and mystery surrounding the inner workings of RTTM-based systems. It is certainly not true that all RTTM-based systems are equivalent, nor is it true that one can easily discern the differences between available systems or determine which might be most appropriate for a particular pipeline. The differences between systems are often confused by supplier claims. Unlike flow meters that can be tested and evaluated in a test facility, RTTM leak detection systems are difficult to test and evaluate in an isolated "test bed" environment.

The performance of an RTTM-based system depends on all of the following:

- The quality, spacing, and placement of pressure and temperature measurements.
- The quality of flow measurements and the size and variability of flows and any flows that are not metered entering or leaving the pipeline system.
- The quality and placement of flow measurements within a pipeline system permitting isolation of portions of the pipeline into independent mass balance sections.

Pipeline Leak Detection Handbook. DOI: http://dx.doi.org/10.1016/B978-0-12-802240-5.00004-2

- The quality and placement of fluid property measurements. By this we mean measurements that provide information about fluid composition, specific gravity, viscosity, and other fluid characteristics that differentiate fluids within the pipeline.
- The quality and placement of any measurements that provide data to facilitate computing heat loss from the pipeline to its surroundings (such as ambient temperature, ground moisture, cloud cover, etc.).
- The completeness and appropriateness of the equations used to represent the transient behavior of the fluid in the pipeline and the interaction of the fluid with the pipeline surroundings (particularly thermal interactions).
- The selection of appropriate or best boundary conditions for the RTTM.
- The fidelity with which the equations defining the transient behavior of the pipeline are solved.
- The ability of the RTTM-based system to deal with noise and errors in the pipeline instrumentation.
- The ability of the RTTM-based system to extract evidence of a leak from other mass imbalances that are due to limitations, errors, and uncertainties in the input data, pipeline equations, and numerical solutions of the pipeline equations.

It is emphatically incorrect to assume that all RTTM-based leak detection systems are the same. There is both a science and an art to the design of an RTTM-based leak detection system. We discuss this further in the upcoming sections.

4.1 THE REAL-TIME TRANSIENT MODEL

By RTTM, we mean the real-time transient model of the pipeline system as it is represented through a software-based mathematical solution of the equations describing the pipeline state. When we are referring to the associated leak detection system of which the RTTM is a part, we call it an RTTM-based leak detection system.

In this section we discuss the RTTM. Let us start by discussing some fundamentals.

The pipeline state is fundamentally transient in that it changes constantly as time progresses. Some aspects of the pipeline state change rapidly, such as pressure, temperature, and flow disturbances that move at the speed of sound. Others propagate with the speed of the fluid, such as the variations in fluid properties or variations in inlet temperatures. Still other effects change very slowly, such as the adjustment of the temperature "profile" in the ground outside of the pipeline resulting from changing ambient conditions, changes in pipeline flow rates, or changes in pipeline inlet temperatures.

In an RTTM-based leak detection system, the real-time transient model portion of the system (the RTTM) plays a fundamental role.

However, different RTTM providers may make very different choices in the ways in which the RTTM is implemented.

Some RTTMs attempt to account for all of the transient effects described. Another RTTM might ignore the slowly changing thermal profile in the pipeline surroundings. Another RTTM may choose to ignore thermal effects altogether. Others may ignore some of the complicating terms in the pipeline equations to reduce the complexity and processing requirements of the RTTM portion of the system. Some systems may choose to treat all of the fluid in the pipeline as identical, whereas others may attempt to track all fluid properties throughout the pipeline system.

RTTM-based leak detection system providers may differ in their selection of measurements to be used as inputs (boundary conditions) to the solution of the RTTM equations.

Finally, each RTTM-based leak detection system provider applies their own set of logic and rules to differentiate between a true leak and other sources of errors in the leak signal signal(s) developed by the RTTM based system.

4.1.1 Fundamental Equations and Physics

The relationship between pressure, temperature, and flow rate in a pipeline is based on a set of mostly partial differential equations that require numerical discretization to be solved. The solutions of these equations depend on initial conditions (the state of the pipeline fluid at the start of the solution) and boundary conditions (individual measurements at discrete points in time). The equations are typically expressed in terms of derivatives in terms of x (distance along the pipeline mid-line) and t (time). RTTMs model the pipeline flow as a one-dimensional (1D) transient phenomena—pressures, velocity, and temperature change with distance along the pipeline and progress forward in time. Often, the RTTM includes a model of transient heat loss to the pipeline surroundings that depends on the radial distance from the pipeline, which is a second spatial dimension for the thermal solution. However, the model of the fluid in the pipeline is virtually always represented using a single spatial dimension, the distance along the pipeline mid-line.

We review the fundamental equations that form a basis for the RTTM. Chapter 1, Introduction includes a table of nomenclature that we use in the equations here.

Continuity Equation (Conservation of Mass)

The continuity equation (Eq. 4.1) is the statement of conservation of mass in the pipeline: mass in minus mass out equals change of mass.

- The first term in the equation, $\partial(\rho v A)/\partial x$, is "mass flow in minus mass flow out" of a slice of the pipeline cross-section.
- The second term, $\partial(\rho A)/\partial t$, is the rate of change of mass of a slice of pipeline cross-section.

$$\frac{\partial(\rho vA)}{\partial x} + \frac{\partial(\rho A)}{\partial t} = 0$$

EQUATION 4.1 Continuity Equation (Mass Balance)

where ρ is mass density, v is the velocity, A is the pipe cross-sectional area, x is the coordinate along the pipe centerline, and t is time.

Momentum Equation (Newton's Second Law of Motion)

The momentum equation (Eq. 4.2), an expression of Newton's second law of motion, represents the transient force balance on the fluid within a slice of the pipeline cross-section.

- The left side, $\rho(\partial v/\partial t + v\partial v/\partial x)$, is mass times acceleration per unit volume of fluid (there is a velocity change in time, t, as well as a change as it moves in distance, x).
- The right side (RHS) represents the forces acting on a unit mass of fluid.
- The first RHS term, $-\partial p/\partial x$, is the net force imposed by the pressure gradient.
- The second RHS term, $-\rho g\partial z/\partial x$, is the force of gravity on the element as it moves in the vertical direction (due to the slope of the pipeline).
- The final term $-\rho fv|v|/2D$ is the frictional force that acts in a direction opposite to the velocity.

$$\rho\left(\frac{\partial v}{\partial t} + v\frac{\partial v}{\partial x}\right) = -\frac{\partial p}{\partial x} - \rho g\frac{\partial z}{\partial x} - \rho\frac{fv|v|}{2D}$$

EQUATION 4.2 Momentum Equation (Momentum Balance)

Energy Equation (Conservation of Energy)

The energy equation (Eq. 4.3) represents conservation of energy of a fluid element.

- The left side represents the rate of change of internal energy of a fluid element. There is a time component and a spatial component because the fluid element under consideration is moving. The spatial component represents the convection of thermal energy with the fluid as it moves through the pipeline.
- The first two RHS terms, $-T(\partial p/\partial T)_\rho(A_t + (vA)_x) + pA_t$, include a thermodynamic derivative $(\partial p/\partial T)_\rho$ that can be computed with the assistance of an equation of state. Inherent in these two terms is the work in the fluid element as it expands or contracts and the Joule-Thomson effect.
- $\rho Af/2D|v^3|$ represents the work of friction on the fluid.
- The final term, $-qA4/D$, represents the loss of heat from the pipeline through conduction and/or convection from the pipe surface,

where q represents the heat flux (heat flow rate per unit surface area) at the inner pipe surface. The multiplier $4/D$ in this term arises from the fact that the ratio of the inside pipe surface area to the internal pipe volume is $\pi D/(\pi D^2/4) = 4/D$.

$$\rho A c_V \left(\frac{\partial T}{\partial t} + v\frac{\partial T}{\partial x}\right) = -T\left(\frac{\partial p}{\partial T}\right)_\rho (A_t + (vA)_x) + pA_t + \rho A \frac{f}{2D}|v^3| - qA\frac{4}{D}$$

EQUATION 4.3 Energy Equation

Note that to compute q, one must be able to represent the thermal characteristics of the pipeline surroundings.

For below-ground pipelines, the heat equation, Eq. (4.4), represents the heat transfer in the ground surrounding the pipeline.

$$\frac{\partial T}{\partial t} = \frac{k}{\rho c_P}\nabla^2 T = \alpha \nabla^2 T$$

EQUATION 4.4 Heat Equation for Heat Flow in Ground Surrounding Pipe

One often assumes radial symmetric heat flow from the pipeline to an "ambient" temperature at some distance from the pipeline. In cylindrical coordinates centered on the pipeline cross-section, the heat equation is then written as:

$$\frac{\partial T}{\partial t} = \frac{\alpha}{r}\frac{\partial}{\partial r}\left(r\frac{\partial T}{\partial r}\right) = \alpha\left(\frac{\partial^2 T}{\partial r^2} + \frac{1}{r}\frac{\partial T}{\partial r}\right)$$

EQUATION 4.5 Heat Equation Assuming Radial Symmetry

We note that the actual thermal gradient is quite two-dimensional (2D), and that the radial symmetry is an approximation. However, it has been shown to work well in RTTM applications. It also has the advantage of being much easier to solve numerically than the 2D form of the equation.

This equation can be solved in conjunction with the pipeline energy equation, where the heat flux at the pipeline surface is the inside boundary condition for this equation (at $r = r_S$) and the ground temperature at some distance from the pipeline is the outer boundary condition.

Energy Equation and Ground Thermal Modeling: Discussion

The best attempts to simulate below-ground pipeline heat transfer are only approximations to reality. In the real world, the following distort this simplified representation:

- The ground temperature profile is dependent on the distance below the surface; it is not radial, but rather 2D. It is influenced by the temperature of the fluid in the pipe (assumed well-mixed and uniform across the cross-section), the above-ground temperature, the depth below the ground surface, and the distance from the pipe wall.

- The thermal properties of the pipeline surroundings are unlikely to be uniform.
- The thermal properties of the pipeline surroundings are influenced by ground moisture content. In some cases, perhaps even convective heat transfer effects could dominate.

These variables are also dynamic and dependent on time or year and weather conditions.

There are probably as many variations on the representation of the energy equation and the associated ground thermal model as there are RTTM vendors. In some cases, the final term in the energy equation may be replaced with a term representing the heat transfer through a fluid "boundary layer" between the bulk pipeline fluid and a thin boundary layer of nearly stagnant fluid near the pipe wall represented by a "film" heat transfer coefficient. In other cases, the pipe wall may be ignored and the final term may be used to represent the heat transfer in a thin "shell" in the ground outside of the pipeline.

In the grossest case, and unfortunately perhaps the most common, the transient heat transfer in the ground is ignored. In this case, q in the energy equation is replaced with a simple term representing the steady-state heat transfer rate between the pipe and ground.

Unfortunately, an RTTM using a steady-state representation of heat flow in the ground surrounding the pipeline will poorly simulate the temperature changes in the pipeline.

As we have described, there are a number of difficulties inherent in trying to accurately represent the transient heating and cooling of the ground surrounding the pipeline. However, the authors have concluded the following from many years of experience with real-time transient models for below-ground pipelines.

1. The ground surrounding the pipeline has a significant amount of thermal mass.
2. From a leak detection perspective, one of the predominant influences of the ground is its ability to dampen the rate of change of temperatures in the pipeline. As the pipeline temperature starts to change, heat immediately starts transferring between the pipe and the ground near the pipe. The ground surrounding the pipeline effectively adds a great deal of thermal mass to that of the fluid itself, reducing the pipeline fluid rate of temperature change.
3. The time constant for the ground outside of the pipeline to reach a new thermal equilibrium following a change in the pipeline temperature is many hours or days.

4. In contrast, a simple HTC formulation ignores the entire thermal mass of the pipeline surroundings. Instead, the ground is treated as an insulating layer whose heat capacity is zero.
5. Consequently, a transient thermal model of the pipeline fluid coupled with a steady-state model of the pipeline surroundings is substantially worse than assuming that the pipeline temperature is constant, at least from the perspective of leak detection.

Pipe Wall Expansion Equation

The pipe cross-sectional area changes with pressure and temperature. For steel pipe, the expansion is fairly small. However, despite that fact, it is enough to have a noticeable effect on pipeline transients, particularly in liquid pipelines.

$$A = A_0 + \left.\frac{\partial A}{\partial p}\right)_T (p - p_0) + \left.\frac{\partial A}{\partial T}\right)_p (T - T_0)$$

EQUATION 4.6 Pipe Cross-Sectional Area Equation

where

$$\left.\frac{\partial A}{\partial T}\right)_p = 2\alpha_{\text{pipe}} A_0$$

EQUATION 4.7 Variation of Cross-Sectional Area with Temperature

$$\left.\frac{\partial A}{\partial p}\right)_T = \begin{cases} \dfrac{D}{Ew}(1 - \nu^2)A_0 & \text{Ends fixed} \\ \dfrac{D}{Ew}(1.25 - \nu)A_0 & \text{Ends free} \end{cases}$$

EQUATION 4.8 Variation of Pipe Cross-Sectional Area with Pressure

α_{pipe} is the coefficient of thermal expansion of the pipe material, ν is its Poisson's ratio, D is the pipe diameter, E is Young's modulus, and w is wall thickness.

In this equation, the subscript 0 stands for conditions at standard temperature and pressure (STP), so A_0 is the pipeline cross-sectional area at STP. Note that the change of area with pressure is different for a pipe with fixed ends as compared to a pipe with ends that are free to move. We believe that in above-ground pipes, which do not use constrained supports, the ends are effectively free to move, as is the also the case for pipes lying on the sea floor. However, pipes that are buried are effectively constrained by the fill and ground around them, and therefore are likely to behave as if their ends are fixed.

Equation of State

An equation of state provides a relationship between density, pressure, and temperature, which we can express generally as $\rho = f(p, T)$. The equation of state can be quite complex, and typically it will have dependencies on real-time variations in fluid properties. For crude oil and petroleum products, API/ASTM equations are dependent on the API or specific gravity, which is often measured in real time. For natural gas, AGA-8 is often used; in its detailed form, it is dependent on the composition of the gas. Precise calculation of the packing rate for leak detection purposes requires an accurate equation of state.

Viscosity Equation

To compute the friction factor, one must know the viscosity of the fluid. Viscosity is usually dependent at least on temperature, and fluid composition. A viscosity equation or tabulated data for the viscosity of the pipeline fluid is required. Either dynamic or kinematic viscosity is sufficient because the two are related through the density of the fluid.

Darcy-Weisbach Friction Factor

The momentum equation is dependent on f, the Darcy-Weisbach friction factor that is dependent on Reynolds number (Re) and relative roughness. The Moody diagram provides a chart of friction factor compared with Re for different relative roughness levels. An RTTM must include a computation of the friction factor at actual pipeline conditions at each point in the pipeline.

Batch or Composition Tracking

It is necessary for the RTTM to keep track of the changing pipeline fluid characteristics at every point in the pipeline. This requires tracking composition or fluid property changes as they move through the pipeline at the speed of the flow. For instance, in an oil pipeline, the RTTM will likely track specific gravity, DRA (drag-reducing additive), concentration (if used), and perhaps water content of the oil as it moves through the pipeline. For a natural gas pipeline, an RTTM will typically track composition changes moving through the pipeline. Alternatively, it may track the gas specific gravity and heating value changes. Unless the pipeline fluid characteristics are uniform in space and time, fluid property tracking is necessary for accurate RTTM calculations. If there is significant mixing of fluids in the pipeline, then the mixing must be calculated as well.

Check Valve and Block Valve Equations

An RTTM must generally provide a mathematical representation of the automatic opening and closing of check valves along the pipeline as flow changes direction. Even for instrumented valves, whose position may be known, an RTTM must calculate the resistance created by a valve moving from full open to full closed and vice versa. For a check valve, it is useful to model the frictional effect of a floating check valve clapper. At minimum, the RTTM must be able to represent a valve as fully open (very low resistance) or fully closed (nearly infinite resistance).

Other Equations

In addition, depending on the RTTM provider's methodology and available instrumentation, the RTTM may mathematically represent the operation of other types of equipment (such as pumps and compressors).

4.2 NUMERICAL METHODS

The preceding section provides the equations governing the motion of fluids in a pipeline. They include several partial differential equations with independent variables x and t. All of the equations defined in Section 4.1 must be solved at every time t and distance x. In our discussion, we assume that the partial differential equations are solved using finite difference methods. We note that other discretization methods that are commonly used include finite volume and finite element methods, which are similar but with important differences. These methods are not discussed here.

An RTTM is designed to simulate the pipeline transients moving through space and time. The four partial differential equations cannot be solved analytically. Instead, their solution requires discretization in both space and time and a numerical solution to solve all of these, along with the other equations, at every discrete point in space and time.

From a mathematical perspective, there are a great number of details to developing an effective solution to the more than eight pipeline fluid flow equations. The fundamentals of the solution approaches are substantially influenced by the numerical technique selected for the solution of the partial differential equations. Three different numerical techniques are used to solve the partial different equations:

1. Explicit integration
2. Implicit integration
3. Method of characteristics

We briefly describe the fundamentals of the three approaches here. A further discussion elsewhere [1] provides examples of differences between the three.

Each technique solves a numerical approximation of the differential equations. Any one of these can provide useful results, but there are advantages and drawbacks to each.

Fig. 4.1 illustrates the approach that an RTTM takes to solve the equations describing the pipeline state.

1. The pipeline is subdivided along its length into a series of discrete points at locations denoted as x_i, where i is an index spanning 0 to $N-1$, where N is the number of discrete points at which the equations are solved. The pipeline length between x_i and x_{i+1} is referred to as the distance step, Δx_i. The discrete computational points $x_{i,i=0...N-1}$ are referred to as knots.
2. The value for every pipeline variable, represented generally as χ, is required at the start of the simulation. The collection of these values is referred to as the initial state. The initial state may be obtained from a steady-state solution to the pipeline equations by setting all variables to a suitable constant value, from a previous simulation, or perhaps by other means.
3. The state of every pipeline variable is computed for successive discrete points in time. Assuming that one has a value for the pipeline variables at time t_j, the task of the numerical solution of the RTTM is to compute the values of the pipeline variables at time $t_{j+1} = t_j + \Delta t_j$, where Δt_j is referred to as the time step. We use $\chi_{i,j}$ to represent the value of a generic variable χ (such as fluid pressure, temperature, velocity, density, pipe area, temperatures in the pipeline surroundings, etc.) at location x_i and time t_j. The collection of these values is referred to as the pipeline state at time t_j.

Referencing Fig. 4.1, assume that pipeline state is known at t_j. The task of the RTTM is to compute the pipeline state at time t_{j+1}. Then, time step by time step, the numerical solution of the RTTM moves the pipeline state forward in time from t_j to t_{j+1}, to t_{j+2}, and on and on. Because inputs to the solution are real-time pipeline measurements, the RTTM pauses at the latest time, waits for more input data, and then steps forward as time progresses and as updated pipeline measurements are available.

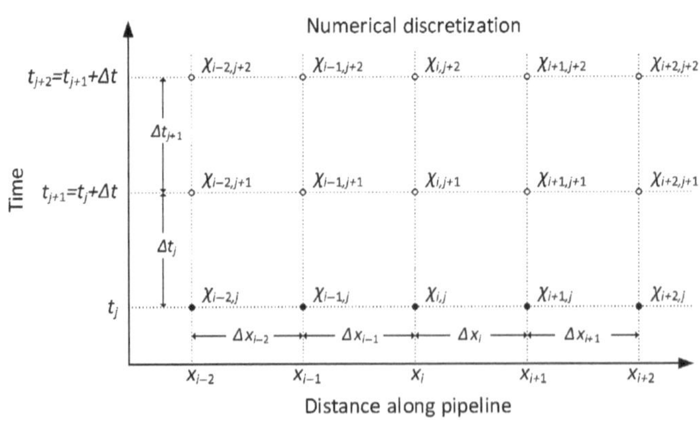

FIGURE 4.1 Numerical discretization illustration.

The differential equations include a number of partial derivatives in space and time (ie, $\partial\chi/\partial t$ and $\partial\chi/\partial x$). The RTTM must have a method for representing these derivatives numerically so that the collection of variables at time t_{j+1} can be expressed algebraically in terms of the values of variables at time t_j. In general, each variable $\chi_{i=0...N-1,\,j}$ is assumed known because it was previously calculated at the earlier time step, and each variable at the new time, $\chi_{i=0...N-1,\,j+1}$ is unknown.

4.2.1 Explicit Numerical Solution

The simplest solution method is the explicit approach (refer to Fig. 4.2). The heavy dotted lines show the spans in space and time used for expressing the partial derivatives. We represent our partial derivatives using Eqs. (4.9 and 4.10).

$$\frac{\partial\chi}{\partial x_i} = \frac{\chi_{i+1,j} - \chi_{i-1,j}}{x_{i+1} - x_{i-1}}$$

EQUATION 4.9 Explicit Distance Derivative

$$\frac{\partial\chi}{\partial t_i} = \frac{\chi_{i,j+1} - \chi_{i,j}}{\Delta t_j}$$

EQUATION 4.10 Explicit Time Derivative

Note that the explicit distance derivative is written in terms of variables at time t_j. Thus, the explicit time derivative has only a single unknown. The consequence of this is that we can express all of our equations for the unknown values of the variables at x_i and t_{j+1} in terms of each other and the known variables at time t_j. In the explicit approach, the solution proceeds

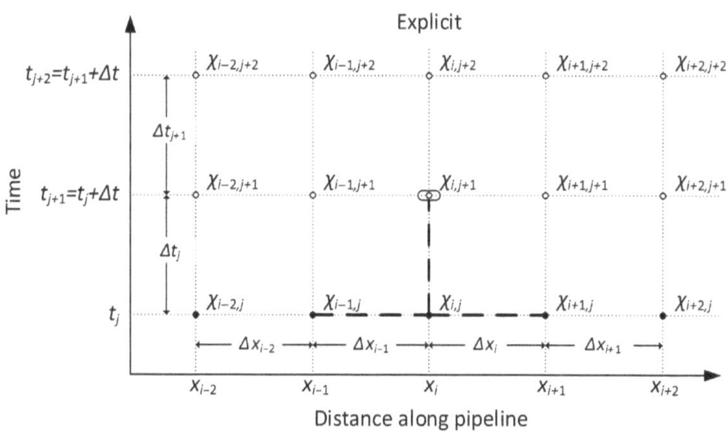

FIGURE 4.2 Illustration of explicit numerical solution.

point-by-point through the pipeline with the values at each point i computed independently of values at any other point in the pipeline at time t_{j+1}. The oval next to unknown $\chi_{i,j+1}$ in Fig. 4.2 shows the unknown variable that is expressed in terms of the known values at the other ends of the dashed lines. In the finite difference paradigm, the dashed lines are commonly referred to as a *stencil*.

There are several nonlinear terms in the transient flow equations. One may linearize the equations in terms of the changes in the variables. Linearization involves writing the equations in terms of the known values at time t_j (eg, $\chi_{i,j}$) and the change in the value over the time step (eg, $\Delta\chi_i$), and then discarding higher-order terms. For example, the linearization of ρv is shown in Eq. (4.11). The second-order term, $\Delta\rho_i\Delta v_i$, is discarded in the linearization.

$$\begin{aligned}\rho_{i,j+1}v_{i,j+1} &= (\rho_{i,j} + \Delta\rho_i)(v_{i,j} + \Delta v_i)\\ &= \rho_{i,j}v_{i,j} + \Delta\rho_i v_{i,j} + \rho_{i,j}\Delta v_i + \Delta\rho_i\Delta v_i\\ &\approx \rho_{i,j}v_{i,j} + \Delta\rho_i v_{i,j} + \rho_{i,j}\Delta v_i\end{aligned}$$

EQUATION 4.11 Linearization Example

The combined set of discretizations and other assumptions result in a set of equations for each knot that are equal to the number of unknowns, allowing one to solve for the new state at every knot at time t_{j+1}. All unknown variables at the current time step are *explicitly* specified in terms of variables at the previous time step. Consequently, explicit methods are computationally easy to set up.

A significant issue is that explicit methods are subject to a very restrictive maximum space and time step limit, often referred to as the Courant limit. Under this limit, our maximum time step is constrained by:

$$\Delta t_{\max} \leq \frac{1}{2}\frac{\min(\Delta x)}{a}$$

EQUATION 4.12 Explicit Finite Differences Time Step Requirement

where $\min(\Delta x)$ is the shortest distance step and a is the commodity speed of sound. If this constraint is violated, then the solution will become unstable and unusable.

In conclusion, explicit methods present the following pros and cons:

Advantages:

- Equations are easily expressed.
- The explicit solution is straightforward and does not require matrix methods.
- They permit variable space sizes within the Courant limits.

Disadvantages:

- Courant condition constraints are very restrictive and limit grid spacing and time steps.
- Many problems cannot be solved in reasonable time using explicit methods.
- For a given time step and distance step, the explicit method is only accurate for the first-order terms (all second-order and higher terms show up as errors).
- May be subject to significant numerical dispersion problems.

4.2.2 Method of Characteristics Solution

The method of characteristics (MOC) approach has been favored by textbook authors [2,3] because it is explicit and thus easy to set up and solve. It also tends to preserve wave behavior that is very important for fast, brief, transient calculations that are required for pipeline surge calculations.

The approach is based on the assumption that pressure and velocity transients propagate with the speed of sound within a pipeline system. In the $x-t$ domain, these paths are called characteristic curves. Because (under the assumptions upon which MOC is based) all transients move on these paths, the partial differential equations are converted to ordinary differential equations that express the propagation along these characteristic curves. These curves define wave propagation in both the upstream and downstream directions and allow us to algebraically calculate our unknown variables at points where these characteristic curves intersect each other.

Fig. 4.3 shows a typical MOC solution grid in the $x-t$ plane. The diagonal dashed lines show the characteristic paths along which a solution would

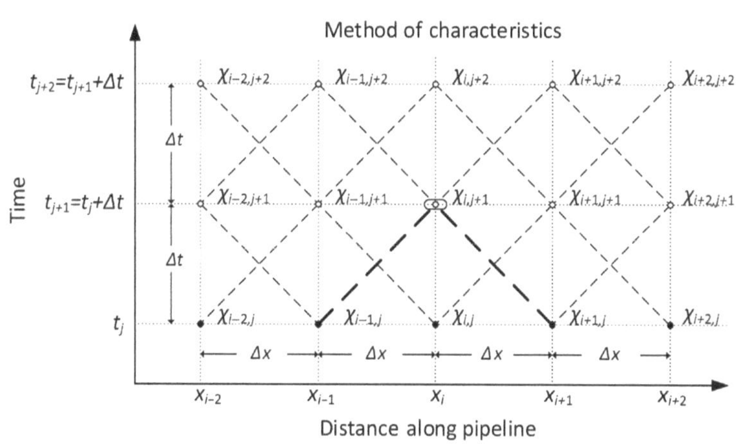

FIGURE 4.3 Illustration of method of characteristics solution.

propagate at the speed of sound. Because waves may propagate in both directions along the pipeline, changes in variable $\chi_{i-1,j}$ may propagate along the diagonal to the right from $\chi_{i-1,j}$ to $\chi_{i,j+1}$ and along the diagonal to the left from $\chi_{i-1,j}$ to $\chi_{i-2,j+1}$. As illustrated by the heavy dashed lines, the variables at $(i,j+1)$ result from values propagating downstream from $(i-1,j)$ and upstream from $(i+1,j)$. A necessary condition for the MOC solution is that the ratio of the distance step to time step must be equal to the speed of sound a (ie, $a = \Delta x/\Delta t$), which presents severe restrictions on the construction of the grid.

Wiley and Streeter [2] and Chaudhry [3] provide much greater detail on this method, but each ignore the energy equation (Eq. 4.3) and do not consider heat transfer to the surroundings. The implicit assumption of the textbook solutions is that temperature is unimportant and can be treated as constant. However, for real-time transient modeling aimed at performing leak detection, both temperature variations and the energy equation are important.

A separate complication is the movement of fluid property changes through the pipeline. As an illustration, a change in fluid density or bulk modulus moving through the pipeline will impose a "kink" in the characteristic curve that slowly moves through the pipeline at the speed of the fluid.

Even without these complications, the requirement that the ratio of distance step to time step must be equal to the speed of sound is challenging to deal with on a real pipeline for several reasons including the following:

1. The speed of sound is unlikely to be uniform in the pipeline. In gas, the speed of sound is proportional to the square root of the temperature. In liquids, the speed of sound will vary with variations in fluid composition. Once one has selected a grid spacing, the nonuniform speed of sound makes it impossible to exactly satisfy $a = \Delta x/\Delta t$ throughout the entire pipeline.
2. There must be an integer number of discrete intervals in each unbranching uniform section of the pipeline. In an RTTM, it is likely to be important to place knots at wall thickness changes, elevation peaks, every valve, and so on. Without resorting to very small distance steps, the distance step cannot be made uniform throughout the pipeline.
3. The uniform grid spacing requirement makes it difficult to match up the various branches in a networked pipeline, each of which may have a different length. Each segment is very unlikely to be divisible by an integer number of grid segments.

Furthermore, varying fluid compressibility, viscosity, and friction factor all require special treatment in the formulation of an MOC solution for an RTTM.

In conclusion, MOC methods offer the following pros and cons:

Advantages:

- Because this is a variant of an explicit method, the equations are easily set up and easy to solve.

- Reliable preservation of wave behavior makes MOC very useful for surge analysis. This behavior is not necessarily as valuable for leak detection.

Disadvantages:

- Courant condition constraints are again very restrictive and limit grid spacing and time steps.
- Many problems cannot be solved in reasonable time using the method of characteristics.
- For a given time step and distance step, MOC is only accurate for the first-order term.
- This approach is not amenable to variable grid spacing.
- The constant speed of sound is restrictive and cannot generally be satisfied everywhere in the pipeline. This may affect the accuracy of the packing calculation.
- Uniform grid spacing may be impossible achieve.

4.2.3 Implicit Numerical Solution

For a given time and distance step, an implicit approach can provide a more accurate representation of the differential equations. Unknowns at the current time step are expressed in terms of values at the previous time step (as in the explicit method) combined with values of neighboring and related values at the current time step.

Refer to Fig. 4.4. This figure illustrates one form of implicit solution sometimes referred to as the box method. Note the change in the stencil. The equations are now written in terms of distance variable $x_{i+1/2}$, time $t_{j+\theta}$, and user-specified parameter θ (with a value between 0 and 1) in terms of the values at the "box" corners surrounding this point, $\chi_{i,j}$, $\chi_{i+1,j}$, $\chi_{i,j+1}$, and

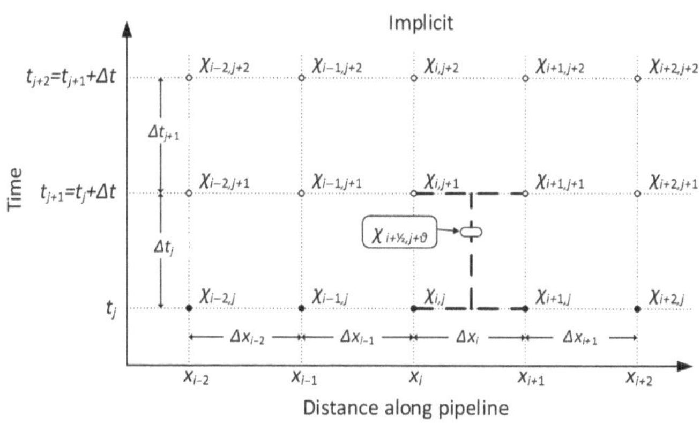

FIGURE 4.4 Illustration of implicit numerical solution.

$\chi_{i+1,j+1}$. These derivatives are expressed by Eqs. (4.13 and 4.14). Pictorially, the oval labeled $\chi_{i+1/2,j+\theta}$ shows the point at which the equations are written, with the dashed lines showing the dependencies on the adjoining variables.

$$\left.\frac{\partial\chi}{\partial x}\right|_{i+1/2,\,j+\theta} = (1-\theta)\frac{\chi_{i+1,j}-\chi_{i,j}}{x_{i+1}-x_i} + \theta\frac{\chi_{i+1,j+1}-\chi_{i,j+1}}{x_{i+1}-x_i}$$

$$= (1-\theta)\frac{\chi_{i+1,j}-\chi_{i,j}}{x_{i+1}-x_i} + \theta\frac{\chi_{i+1,j}+\Delta\chi_{i+1}-(\chi_{i,j}+\Delta\chi_i)}{x_{i+1}-x_i}$$

$$= \frac{\chi_{i+1,j}-\chi_{i,j}}{x_{i+1}-x_i} + \theta\frac{\Delta\chi_{i+1}-\Delta\chi_i)}{x_{i+1}-x_i}$$

EQUATION 4.13 Implicit "Box Scheme" Distance Derivative

$$\left.\frac{\partial\chi}{\partial t}\right|_{i+1/2,\,j+\theta} = \frac{\chi_{i,j+1}+\chi_{i+1,j+1}-(\chi_{i,j}+\chi_{i+1,j})}{2\Delta t_j}$$

$$= \frac{\Delta\chi_i+\Delta\chi_{i+1}}{2\Delta t_j}$$

EQUATION 4.14 Implicit "Box Scheme" Time Derivative

We linearize our equations in a fashion similar to that described for the explicit method in Section 4.2.1. However, the linearization in the implicit method is slightly more complex. For example, in the box scheme illustrated here, because the equations are being written at a nongrid point ($x_{i+1/2,j+\theta}$), the linearizations are also written at this point, resulting in contributions from terms at each of the "box" corners.

The combination of the numerical discretization of the partial derivatives and the linearization of the nonlinear terms results in a series of equations that are expressed in terms of variables at more than one knot. In the box scheme approach illustrated here, each equation is written in terms of variables at two neighboring knots. Therefore, the implicit RTTM numerical solution requires a simultaneous solution of all of the equations for every "box" using matrix methods.

In terms of computational overhead, the implicit model imposes a substantially higher load than either of the other approaches for a single step forward in time because it always requires a matrix solution of some sort. The matrix is typically a band or sparse matrix that requires far less CPU and memory than a full matrix solution but still is somewhat greater than the simpler computational requirements of the other approaches.

However, compared to an explicit model, the higher-order accuracy of the solution offsets this cost, because longer time and distance steps may be taken without sacrificing accuracy. In addition, the method of characteristics has challenges of its own in that the distance steps all need to be nearly identical. This means that the size of the distance steps is limited by the portion of the pipeline requiring the highest level of distance detail (eg, the distance between two pipe wall thickness or pipe diameter changes).

Despite these mitigating factors, the implicit finite difference approach is likely to impose somewhat higher computational load, especially during times of very dynamic pipeline behavior. One might expect this to be a small multiplier of the computational load of the other methods. During times of stability, however, the longer time steps permitted by this method can permit much faster execution than is possible for explicit methods.

Implicit methods offer the following pros and cons:

Advantages:

- The implicit method often has approximately second-order accuracy, which outperforms the explicit and MOC approaches.
- Solutions are often very stable or even unconditionally stable, permitting much longer time steps.
- Computations in large pipeline systems become more tractable for given computer resources because the Courant condition is much less restrictive.

Disadvantages:

- Solution methods require complex matrix and sparse matrix calculations.
- Debugging solutions can be more difficult.
- May be subject to significant numerical dispersion problems depending on spacing and time step issues.

4.2.4 A Comparison of Numerical Methods

When implemented with care and as appropriate to the problem at hand, any one of the different numerical methods can provide the level of pipeline simulation required for the RTTM. More important than the actual numerical method is the completeness of the model (ie, how well the equations being solved represents the actual pipeline conditions), the approach to using measurements as inputs to the RTTM (see Section 4.2.3), and the approach used to extract a leak signal from errors and "noise" (discussed in detail in Chapter 5: Statistical Processing and Leak Detection). There are other numerical methods that could be used, but the explicit finite difference, implicit finite difference, and method of characteristics approaches discussed here are predominantly being used.

However, when comparing explicit and implicit finite difference approaches, the clear winner seems to be implicit methods because of the greater solution stability and the higher solution accuracy for a given time and distance step. Although an implicit modeling solution is more challenging to implement, the flexibility and higher accuracy are gradually making this approach the method of choice in many RTTM leak detection applications.

From the point of view of the end user, the numerical method used in the RTTM should not be a deciding factor. However, in the long-term, we expect that implicit finite difference methods will become the norm, if they are not already.

4.3 MEASUREMENTS AND BOUNDARY CONDITIONS

In this section, we discuss the choices that an RTTM must make in the design of the product in relation to how measurements are used in the RTTM and how discrepancies between measured and modeled values are handled.

4.3.1 Measurement Placement, Availability, and Reliability

Real-time transient models are ultimately dependent on the availability of timely real-time data that are used as inputs to the RTTM and the RTTM analyses. Ideally, the following would be made available to the RTTM within seconds of real time through a Supervisory Control and Data Acquisition (SCADA) system.

- Accurate measurement of every flow entering or leaving the system
- Intermediate flow measurements sufficient to divide the pipeline into useful sub-systems for leak detection analysis
- Pressure and temperature measurements at all branch points upstream and downstream of every active device that might change the pressure in the pipeline (eg, valves (but not necessarily check valves), compressors, pumps)
- Pressure and temperature measurements at intermediate points limiting the total span between pressure measurements. The appropriate span will depend on specific needs and pipeline characteristics, but shorter spans are always better—from a leak detection and leak location perspective.
- Real-time measurement of fluid properties, specific gravity, or fluid composition. The RTTM needs enough data to be able to compute the following at pipeline conditions:
  - Fluid viscosity
  - Fluid density as a function of pressure and temperature
  - Fluid heat capacity
- Valve positions for any valves that may isolate other instrumentation from the monitored pipeline (eg, valves that might isolate pressure measurements from the pipeline) or that might isolate one portion of the pipeline from another.

As an illustration, consider Figs. 4.5 and 4.6. Fig. 4.5 illustrates a pipeline well instrumented for leak detection with the following salient details

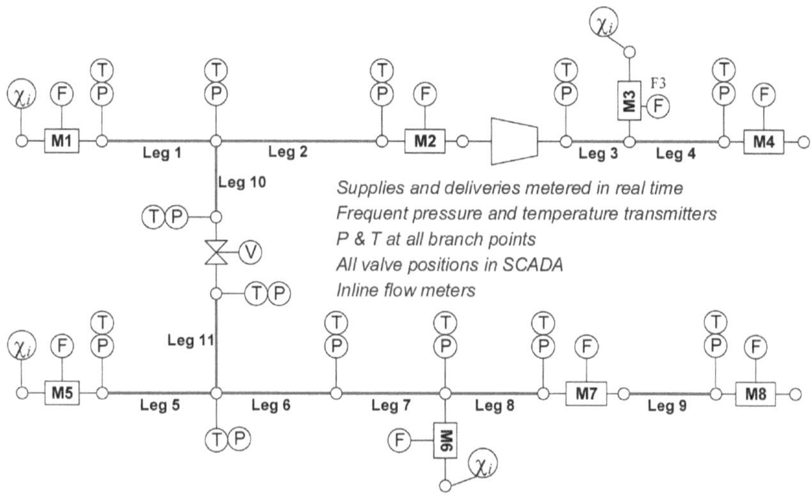

FIGURE 4.5 Pipeline well-instrumented for leak detection.

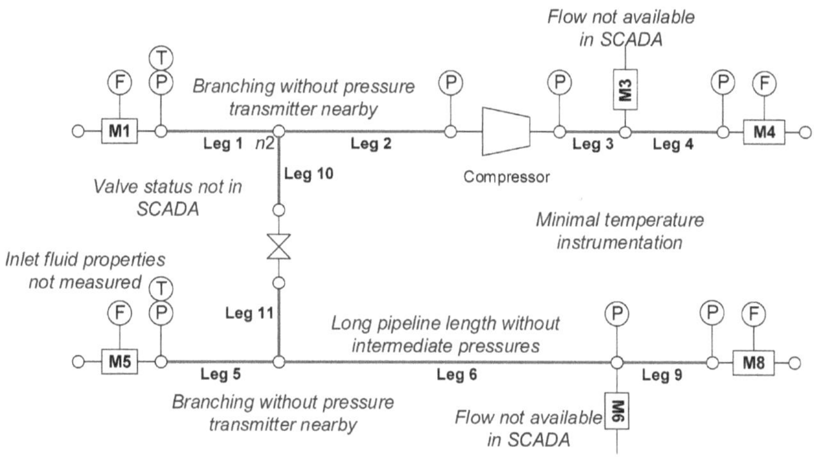

FIGURE 4.6 Pipeline poorly instrumented for leak detection.

(measurements are denoted by symbols within circles with F, T, P, V, and χ_i representing flow rate, temperature, pressure, valve position, and fluid properties).

1. All flows into and out of the pipeline are metered, allowing one to calculate a flow balance across the pipeline system.
2. At some intermediate points, inline flow measurements are provided (M2 and M7), allowing one to do flow balances across smaller portions of the pipeline system.

3. Pressure and temperature measurements are available at the extremes of the pipeline system, at all branch points, and on both sides of all active devices (eg, valves and compressors). They are also available at additional locations along the pipeline, such as between Leg 6 and Leg 7, Leg 7 and Leg 8, and Leg 8 and Leg 9.
4. Fluid properties are measured for all flows entering the pipeline (and preferably at pipeline outlets).

In contrast, Fig. 4.6 shows a pipeline poorly instrumented for leak detection. Notable deficiencies include:

1. Some flows entering or leaving the pipeline are not metered (M3 and M6).
2. Inlet fluid properties are not measured.
3. Pressures are not available at a valve that may close.
4. Valve position not available for valve.
5. Long leg (Leg 6) without any intermediate pressure measurement

That is not to say that an RTTM requires perfect instrumentation or, in particular, that there is no benefit to installing an RTTM on a poorly instrumented pipeline. Useful data and leak detection can be provided with less than ideal instrumentation. However, the robustness of the system is directly influenced by the accuracy, reliability, and availability of real-time pipeline measurements. Robustness, in usage, incorporates a combination of the following:

- High availability
- Low rate of false alarms
- High-sensitivity leak detection
- Fast response time

The robustness of an RTTM-based leak detection system is directly dependent on the quality, placement, availability, and completeness of the pipeline instrumentation that is available in real time (through SCADA) for timely use by the RTTM-based leak detection system.

Data acquisition rates and time skew also affect the ability of an RTTM to accurately represent fast transients in a pipeline system. In a liquid pipeline, periodicity between data updates of a few seconds is desirable, whereas on a gas pipeline 30 second periodicity can be acceptable. Still, an RTTM can provide benefit even with substantially longer acquisition rates, although speed of response may suffer. The Nyquist-Shannon criteria states that the highest frequency that can be represented in sampled data is one-half the sampling rate [4]. If data are sampled every 30 seconds, then the highest frequency that can be represented by that data are transients with a rise time of 60 s.

Furthermore, if there are higher stable frequencies in the underlying data, then a phenomenon called "aliasing" can result in those higher frequencies observed as much longer frequency transients in the sampled data. Consequently, one should not expect an RTTM to realistically simulate transients whose rise time is faster than a few times the sampling rate. For example, if data are sampled every 30 s, then one might expect good modeling fidelity for transients with a rise time of 2 min or longer. Recommendations related to sampling of field data, reduction of noise in the data, and the problem of time skew are discussed in more detail in Chapter 8, Leak Detection System Infrastructure.

4.3.2 Selection of Boundary Conditions

One defining characteristic that separates different RTTM implementations is the selection of boundary conditions. Boundary conditions are the inputs that bind the RTTM solution to the real pipeline. The following boundary conditions are required for an RTTM.

1. Temperature and pressure or flow rate for every flow into the pipeline.
2. Pressure or flow rate for every flow out of the pipeline. Note that temperature cannot be imposed as a boundary condition for flows leaving the pipeline.
3. Fluid properties sufficient to specify viscosity, equation of state (density as a function of pressure and temperature), and heat capacity for every flow entering the pipeline.
4. Pressure on both sides of every active device and temperature on the downstream side (eg, valves that may open or close, compressors, pumps, throttling devices, etc.). Alternatively, one might substitute a model of the device's operation, but this is less desirable than actual measurements.

In addition, intermediate pressure and temperature measurements and internal flow meters are helpful.

Note that if one imposes a pressure boundary condition, the computed flow rate at that point may not match the measured flow rate (and vice versa). Some implementations allow for a linear combination of flow and pressures as boundary conditions where neither is imposed directly. Each is adjusted in a fashion that provides some minimization of the discrepancy between measured flow and computed flow and between measured pressure and computed pressure.

There are almost always more measurements available than can be imposed as direct inputs to a real-time model, and there are many ways in which the measurements can be imposed as boundary conditions on the RTTM. Various RTTM vendors have arrived at different approaches to applying the available measurements as boundary conditions on the pipeline. The approach used to apply measurements as boundary conditions is a defining characteristic of specific RTTM.

4.3.3 Boundary Condition Strategies

In this section, we consider different approaches to imposing measurements as boundary conditions on the RTTM. Consider the very simple but well-instrumented pipeline of Fig. 4.7.

One approach to applying boundary conditions is to impose all of the available pressures as boundary conditions, segmenting the pipeline into separately modeled elements. Therefore, for our sample pipeline, we might model it as illustrated in Fig. 4.8.

In this example, Leg 1 is modeled with upstream pressure and temperature and downstream pressure as boundary conditions. The RTTM for this leg can therefore be solved as a single entity. Similarly, the equations describing the RTTM state for Legs 2, 3, and 4 can be solved for independently of the other legs. Because each leg is modeled independently of other legs, we refer to this as a segmented model.

Variations of this approach allow for branching to occur without pressure measurement at the branch point. In that case, multiple legs would be solved as a unit. However, the fundamental characteristic of our setup is that maximum use is made of measured pressures and temperatures as boundary conditions on the pipeline. By imposing measured pressures and temperatures wherever possible, we hope to obtain the best representation of the packing rate. Note, however, that nothing constrains the flows, leaving one leg to match those entering the next. When this paradigm is adopted, other methods must be used to ensure that fluid properties move through the pipeline at an appropriate rate and to ensure that batches do not grow or shrink artificially as they move from one leg into the next (eg, state estimation designed specifically for this purpose; see Section 4.4).

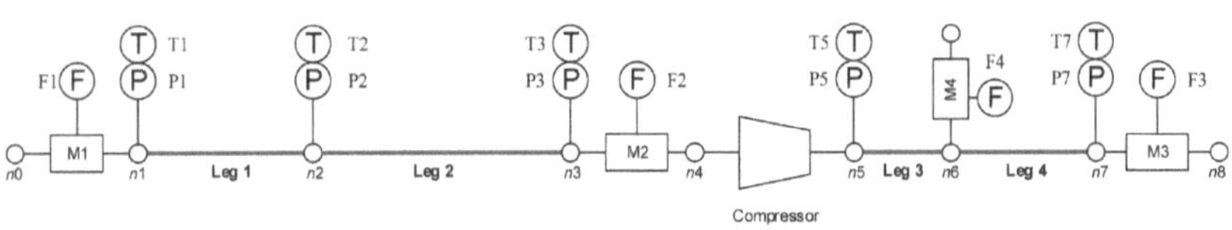

FIGURE 4.7 Example pipeline.

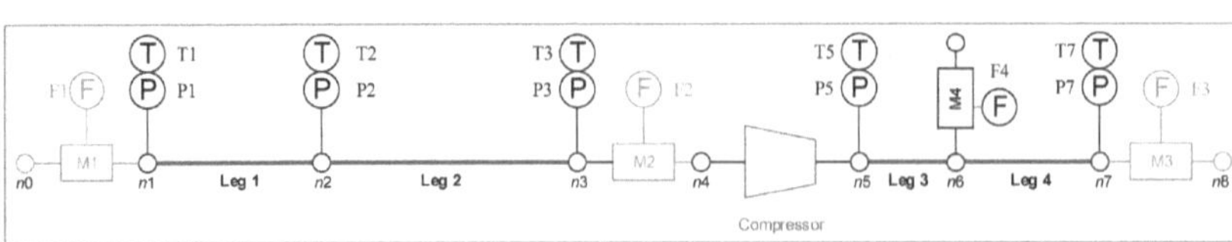

FIGURE 4.8 Example pipeline: maximum use of pressures and temperatures as boundary conditions.

Other approaches for applying measurements as boundary conditions in commercially available RTTMs include:

- Imposing a minimal set of boundary conditions and modeling the pipeline as a unified whole rather than as a series of relatively independent "legs." We refer to this as a unified model.
- Modeling the pipeline as a unified whole, but with the specification of an extra flow into or out of the pipeline at each pressure measurement location, Then, an adjustment is made to every measured pressure such that some totalization of the extra flows is minimized, subject to an objective function that attempts to develop a best fit between the model and measured flows and pressures. We refer to this approach as a modified unified model.

Each of these approaches has its proponents. There are other variations that can be envisioned as well. From the perspective of leak detection, one desires an approach that provides the best leak signal or set of leak signals. Unfortunately, the industry has not reached a consensus on this issue.

4.4 STATE ESTIMATION AND RELATED SUBJECTS

State estimation is the process of developing a "best" estimate of the state of a process by evaluating a collection of measurements representing portions of the process state along with a mathematical representation of the overall process, with the goal of developing a "best" possible representation of the true process state. Take, for example, the segmented pipeline model of Fig. 4.8 and consider the RTTM solution for the state of Leg 1 and Leg 2. We use "left" and "right" to identify the ends of the legs. Using P1 and P2 as the pressure boundary conditions at the left and right ends of Leg 1, respectively, provides a calculation of the flow rates at the left and right ends of Leg 1 (because this is a transient model, these flows will not likely be the same). Similarly, imposing P2 and P3 as boundary conditions on Leg 2 provides a calculation of the flow rates at the left and right ends of Leg 2. However, nothing in the modeling equations constrains the flows to balance between legs or between legs and flow meters.

- Measured flow F1 will not precisely match the left end flow of Leg 1
- Measured flow F2 will not precisely match the right end flow of Leg 2
- The right end flow of Leg 1 will not match the left end flow of Leg 2

This issue is not unique to the segmented model. The same issue arises in the unified model. However, in that case, the inconsistencies are mostly differences between measured and modeled pressures.

State estimation provides a way of accommodating these inconsistencies. The goals of state estimation will vary based on the technical assessments of the RTTM provider and are certainly very dependent on the boundary conditions selected for use for the RTTM. Using the segmented modeling paradigm as an example, because the flows are not forced to balance between separately modeled elements or between legs and flow measurements, it is useful, and for some purposes necessary, to develop a set of flow corrections, one for each independent leg and one for each flow meter that balances flows across every node. The state estimation formulation for this objective could be stated as follows:

> *Compute an additive flow rate correction for every leg and every flow measurement such that when each is normalized by its expected uncertainty, the sums of the squares of the normalized corrections are minimized subject to the constraints that the corrected flows balance across each node in the pipeline system.*

Although it is essential for the RTTM to provide a computation of the pipeline packing rate to use for leak detection, it is also important for the RTTM to move fluid properties and convect temperature changes at the proper speed through the pipeline system. An RTTM based on segmented models might rely on state estimation to compute a "best" set of flow corrections. Then, instead of propagating fluid properties and thermal effects down the pipeline at the computed velocities of the RTTM, the RTTM could use the estimated corrected flow rates of the state to ensure that properties and thermal effects propagate at an appropriate velocity and that batch sizes are preserved as they move through the pipeline.

Therefore, state estimation might be an integral component of an RTTM. For an RTTM based on a segmented modeling approach, it may be essential to ensure that fluid properties and thermal effects propagate at appropriate and self-consistent rates through the pipeline system.

Note, also, that the modified unified model applies state estimation in a different way. Instead of applying corrections to the RTTM results at the end of a time step, every measurement is corrected before being imposed as a boundary condition. This is a form of state estimation applied in a different way.

There is no one perfect solution to the issue of inconsistencies and the fact that generally the system is overconstrained. Imposing pressure at a point requires releasing a flow constraint and vice versa. It is important to move fluids through the pipeline system at appropriate and internally consistent flow rates to avoid artificial growth or shrinkage of batch sizes that will occur if flows are not consistent between one portion of the pipeline and the next. However, one wishes to make best use of the available pressure and temperature measurements to obtain the best possible calculation of the packing rate. State estimation provides a tool for navigating these conflicting demands. However, there are possibly as many different approaches to dealing with these issues as there are vendors selling RTTM-based leak detection systems.

A related subject that we do not deal with directly here is automatic tuning of modeling parameters within an RTTM. Automatic tuning involves the adjustment of modeling parameters to reduce the discrepancies between modeled and measured data. RTTM vendors address this to various degrees of sophistication. The appropriate automatic tuning approaches are dependent on the vendor's approach to selection of boundary conditions and state estimation.

4.5 LEAK DETECTION SIGNALS

An RTTM represents the pipeline as if there is no leak present. Eq. (4.1), the continuity equation (mass balance), imposes the constraint that mass is conserved within the RTTM of the pipeline components and, most fundamentally, within the lengths of pipe between measurement locations. A leak violates this principle. Therefore, a leak causes the RTTM calculations to diverge from the actual pipeline conditions. The actual divergence observed by the RTTM resulting from a leak depends on the boundary conditions that are imposed as inputs to the pipeline model.

The most straightforward way to conceptualize the purpose of an RTTM in leak detection is to consider it to be a packing rate calculator. Recall Eq. (3.5) from Chapter 3, Mass Balance Leak Detection:

$$MB_{\mathrm{MBS,Observable}} = MFB_{\mathrm{MBS}} - MPR_{\mathrm{MBS}}$$

EQUATION 3.5 Observable Mass Balance Definition

Assume that all flows are metered into and out of a section of the pipeline (which we refer to as an MBS), the flow measurements themselves provide all of the data required to compute the flow balance. Combined with the packing rate computed by the RTTM, the volume balance can provide our primary leak detection signal:

$$\text{If } MB_{\mathrm{MBS,Observable}} > Threshold(U(MB_{\mathrm{MBS,Observable}}))$$
$$\text{then}$$
$$MB_{\mathrm{MBS,Observable}} = \text{Leak Size Estimate}$$

EQUATION 3.13 Practical Statement of Mass Balance Leak Detection

We use the segmented model of Fig. 4.8 as an example. We can compute the mass or standard volume balance between M1 and M2, M2 and M3, and for the entire pipeline between M1 and M3. The equations for the volume balance for these three sections are listed in Eq. (4.15) (Fig. 4.9).

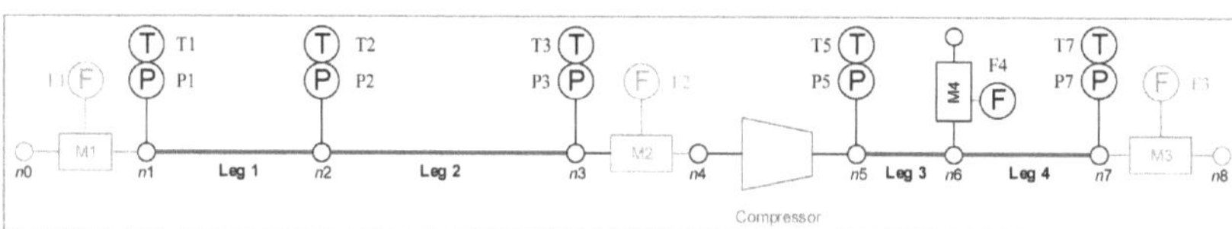

FIGURE 4.9 Example pipeline: segmented pipeline model.

$$VB_1 = F1 - F2 - (PR_{\text{Leg 1}} + PR_{\text{Leg 2}})$$
$$VB_2 = F2 - F3 - F4 - (PR_{\text{Leg 3}} + PR_{\text{Leg 4}})$$
$$VB_{\text{Pipeline}} = F1 - F3 - F4 - (PR_{\text{Leg 1}} + PR_{\text{Leg 2}} + PR_{\text{Leg 3}} + PR_{\text{Leg 4}})$$

EQUATION 4.15 Volume Imbalance Leak Signals for Example Pipeline Segmented Model

Note, however, that there are other signals that can be used for leak detection. For each node or collection of nodes bounded by legs and/or flow meters, we can define the flow discrepancy as flow in minus flow out. For the example pipeline, we can define the flow discrepancies as:

$$FD_{n1} = F1 - F_{\text{Leg 1,Left}}$$
$$FD_{n2} = F_{\text{Leg 1,Right}} - F_{\text{Leg 2,Left}}$$
$$FD_{n3} = F_{\text{Leg 2,Right}} - F2$$
$$FD_{n4,n5} = F2 - F_{\text{Leg 3,Left}}$$
$$FD_{n6} = F_{\text{Leg 3, Right}} - F4 - F_{\text{Leg 4, Left}}$$
$$FD_{n7} = F_{\text{Leg 4, Right}} - F3$$

EQUATION 4.16 Flow Discrepancy Signals for Example Pipeline Segmented Model

where $F_{\text{Leg }x\text{,Left}}$ and $F_{\text{Leg }x\text{,Right}}$ are the RTTM computed flows at the left and right ends of Leg x.

4.6 USING THE LEAK SIGNALS TO DETECT LEAKS

In our segmented model that we have illustrated by the example here, the RTTM provides a real-time calculation of the volume balance of a pipeline mass balance section, $VB_{\text{MBS}}(t_i)$, where t_i is the time of the calculation. The biggest problem we have is that the signal is noisy, and that the noise may be of the same order or larger than the leak we are trying to detect. A common approach to leak detection is to minimize the noise by averaging this signal over one or more time periods. For example, one might have a series of averaging periods over which the signal is averaged, such as 5 min, 15 min, 1 h, and 4 h. Presumably, the leak threshold would be higher for short periods and lower for longer periods, whereas response time would be fastest for short periods and slowest for long periods. To remove any long-term bias or time-correlated noise from the signal, we might also compute a long-term average of the volume balance or use some other method to decorrelate the signal. Other issues that might need to be addressed would involve determining just the right threshold values to minimize the chances of having a false alarm while also maximizing the leak detection sensitivity of the system.

The issue of reliably identifying a leak signal in the midst of errors from a number of sources (commonly called noise) is a fundamental component of any leak detection system. The simple example presented here is one approach. We discuss this issue in more detail in Chapter 5, Statistical Processing and Leak Detection.

4.7 ESTIMATING LEAK LOCATION

Pinpointing the location of a leak is substantially more difficult than detecting a leak. By that, we mean that the leak location calculations are more sensitive to uncertainties in the pipeline model than are the leak detection calculations. Leak location can be found using two approaches: (1) detection of a rarefaction wave generated by the leak or (2) the pattern of inconsistencies in the RTTM results. The former can only be applied when the scan rate is sufficiently high and measurement locations are sufficiently close together. Therefore, it is often not appropriate for RTTM-based leak detection and location calculations. We leave the discussion of the rarefaction wave leak location to Chapter 6, Rarefaction Wave and Deviation Alarm Systems.

In particular, we focus on estimating leak location with the segmented RTTM approach using the pattern of flow discrepancies discussed in Section 4.3.3. This approach also applies to the unified model with an extra flows approach because each extra flow corresponds to a flow discrepancy of the segmented model.

Let us first consider what the pattern of flow discrepancies would be in a perfect world in which:

1. The pressure and flow measurements have no error
2. The RTTM assumptions are precisely correct; fluid properties are known precisely and there is no error in any modeling parameter

In this case, the RTTM results would be perfect in every leg of the pipeline, except for the leg containing the leak. For example, consider the pipeline of Fig. 4.8. If the leak were within Leg 2, then the only nonzero flow discrepancies would be FD_{n2} and FD_{n3}. Once the effect of the leak has fully propagated to the leg ends, the leak size can be estimated as the sum of these two flow discrepancies. One can demonstrate that if the leak is close to the upstream end ($n2$), then FD_{n2} will be largest; if it is close to the downstream end, then FD_{n3} will be largest. In fact, if it is precisely at the end of the leg, then all of the leak will be observed in that end's flow discrepancy and none will be observed at the other end. Note that if the leak is very near the upstream end, $n2$, then it will be impossible to determine whether the leak is in Leg 1 or Leg 2, it will only be possible to determine that the leak is near $n2$.

Even in a perfect world, pinpointing the leak location within a leg is complicated by the fact that the leg flow equations are nonlinear. However, when the leak flow rate is small compared to the actual leg flow rate, one can obtain the following linearized estimate of the leak location expressed as a fraction of the length between the upstream and downstream node.

$$\text{Fractional leak location} = \frac{FD_{\text{down}}}{FD_{\text{up}} + FD_{\text{down}}}$$

EQUATION 4.17 Linearized Leak Location Estimator

Estimating leak location is further complicated when more than two legs join together without a pressure measurement at the interconnection.

Of course, errors in modeling assumptions and measurements further complicate leak location estimation. Certainly, as the leak becomes larger, the impact of the leak becomes greater when compared to the noise. We can generally state the following:

At the limits of detectability of a leak, the leak location error is high and it may be impossible to estimate with any reasonable degree of confidence. There is a direct correlation between the size of the leak and the magnitude of the signals available for estimating the leak location. Therefore, larger leaks will have more easily discernible location signals. However, the nonlinearity in the relationship between location signals (eg, node flow discrepancies) and estimated leak location increases with leak size and requires more complex leak location calculations than the linearized expression of Eq. (4.17).

Desirable characteristics of a leak detection system are the ability to estimate the uncertainty of the leak location and the ability to estimate the leak location taking into full account the nonlinearities inherent in the flow equations.

4.8 IMPACT OF FLUID TYPE: LIQUIDS, GASES, AND MULTIPHASE FLOWS

RTTM-based leak detection operates on fundamentally the same principles regardless of the pipeline fluid. However, fluid characteristics have obvious impacts on the response of the pipeline to fluid transients, which must be simulated in an RTTM to provide sensitive leak detection.

4.8.1 Liquid Pipeline Leak Detection

RTTM-based leak detection systems for liquid pipelines have been a significant focus in the leak detection industry. That may be due, in part, to the fact that leak detection for liquid pipelines is easier than for gas pipelines because the liquids are much less compressible. This is certainly true if the liquid pipeline remains full of liquid at all times. However, it is usual for liquid pipelines spanning any significant elevation changes to exhibit slack line flow (a form of two phase flow) during some operating conditions, particularly during pipeline shutdown and restart.

4.8.2 Gas Pipeline Leak Detection

Because gases are orders of magnitude more compressible than liquids, issues related to the compressibility of the gas are a predominant factor in gas RTTMs. However, the concepts behind a gas RTTM-based leak detection system are precisely the same as those of a liquid pipeline RTTM. Consequently, a well-designed RTTM for gas pipelines and liquid pipelines can work equally well, with the only fundamental difference being the fluid equation of state. All other RTTM equations described in Section 4.1.1 and the leak detection signals generated by the RTTM described in Section 4.5 can be identical for both gas and liquid RTTMs.

Natural gas volume changes are approximately seven-times more sensitive to temperature variations and 300-times more sensitive to pressure variations than crude oil. This has a direct impact on the packing rates experienced by a gas pipeline. As a consequence, an RTTM-based leak detection is the only mass balance leak detection approach that can work well for most gas pipeline.

There are other differences in the behavior of leaks in gas pipelines compared to liquid pipelines.

The speed of sound is somewhat slower in gas pipelines than in liquid pipelines. Therefore, the effects of leaks propagate somewhat more slowly

Because of the high compressibility of gas, a much larger pressure gradient can persist in the region surrounding a leak than can exist in a liquid pipeline. This occurs if the pressure drops substantially in the area of the leak and the expansion of the gas results in much higher velocities near the leak than elsewhere in the pipeline. Because frictional losses are proportional to nearly the square of the velocity, the pressure gradients can be very high.

The density of the gas near a leak site can be very different than that elsewhere in the pipeline. In contrast, this is generally not the case in liquid pipelines. This results in higher nonlinear hydraulic effects in gas pipelines than in liquid pipelines. Among other impacts, leak location calculations based on linearized approximations can be significantly degraded.

An expansion wave propagating from a leak attenuates faster in a gas pipeline than in a liquid pipeline.

Further discussion of gas pipeline RTTM-based leak detection is available elsewhere [5].

4.8.3 Liquid Pipelines With Slack Line Flow

When a pipeline transporting liquids (such as crude oil) transverses an elevation peak, the pressure in the pipeline drops as the elevation rises. The head

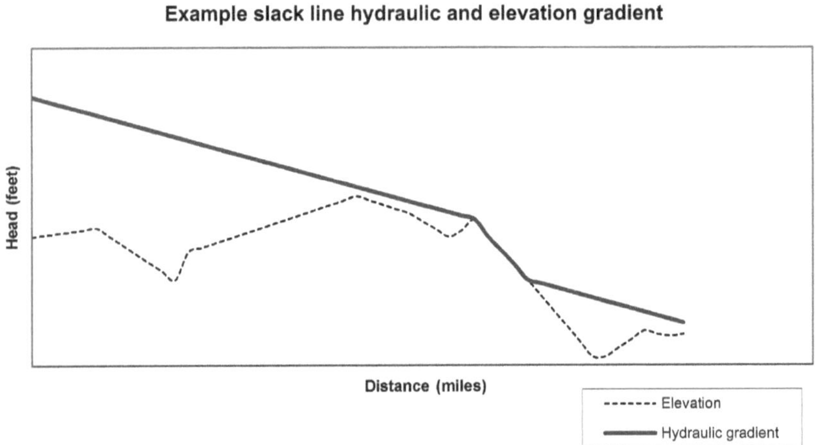

FIGURE 4.10 Slack flow head gradient example.

gradient downstream of the peak is governed by frictional losses. If the pressure at the downstream end (or next pump station suction) is low enough, then the pressure at the peak may drop to vapor pressure and the liquid will vaporize. If the pressure drops even lower at the downstream end, then the fluid will be partially liquid and partially gas downstream of the elevation peak until the elevation drops enough that the pressure in the pipeline exceeds the vapor pressure and the fluid becomes all liquid again. The pressure of the gas/liquid flow regime downstream of the peak is equal to the vapor pressure of the fluid, and this type of flow is referred to as slack line flow. In steady conditions, the slack/tight intercepts are at the peak and at the downstream elevation at which the tight flow head gradient (also referred to as the hydraulic gradient) intercepts the elevation profile, as shown in Fig. 4.10. The slack line flow region is the region in which the head gradient parallels the elevation gradient. Fig. 4.10 is drawn as if the vapor pressure is equal to atmospheric pressure. If it is higher or lower, then the hydraulic gradient in the slack region would be offset from the elevation gradient by the pressure difference (converted to head) between vapor pressure and atmospheric pressure. Because all pipeline liquids vaporize at some point before 0 psia, the maximum negative offset of the head gradient from the elevation gradient would be equal to the fluid head corresponding to 14.7 psia. Slack line flow is discussed in more detail elsewhere [6].

Slack line flow is much more difficult to model than either gas or liquid flow; therefore, it poses special challenges to an RTTM. The modeling challenges include the following:

1. Vapor pressure is a function of fluid composition and temperature. It is difficult to know it precisely at pipeline conditions unless it is a pure fluid.

2. Temperature of the fluid is difficult to model to a high degree of accuracy due to varying soil conditions and ambient temperatures. Therefore, even for a pure fluid, vapor pressure is uncertain.
3. The slack/tight intercept is directly impacted by frictional head losses that control the slope of the head gradient. Therefore, the length of the slack region has uncertainties directly dependent on uncertainties in the friction factor.
4. The vapor/liquid fraction in the slack region has a complex dependency on the upstream flow rate and the frictional losses in the slack region. It is difficult to estimate the vapor/liquid fraction accurately.
5. There may be several slack/tight transitions downstream of an elevation peak.

In addition to these real physical effects that impose uncertainties in the line pack and packing rate, there are numerical challenges involved in modifying the equations of motion presented in Section 4.1.1 to realistically represent the slack conditions and to solve them simultaneously with the adjoining tight line equations. The slack region behaves as if it has nearly infinite compressibility compared to the very "stiff" adjoining tight sections.

Consequently, leak detection in pipeline sections that flow constantly or intermittently slack is subject to many more errors than either liquid or gas leak detection. RTTM leak detection is feasible, but either sensitivity or detection time will suffer. However, other than direct detection of fluid outside of the pipeline, there are no better techniques for leak detection in slack regions.

4.8.4 Multiphase Flow–Based RTTMs

Multiphase flows pose many more modeling uncertainties than either liquid or gas RTTMs. A general multiphase flow model is far more complex than a single phase fluid model. Transient multiphase models designed for engineering simulations have been modified to be driven with real-time measurements as an RTTM. However, this remains an area of challenge for pipeline leak detection.

4.8.5 Dense Phase Fluids

Fluids that have a critical temperature in the range of normal pipeline temperatures will likely have a very strong dependence on temperature at pipeline conditions. For example, the density of ethane (critical temperature = 90°F) and ethylene (critical temperature = 48.6°F) can be very dependent on temperature. Because of this, models of these pipelines will be more subject uncertainties resulting from thermal effects than others. Still, an RTTM-based system is likely to provide significant benefit.

4.9 RTTM UNCERTAINTY RECAP

We have discussed the fact that even an RTTM is subject to a great number of uncertainties due to a variety of factors, including:

1. Errors in input measurements
2. Errors resulting from time skew of the input measurements (see Chapter 8: Leak Detection System Infrastructure)
3. Real physical pipeline unknowns such as roughness and thermal properties of pipeline surroundings
4. Uncertainty in fluid properties such as viscosity (at all pipeline conditions), heat capacity, and equation of state
5. Numerical errors in the solution of the pipeline equations
6. Physical approximations on the pipeline equations
7. The approach used by the RTTM developer to deal with inconsistencies between measurements, noisy data, and other factors

Note that API 1149 [7] is devoted to a discussion of many of these factors. These factors are also discussed in numerous technical papers such as [8], [9], [10], and [11] to list but a few.

RTTM-based systems are certainly the most reliable type of mass balance system available. However, uncertainties limit the sensitivity and speed of response of the leak detection system. Even over the course of long leak detection times, the sensitivity is likely to never better 0.1% of the pipeline flow rate and can be somewhat worse.

REFERENCES

[1] Modisette J, Nicholas E, Whaley R. A comparison of transient pipeline flow models and features. In: PSIG annual meeting; October 18–19, 1984. Available at https://www.onepetro.org/conference-paper/PSIG-8401.

[2] Wylie EB, Streeter VL. Fluid transients in systems. Prentice Hall; 1993.

[3] Chaudhry MH. Applied hydraulic transients. 3rd ed. Springer; 2014.

[4] Wikipedia Contributors. Nyquist–Shannon sampling theorem. Wikipedia, The Free Encyclopedia; 1 Apr. 2016. Web. 2 Apr. 2016.

[5] Nicholas E, Carpenter P, Henrie M. RTTM-based gas pipeline leak detection: a tutorial. In: PSIG annual meeting; May 12–15, 2015. Available at https://www.onepetro.org/conference-paper/PSIG-1504.

[6] Nicholas E. Simulation of slack line flow – a tutorial. In: PSIG annual meeting; October 19–20, 1995. Available at https://www.onepetro.org/conference-paper/PSIG-9505.

[7] API Technical Report 1149. Pipeline variable uncertainties and their effects on leak detectability. 2nd ed.; September 2015.

[8] Nicholas RE. Leak detection and location sensitivity analysis. In: Pipeline engineering symposium; 1992, PD, vol. 46. New York: The American Society of Mechanical Engineers.

[9] Nicholas RE. Leak detection by model compensated volume balance. In: Pipeline engineering symposium; 1987, PD, vol. 6. New York: The American Society of Mechanical Engineers. p. 13–20.

[10] Nicholas RE. Leak detection on pipelines in unsteady flow. In: Forum on unsteady flow; 1990, FED, vol. 102. New York: The American Society of Mechanical Engineers. p. 13–20.

[11] Whaley RS, Mailloux JL, McDonnold J. Model based leak detection on gas pipelines – results from the field. In: PSIG annual meeting; 1991.

Chapter 5

Statistical Processing and Leak Detection

Leak detection systems are time-series, decision-making, information-based systems. Over a given set of time, the leak detection system receives a series of information inputs, calculations occur, and decisions are derived. Furthermore, and especially for an internal leak detection system (LDS), the information inputs are time series of numerical quantities (such as pressures and flows sampled at discrete points in time). The leak signals developed by the LDS are also usually a time series of values (such as mass balance for each processing step).

Numeric uncertainty and noise are always present in the input data and, consequently, in the leak signals produced by the leak detection system. This chapter discusses approaches used to deal with uncertainties that are inherent in the leak signal. These approaches are potentially applicable to time-varying leak signals for any type of leak detection system. Dealing with noise and errors in the LDS inputs is discussed separately in Chapter 8, Leak Detection System Infrastructure.

Leak signal uncertainties result from noise in the data inputs to the leak detection system, data conditioning within the system, and errors inherent in the leak signal computation. We refer to the LDS processing that converts the input data into one or more leak signals or alarms as the leak signal computation.

For example, the process of calculating the observable mass balance from the time changing pipeline inputs is a leak signal computation. Uncertainties inherent in the measurement data, their processing (data conditioning), and calculation errors inherent in the model that are used to compute the packing rate all contribute to the leak signal noise. Chapter 3, Mass Balance Leak Detection and Chapter 4, Real-Time Transient Model–Based Leak Detection discuss uncertainties inherent in the mass balance and model calculations.

In the following sections, we delve into some of the fundamental methods commonly used to process the leak signal. We follow this with a discussion of more rigorous statistical methods that can be applied to these signals.

Pipeline Leak Detection Handbook. DOI: http://dx.doi.org/10.1016/B978-0-12-802240-5.00005-4

5.1 INTRODUCTION TO LEAK SIGNAL PROCESSING

In this section, we describe a very common leak signal processing approach used to determine when a leak signal is large enough to warrant the declaration of a leak alarm: the comparison of the leak signal to a threshold. With this approach, a leak detection threshold is established by some means: it could be fixed in advance, could be set through a manual tuning process, or could be developed automatically through some statistical or other automatic method. When the leak signal violates the threshold, an alarm is declared. Using this paradigm, we illustrate various challenges that one faces when turning a leak signal into a reliable leak alarm.

Let us start by taking a simple approach in which one establishes a fixed leak threshold value that has no dependence on process variables. In this case, absolute accuracy of the leak signal is critically important. That said, the leak signal will have uncertainties that may include bias and noise.

Bias is a difference between the output and the true value that is constant over time. This can result from field data bias that the input data conditioning did not eliminate, or leak signal bias generated by the leak detection algorithm. A fixed bias is often easy to remove. However, leak detection signals may also exhibit slowly changing errors that may appear over shorter periods as a bias. These slowly varying errors are referred to as shifts and drifts. These are much more difficult to remove because they may resemble a leak or could look like a bias. When dealing with a fixed threshold, the LDS has no obvious way to distinguish slowly varying shift or drift from a real leak. In that case, a constant threshold value must be set high to avoid false alarms during the worst case.

Alternatively, a leak detection system may compute a *dynamic* threshold based on process conditions, the history of the leak signal, or other factors. In this case, the dynamic threshold itself may be subject to the input and leak signal uncertainties.

Once a thresholding scheme has been established, the question then becomes when to declare the leak alarm. The simplest approach is to declare an alarm every time the leak signal exceeds the threshold. For a given threshold, this approach results in the fastest alarm declaration.

One negative aspect of this method is that multiple alarms and associated alarm clears will likely occur. This is a result of a noisy leak signal varying around the alarm threshold. As the leak signal fluctuates above the threshold, an alarm is declared. As it fluctuates back below the threshold, the alarm clears.

A second approach that eliminates this intermittent leak alarm problem requires the derived leak value to continuously exceed the leak threshold for some minimum time. Similarly, to clear the alarm, the leak signal must drop below the leak threshold for a minimum period. Note that a downside is that this results in a delay before issuing and clearing alarms.

In the previous discussion, we addressed signals that can be expected to be approximately constant once the leak has occurred. However, there are LDS technology solutions, such as some external spill volume detectors (see Chapter 7: External and Intermittent Leak Detection System Types) or some mass balance systems that utilize cumulative volume imbalances (see Chapter 3: Mass Balance Leak Detection), where the leak signal will grow over time. In this case, one can further refine the declaration logic to require not only the threshold to be violated and persistence to be met (previous paragraph) but also the signal to continue to increase.

Let us consider some examples. Fig. 5.1 demonstrates a fixed threshold of 250 barrels per hour (BPH) with a varying leak rate. At approximately time 46, the derived leak goes above the leak threshold. With no signal conditioning, a leak alarm would be generated. Then, at time 47, the derived leak signal drops below the leak threshold and the alarm would clear. As shown in this figure, this sequence would continue for the duration of the chart. Correlated instrument noise and a flow meter offset could easily produce these results.

Fig. 5.2 provides an example of an approach that helps reduce false alarms. This method utilizes a time delay, or persistence of, for example, five time units, after the time when the derived leak value first exceeds the leak threshold value. In this figure, the derived value first crosses the threshold at approximately time 41 but drops below it again at time 42. No alarm occurs because the persistence requirement is not completed. At approximately time 50, the derived leak finally goes above the threshold and stays. The alarm finally occurs at time 56, as the persistence value is addressed. In this example, the persistence requirement prevented one false alarm sequence, but the time to declare the alarm took longer.

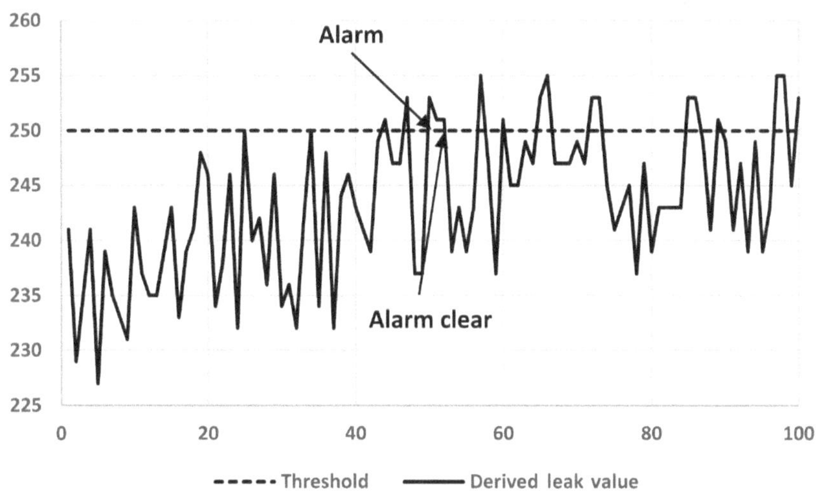

FIGURE 5.1 Fixed threshold example.

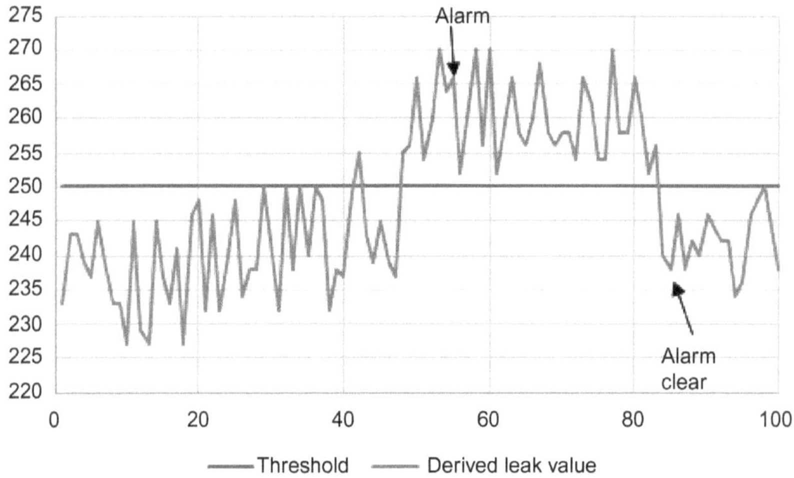

FIGURE 5.2 Delayed alarm example.

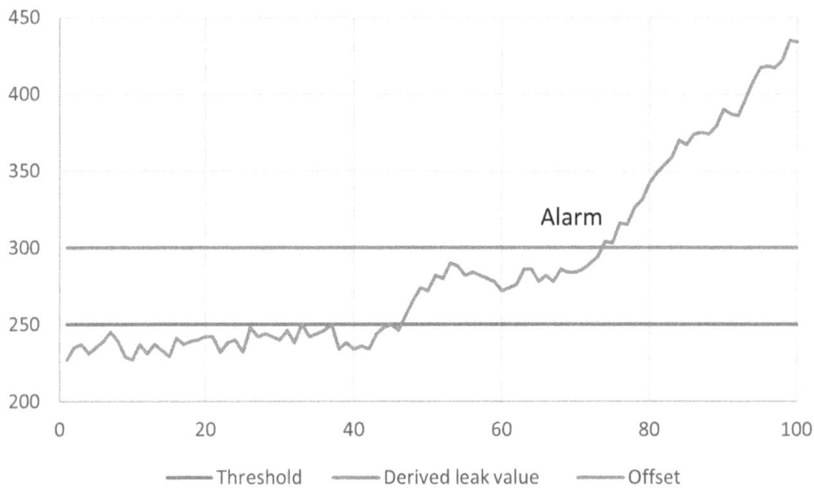

FIGURE 5.3 Delayed alarm with offset example.

Fig. 5.3 shows the impacts of adding a minimum leak volume offset to the leak threshold. This approach is based on the concept that if the leak is real, the derived leak volume will continue to increase over time. Therefore, not only does the derived leak volume have to exceed the leak threshold and maintain a level above the threshold for an established persistence period but it also must acquire enough volume to exceed the offset value. An alarm occurs after meeting all three conditions.

The preceding discussion was developed within the context of a fixed threshold. Although this approach is used, advanced leak detection systems often utilize dynamic thresholds rather than fixed thresholds. Dynamic thresholds change their values based on operational and environmental changes. As an example, starting or stopping a pump will result in a change of threshold because the leak detection algorithm knows that packing will change. The change in packing rate could look like a leak at other times (see Chapter 2: Pipeline Leak Detection Basics, Chapter 3: Mass Balance Leak Detection, and Chapter 4: Real-Time Transient Model–Based Leak Detection for further discussion of packing rates).

Thresholds are often adjusted dynamically, depending on changes in known sources of noise. For example, one might raise the threshold if the packing rate increases because modeling errors are more likely to be introduced when this contribution is large. Alternately, one might raise the threshold if there is a significant increase in leak detection signal noise.

In summary, a fundamental task of the LDS is to develop an appropriate leak signal (such as the observed mass balance), which can be compared to a threshold of some sort. An equally important task is to determine when the signal is significant enough to declare a leak alarm.

The remaining sections of this chapter more rigorously examine the issues associated with converting the leak signal into a reliable leak alarm.

5.2 SIGNAL PROCESSING BASICS

An ideal leak detection system would detect all leaks and would produce no false alarms. This ideal system would also produce an alarm almost instantaneously following the start of the leak. Unfortunately, this ideal does not exist, so we are faced with a situation in which we must deal with LDS errors. Leak detection systems suffer from type I and type II errors: false-positive and false-negative alarms. Note that these errors are not mutually exclusive at the system level: an LDS can kick out both types of errors.

False-negative alarms are the most serious types of classification errors and occur when a leak is present and the leak detection system does not generate an alarm. False-positive alarms are leak alarms with no corresponding leak. To the controller, the impact of these alarms can range from helping to hone their diagnostic skills (if they do not occur too frequently) through increasing stages of irritation to outright contempt for the system (because it is always alarming, rendering the system useless). The human factors consequences of these reactions are discussed in Chapter 10, Human Factor Considerations in Leak Detection.

In the following sections we discuss various techniques that are commonly used to minimize both types of errors.

5.2.1 Outlier Rejection

One type of error that may be observed in leak signals and input data are brief but large excursions from the average signal trend. These may result from brief data outages, pipeline upsets, or other causes. An outlier is a value that deviates significantly from the other nearby values in the data time series. Fig. 5.4 shows a leak signal with an outlier. As noted in American Petroleum Institute (API) Measurement Standards [1] and API TR 1149 [2], signal errors are often considered to be Gaussian in nature. However, in the real world, outliers that are not in line with the assumption of normality occur and can be troublesome to a leak detection system.

To illustrate this, we use a simple pipeline flow balance example. For this pipeline we have an inlet and outlet flow meter. We are also deriving the flow balance based on a simple average of the last 10 data updates. Further, our leak detection threshold is set at 2% of the inlet flow rate.

Let us say that over the last averaging period, flow in averaged 200.14 BPH and flow out averaged 198.77 BPH. This leaves us with a difference of 1.37 BPH, or a difference of approximately 0.6%, of inlet flow. For this example, this flow difference is a historical bias due to the downstream flow meter.

Suppose that during the next data update period the upstream flow jumps to 230 BPH at the upstream site, but then during the next sample it reverts to a value close to an average of 200 BPH. The large excursion for the single sample by approximately 30 BPH increases the 10-sample

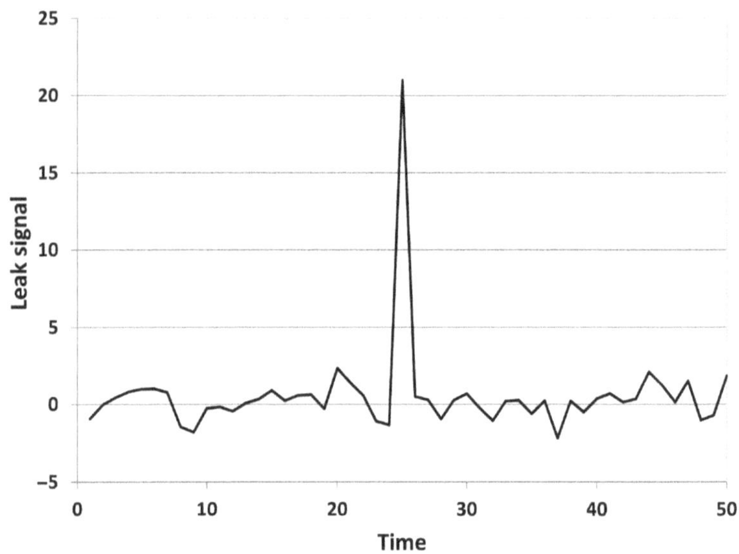

FIGURE 5.4 Leak signal time series with outlier example.

average from 0.37 to approximately 4.37 BPH, or by only approximately 3 BPH. This exceeds the alarm threshold of 4 BPH. At this point, unless a persistence requirement greater than the 10-sample period is imposed, a leak alarm would be generated. Because the outlier remains in the 10-sample average for 10 samples, the impact of the outlier persists for the next 10 samples.

Ideally, our leak detection system would clearly identify outliers and remove them from consideration. We discuss a few ways one can go about doing this. First, we look at a simple, single-outlier, sequential test method for rejecting outliers in input data. This approach applies a standard modified Z-score per Eq. (5.1) to each field measurement as it arrives in the leak detection system.

$$M_z = \frac{0.6745(x_i - \tilde{x})}{MAD}$$

EQUATION 5.1 Modified Z-Score

where M_z is the modified Z-score, x_i is the current leak signal reading, $\tilde{x}$ is the median value, and MAD is median absolute deviation of the data set. If the modified Z-score exceeds an established limit, then the current reading is rejected as an outlier.

Another approach relies on a Grubbs test, which is often referred to as the maximum normed residual test in Eq. (5.2).

$$G = \frac{Y_{\max} - \mu}{\sigma}$$

EQUATION 5.2 Grubbs Equation

where $Y_{\max}$ is the maximum reading in the data set, μ is the sample mean, and σ is the standard deviation of the sample.

In the Grubbs test, the hypothesis of no outliers is rejected according to Eq. (5.3).

$$G > \frac{N-1}{\sqrt{N}} \sqrt{\frac{t^2_{\alpha/(2N),N-2}}{N-2+t^2_{\alpha/(2N),N-2}}}$$

EQUATION 5.3 Grubbs Hypothesis Rejection

where N is the number of samples in our test set and $t_{\alpha/(2N),N-2}$ = critical value of the t-distribution with $N-2$ degrees of freedom and a one-sided significance level of α/N.

Another rather simple approach to eliminating outliers is to incorporate some form of median smoothing, which retains only the median value in the current data set.

5.2.2 Data Averaging/Accumulation

In this section, we discuss the use of the mean as a method to reducing leak signal variability. The mean principle states that as sample size grows, its mean will get closer to the average of the whole population. This can be applied to reduce variability in the leak signal and in the input data.

In this approach, we could use a simple moving average. Let us say you keep track of the last five flow measurements. Each time the system receives a new flow measurement, you discard the oldest and average the new data sample with the previous four values.

Another approach is a time-weighted average. In this method, you still use the current and previous four inputs to derive the modeled value. The difference in this approach is that you weight each value, with the most current value receiving the highest rating and the oldest value assigned the lowest rating. Fig. 5.5 demonstrates the impact of these two approaches on the resulting data.

This figure shows considerable variability in the measured value. Performing a simple five-reading average reduces this variability. A weighted average results in further smoothing of the data. Although the resulting data still have variability, it more closely matches the most recent flow rate measurement.

5.2.3 Use of Multiple Averaging Periods

The previous section discussed the use of averaging as a means of reducing uncertainty. This approach relies on statistical and probability theories of sample size and the differences associated with large and

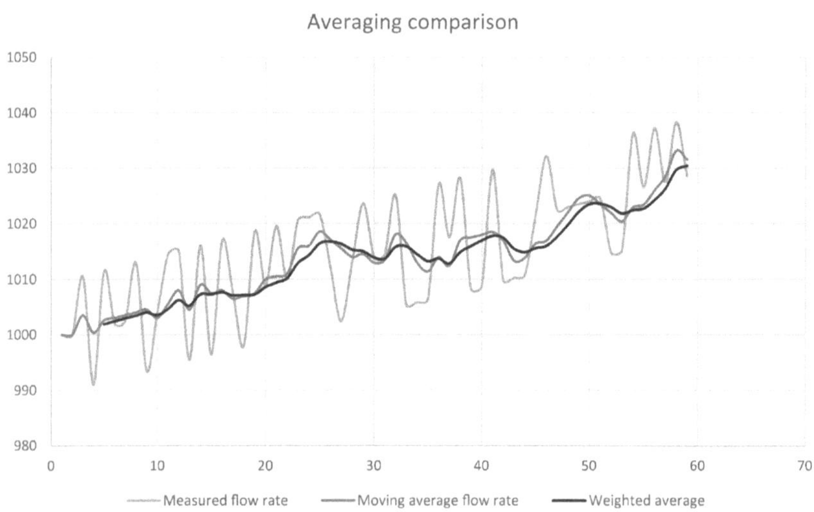

FIGURE 5.5 Data averaging example.

small samples. Computational leak detection systems leverage this by using different averaging periods.

RTTM-based leak detection systems (see Chapter 4: Real-Time Transient Model–Based Leak Detection) often use a range of averaging periods to smooth the mass balance leak detection signal. These might span a few minutes to many hours. The shorter averaging periods utilize smaller sample sizes and target detection of large leaks quickly, whereas longer averaging periods use larger sample sizes and target smaller leaks. The reasons why this is an effective approach are discussed in Section 5.3.

5.2.4 Long-Term Average Analysis

While averaging values over time can reduce variability, it will not eliminate bias, slowly varying shifting or drifting of the signal, or other time correlation. A simple method to remove these artifacts is to calculate a much longer-term average of the signal and then subtract it from the leak or data signal. An important consideration is that the long-term average should be extracted from a data set that is distinct from the data set being evaluated for the presence of a leak.

Take the simple example of a mass balance segment with a flow meter on the inlet and outlet. There is no off-take or inlet flow within this segment. Over time, the system looks for a flow balance by subtracting the average outlet flow rate from the inlet flow rate. Let us further say that the inlet flow rate is 10 BPH higher than the outlet flow rate. By keeping track of this difference and averaging it over a long period, we can derive the indicated inlet flow meter bias. Subtracting this long-term average-derived bias from the incoming meter raw values generates a flow rate that more closely represents the pipeline operating state.

Subtracting a long-term average is a variant of a decorrelation approach, which is discussed further in Section 5.3.

5.3 STATISTICAL PROCESSING AND SIGNIFICANCE TESTING

In the previous section, we examined a number of common techniques that are used to process leak detection signals. These techniques are easily applied but have the downside of lacking a certain amount of technical rigor; they often require tedious hand-tuning to establish thresholds and other parameters that are essential to their operation.

In this section, we investigate the statistical bases that lie behind the techniques discussed previously. Alternately, they can be used rigorously in their own right to detect a leak signature. If applied directly as leak detection signal extraction methods, then they typically require a good degree of understanding and parameterization of the noise in the leak signal. Consequently, they benefit either from a sound analysis of extensive recorded data sets or from online tuning and data collection techniques. A well grounded statistical approach can provide a sound basis for a considerable degree of automatic tuning, and can also be invaluable in supporting performance mapping, which is discussed in Chapter 9, Leak Detection Performance, Testing, and Tuning.

5.3.1 Random Noise, Time Correlation, Probability Distributions, and Significance Testing

Consider a noisy trend of data on which we impose a leak. If the leak starts abruptly, then the leak signature is a simple positive step change that occurs at some time t_{Leak}. The noise is a pre-existing condition that can be characterized according to: (1) its proportionate degree of randomness; (2) the probability distribution associated with that randomness; and (3) the nature of any remaining deterministic time correlation. As an example, consider a trend of "white noise" with a superimposed leak, as shown in Fig. 5.6A. White noise is

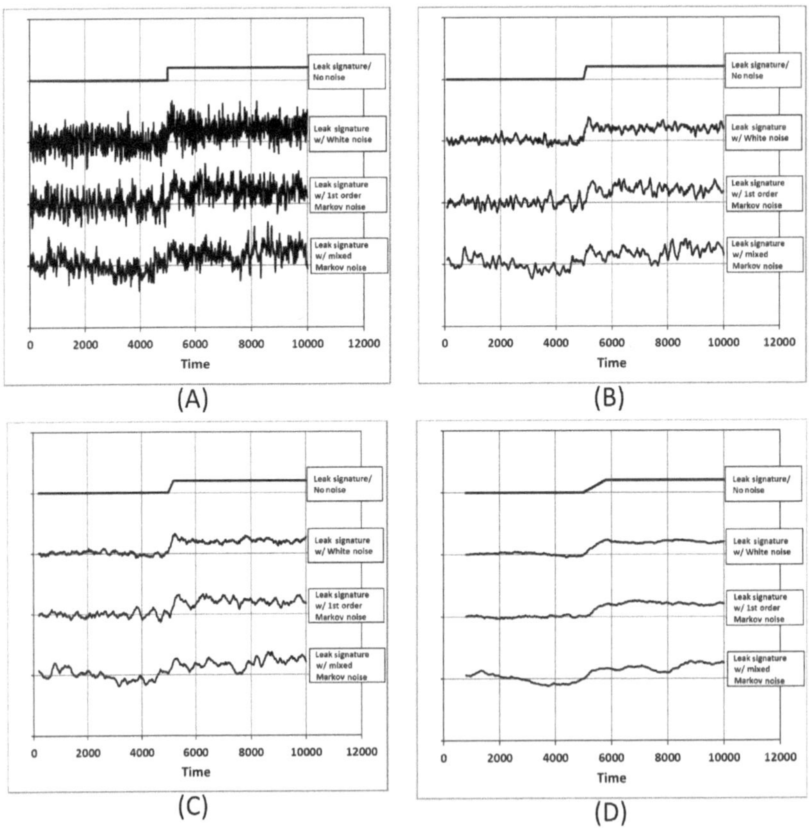

FIGURE 5.6 Noisy data trends with imposed leak (10-s scan) for (A) raw data, (B) 100-s moving average, (C) 200-s moving average, and (D) 800-s moving average.

randomly generated with a Gaussian or normal distribution applied at each scan time and absolutely no time correlation or dependence on previous values. Because there is no time correlation, and because the underlying statistical parameters do not change over time, we also refer to this statistical time series as independent and identically distributed (*iid*).

Consequently, the time series associated with this trend is entirely random, with a well-understood and commonly used probability distribution. Many statistical significance tests are based on collections of data of this type. We note at the outset that many of these tests are inappropriate to use on real-world data without some preprocessing because that data do not fully satisfy the assumptions of randomness and normality.

Fig 5.6A also shows a leak detection signal trend (the third from the top) whereby the data noise is now partially correlated in time. To provide this correlation, this trend is calculated as a first-order autoregressive Markov process (often referred to as an AR(1) process), whereby the data at each scan is given by:

$$VB_i = \varphi VB_{i-1} + \varepsilon_i$$

EQUATION 5.4 First Order Autoregressive Process

where i is the current time step index, φ is the autoregressive factor, and ε_i is random noise (such as Gaussian noise, which applies in this example). This model provides correlation by way of the autoregressive factor, which makes the value of the trend partially dependent on its value at the previous time step. This equation simulates noise that arises from instrument drift, whereby the error has a stochastic tendency to return to its true value. The autoregressive factor typically takes on values between 0 and 1. Values less than -1 or greater than 1 are unstable (ie, the error grows without limit), and values less than zero represent anti-correlation. A value of zero corresponds to white noise and, thus, no time correlation. For the case shown, $\varphi = 0.5$.

Finally, the time correlation can more generally be expressed as the sum of any number n of the previous values. An arbitrary AR(n) process is of the form:

$$VB_i = b + \sum_{j=1}^{n} \varphi_j VB_{i-j} + \varepsilon_i$$

EQUATION 5.5 AR(n) Autoregressive Process

where b is a constant bias. This equation would introduce very long time dependence, so that current values would be influenced by values that potentially occurred at points far in the past.

In our case, we simulated a fourth process (the bottom trend in Fig. 5.6A) with complex long-range autocorrelation by adding together m

individual AR(1) processes with different values of φ for each one, so that for each individual process j:

$$X_{k,i} = \varphi_k X_{k,i-1} + \varepsilon_{k,i}$$

EQUATION 5.6 AR(1) Subcomponent Process

where $k = 1, 2, 3, \ldots, m$, and:

$$VB_i = \sum_{k=1}^{m} X_k$$

EQUATION 5.7 Mixed-multiscale Markov/AR(1) Process

For the case shown, we used a range of values φ_j between 0.33 and 0.95 to ensure a range of significant long-range autocorrelations. This model would apply in the case where there are several AR(1) noise sources, as might occur if we were trying to extract a leak signature from a signal with several autocorrelated error sources, such as a couple of drifting flow meters and a transient model. It is important to note that all of the noisy leak signal trends in Fig. 5.6A are characterized by full Gaussian noise and that the noise for all trends has the same variance. In other words, if you measure each of the noisy signals during the preleak period, then all three trends will exhibit the same standard deviation (set to 75% of the leak size) and the data distribution would be normal. Visually the trends look similar, but they actually behave much differently under statistical averaging, as we shall see.

In addition, be aware that there are several other approaches to modeling time series that can be considered. These include moving average models, in which the current value is a function of the previous error terms, periodic or seasonal models, and mixed models, which combine all contributions. Readers interested in pursuing this further are referred to reference [3].

In Fig. 5.6A, we assumed, in line with reference [4], that the underlying probability distribution of the errors is Gaussian, but in the real world this is not always the case. Real-world distributions can exhibit what are commonly referred to as "fat tails," which are characterized by a heavier set of tails and a shallower central peak when compared to the probability density function for a Normal distribution of the same variance. In short, a fat-tailed distribution is more prone to exhibiting outliers than an equivalent Gaussian distribution. Furthermore, many standard statistical significance tests that assume normality in the underlying data cannot properly be applied to these distributions. An example of this is shown in Fig. 5.7. A t-distribution with a degree of independence $n = 2$ is used to illustrate a fat-tailed distribution. Such distributions can arise as a result multi-mode instrument or modeling noise in the incoming data stream. If we are concerned with minimizing the generation of false alarms, then a fat-tailed distribution is going to imply a far higher threshold than will a normal distribution.

One way to deal with heavy tails is to use some means to reject low-frequency outliers in leak signal data collection, as discussed in Section 5.2.

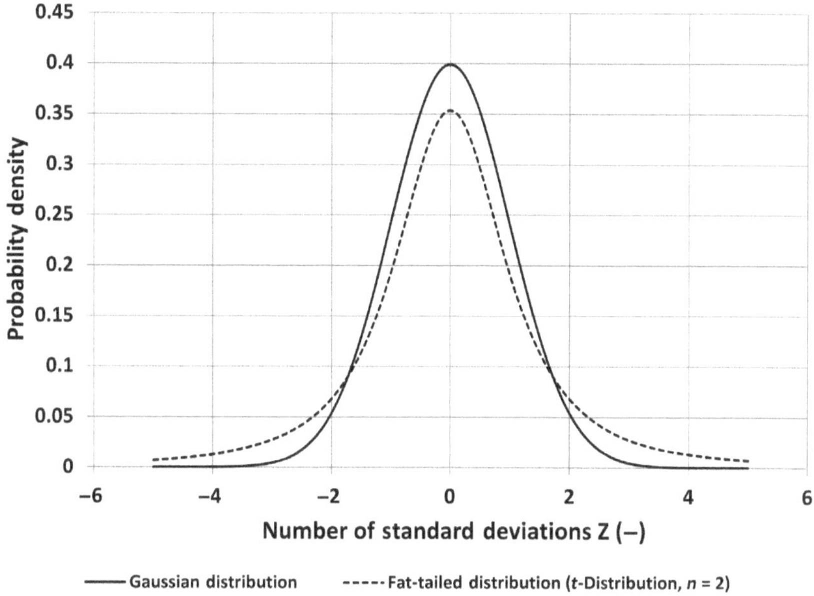

FIGURE 5.7 Leak signal random noise distributions.

Another method is to use the central limit theorem, which states that the arithmetic mean of a sufficiently large number of independent random variables will be approximately normally distributed, regardless of the underlying distribution, as long as the distribution has a well-defined expected value and variance. Thus, if we collect a sufficiently large number of samples, then we should be able to use the assumption that we have normal distribution, along with the variance (presumed known based on an earlier analysis of the data) in the unbiased nonleaking data stream, to set a useful threshold. To use this approach, however, a sufficiently large sample size containing the leak must be accumulated, resulting in a potential delay in the detection of a leak.

Use of summation or data aggregation to minimize the chances of having a type I or type II error leads us into the topic of significance testing, which is the subject of our next section.

5.3.2 Fixed Sample Size Significance Tests

Assume that at any time we can aggregate a fixed quantity of the previous n leak detection signal points. We also assume that the data points are normal because they simply are, or because we have weeded out any unusual outliers, or because n is large enough that the central limit theorem applies. In addition, we assume that the noise is white and that there is no time correlation.

What we would like to do is calculate a threshold value that will be high enough to minimize the type I probability that we will have a false alarm, but not so high that we will experience a type II error and fail to catch a leak. We assume:

$$P_{\text{Type I}} = \alpha; \quad P_{\text{Type II}} = \beta$$

EQUATION 5.8 Type I and II Error Probabilities

where $P_{\text{Type I}}$ is the probability of experiencing a false alarm and $P_{\text{Type II}}$ is the probability of failing to detect a real leak. We are fundamentally performing a hypothesis test on two Gaussian distributions H_0 and H_1:

$$H_0|(1-\alpha): \quad q_{\text{Leak}} \leq 0; \quad H_1|(1-\beta): \quad q_{\text{Leak}} > 0$$

EQUATION 5.9 No Leak/Leak Hypotheses

where H_0 is our null hypothesis (there is no leak) and H_1 is our alternate hypothesis (there is a leak). This set of equations states that when we perform the test, we know that if we accept the null hypothesis we can be confident that we have done so correctly with a confidence of $(1-\alpha)$. Similarly, we accept the alternate hypothesis with a confidence level of $(1-\beta)$. We can further assume that the null and alternate hypothesis volume balance standard deviations σ_{VB} are the same because the presence of the leak should shift the volume balance upward but otherwise will not change the nature of the noise.

According to another report [3], we can show that if we are comparing two Gaussian or normal distributions, then:

$$q_{\text{Leak}}(H_1) = \frac{(Z_\alpha + Z_\beta)\sigma_{VB}}{\sqrt{n_{\text{Fixed}}}}$$

EQUATION 5.10 Fixed Aggregator Leak Rate

Here, n_{Fixed} is the number of discrete independent samples, VB_T is the implied leak detection threshold, and Z_x is the number of standard deviations required to achieve a one-tailed confidence of x. The implied threshold VB_T is:

$$VB_T = \frac{Z_\alpha \sigma_{VB}}{\sqrt{n_{\text{Fixed}}}}$$

EQUATION 5.11 Fixed Aggregator Threshold

At every update of our pooled data set consisting of the last n_{Fixed} points, and assuming no persistence requirement, we will issue an alarm if the volume balance exceeds VB_T. If we set $\alpha = \beta$, then $Z_\alpha = Z_\beta$ and the detectable leak size with confidence $1-\alpha$ is:

$$q_{\text{Leak}}(H_1|(1-\beta)) = \frac{2Z_\alpha \sigma_{VB}}{\sqrt{n_{\text{Fixed}}}}$$

EQUATION 5.12 Fixed Aggregator Detectable Leak Size

Clearly, increasing the number of samples has the benefit of reducing the threshold as well as the size of the detectable leak. If we return to Fig. 5.6 and examine the impact of the sample size on the leak trends with imposed white noise, then we can see the clear decline in the standard deviation of the noise as the sampling size increases from 1 to 80. This means that we can find smaller leaks at the cost of having to wait longer to achieve statistical significance and minimize the chances of having a false alarm.

5.3.3 Colored Noise, Whitening Filters, and Decorrelation

We just saw the significant reduction in white noise achieved by aggregating or averaging the data stream. Fig. 5.6 also shows the impact of averaging on the leak trends that include autocorrelated first-order Markov noise and mixed Markov noise. It is visually clear that aggregating the raw data reduces the auto-correlated noise much less than it did the imposed white noise. We can show this more effectively if we examine the impact of changing the averaging period or aggregation size on the aggregate standard deviation. Results are shown in Fig. 5.8.

The figure clearly shows that the reduction in the error of the mean that occurs with increase in sample size n is much weaker for the first-order and mixed Markov noise cases than it is for the white noise case. The white noise decreases approximately $1/n^{0.5}$, as we would expect from Eq. (5.12). The rate of decrease is much slower for the cases that exhibit time correlation. The power function decrease in the error signal with increasing n

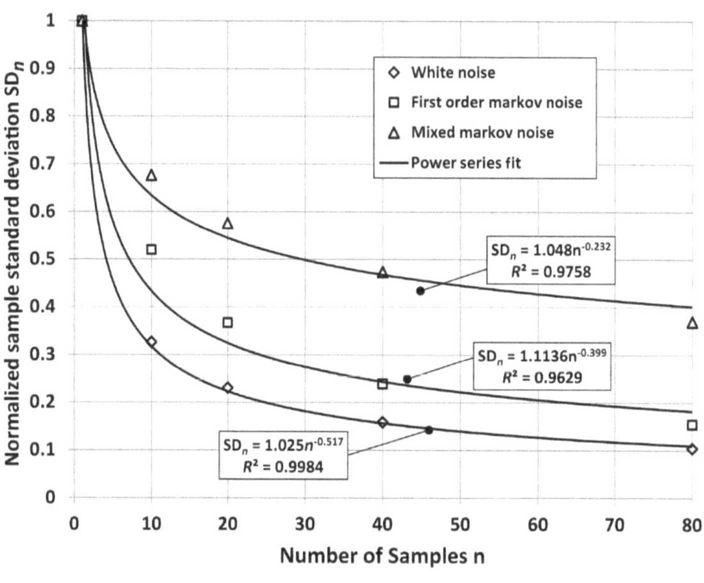

FIGURE 5.8 Impact of sample size on aggregate noise.

(where the exponent is now less than half) is often noted in real-world data trends and is commonly referred to as Hurst or fractal noise. The presence of autocorrelated noise has a general tendency to invalidate standard statistical significance tests [5], which usually assume statistical independence. All of these outcomes are obviously highly undesirable for our LDS signal analyzer.

Noise that is correlated in time is often referred to as colored noise. White noise has a power spectrum that is constant over the range of time frequencies, which implies no autocorrelation. For colored noise, however, the power spectrum is not constant. Therefore, the level of noise occurring at a given instant of time is correlated with the level of noise occurring at some other instant of time.

Colored noise can be handled by processing the signal through a whitening filter or decorrelator. In general, the form of the filter utilizes a decorrelation matrix $\mathbf{W_D}$. Application of the decorrelation matrix to an aggregated vector of correlated samples $\mathbf{X_i}$ produces a fully random whitened output array $\mathbf{Y_o}$:

$$\mathbf{Y_o} = \mathbf{W_D X_i}$$

EQUATION 5.13 Signal Decorrelation Equation

where $\mathbf{W_D}$ is dependent on the covariance matrix $\mathbf{M_C}$ for $\mathbf{X_i}$ such that:

$$\mathbf{W_D^T W_D} = \mathbf{M_C^{-1}}$$

EQUATION 5.14 Decorrelation Matrix Equation

The output array (or any portion of it) can then be more easily analyzed using standard statistical significance tests. The element in the i, j position of $\mathbf{M_C}$ specifies the covariance between the ith and jth elements of random vector $\mathbf{X_i}$, where the covariance in turn specifies the degree to which two random variables change together. In principle, if the input vector includes but is larger than the n aggregated components we wish to analyze, then the decorrelation process strips out the deterministic and nonrandom components of the inputs based on their dependence on other components in $\mathbf{X_i}$. The covariance matrix is usually determined based on analysis of a large sample of the leak signal assuming no leak present.

Another option that can be used to whiten the signal is to use linear predictive coding. A linear predictive coder is an analytical model that assumes that the deterministic components in the data are in the form of an AR(n) process (see Eq. 5.5). Assuming the availability of a significant amount of prerecorded and preprocessed leak detection signal data, the coefficients φ_j can be precalculated via a regression process. At each point in time, the whitened signal can be calculated by subtracting the predicted value of the signal based on Eq. (5.5).

Uncomplicated methods of partially decorrelating the input aggregate include subtracting a long-term average of the data that are outside of or prior to the range being analyzed, as discussed previously. Keep in mind that this approach, although simple and straightforward, should be used with care because it can actually increase the noise in the processed sample if the original aggregated data are already white, exhibit minimal autocorrelation, or are anti-correlated.

5.3.4 Sequential Probability Ratio Testing

Sequential probability ratio testing (SPRT) originated in the 1930s to assist in quality control studies in the manufacturing process control field. SPRT applies statistical theory and methods to a data set in which the number of observations n_{SPRT} is not fixed in advance [6]. The data are again assumed to consist of *iid* observations (white noise). The test has since been applied in a number of fields, including drug efficacy and other health studies, and is also used in real-world leak detection systems [7] to statistically select between the null hypothesis (no leak) or the alternative hypothesis (a leak of some size based on the desired confidence of detection).

SPRT is based on minimizing the number of observations k based on the sequence of likelihood ratios calculated using probabilities of observations X_i under the null (H_0) and alternate (H_1) hypotheses. We define the likelihood ratio:

$$\Lambda_i = \frac{f_1(X_i)}{f_0(X_i)}$$

EQUATION 5.15 SPRT Likelihood Ratio Equation

where $f_1(x)$ is the probability density function applicable under the alternate hypothesis and $f_0(x)$ is the probability density for the null hypothesis. To accomplish this, we collect incoming data one-at-a-time and calculate the following log-sum:

$$S_i = S_{i-1} + \ln\left(\frac{f_1(X_i)}{f_0(X_i)}\right) = S_{i-1} + \ln(\Lambda_i)$$

EQUATION 5.16 SPRT Log-sum Equation

subject to the following threshold-based stopping rules:

- If $S_i > b$, accept H_1
- If $S_i < a$, accept H_0
- Otherwise continue collecting data

The threshold constants a and b are calculated from:

$$a \approx \ln\left(\frac{\beta}{1-\alpha}\right) \quad \text{and} \quad b \approx \ln\left(\frac{1-\beta}{\alpha}\right)$$

EQUATION 5.17 SPRT Threshold Constants

where α and β were Type I and Type II error probabilities defined previously in Eq. (5.8). If the noise is Gaussian, then the sequence S_i is given by:

$$S_i = S_{i-1} + \ln\left(\frac{f_1(X_i)}{f_0(X_i)}\right) = S_{i-1} + \frac{\mu_1}{2\sigma_{VB}^2}(2X_i - \mu_1)$$

EQUATION 5.18 SPRT Gaussian White Noise Log-sum Equation

Where σ_{VB} was previously defined as the standard deviation of the noise in the leak detection signal and μ_1 is the H_1 hypothesis leak rate assuming a detection confidence $1 - \beta$:

$$\mu_1 = q_{\text{Leak}}(H_1)$$

EQUATION 5.19 SPRT Gaussian White Noise Leak Rate Hypothesis

Under H_0, the expected stopping time $E_0(n_{\text{SPRT}})$ (or estimated number of data points n_{SPRT}) for SPRT is given by:

$$E_0(n_{\text{SPRT}}) \geq \frac{(1-\alpha)\ln\left(\frac{1-\alpha}{\beta}\right) - \alpha\ln\left(\frac{1-\beta}{\alpha}\right)}{D(f_0 \| f_1)}$$

EQUATION 5.20 SPRT Null Hypothesis Stopping Time

Similarly, the expected stopping time under H_1 is given by:

$$E_1(n_{\text{SPRT}}) \geq \frac{(1-\beta)\ln\left(\frac{1-\beta}{\alpha}\right) + \beta\ln\left(\frac{\beta}{1-\alpha}\right)}{D(f_1 \| f_0)}$$

EQUATION 5.21 SPRT Alternate Hypothesis Stopping Time

The parameter $D(f_i \| f_j)$ is the Kullback-Leibler (K-L) divergence, defined by:

$$D(f_i \| f_j) = \int_{-\infty}^{\infty} f_i(x)\ln\left(\frac{f_i(x)}{f_j(x)}\right)dx$$

EQUATION 5.22 Kullback-Leibler (K-L) Divergence Equation

If the noise is Gaussian and the sequence is *iid* or white (we assume we previously ran our raw signal through a decorrelator), then the estimated stopping time (assuming that the standard deviation for both f_1 and f_2 is σ_{VB} and the mean of the distribution under $H_1 = \mu_1$) is:

$$E_0(n_{\text{SPRT}}) = E_1(n_{\text{SPRT}}) \geq \frac{2\sigma_{VB}^2(1-2\alpha)}{\mu_1^2}\ln\left(\frac{1-\alpha}{\alpha}\right)$$

EQUATION 5.23 SPRT Gaussian White Noise Stopping Time Equation

In general, SPRT runs until either the null or alternate hypothesis is accepted. If we compare the stopping time n_{SPRT} to the fixed sample size n_{Fixed} required for the same values of $\mu_1 = q_{\mathrm{Leak}}$, σ_{VB}, and α/β, then we typically find that SPRT requires only 40% to 50% of the number of samples required under a fixed sample size approach [8]. Consequently, sequential approaches have a significant advantage over fixed sample size approaches in terms of requiring a shorter time to detect a leak.

Application of SPRT in a continuous process situation in which false alarms are to be expected on an occasional basis requires some modification. One approach is to restart the process every time a hypothesis is accepted. Another approach is to use the CUSUM (cumulative sum control chart) approach [9], which modifies the stopping rule by ignoring the lower threshold. Under such an approach, the cumulative sum is simply updated and the effective window size grows without bound. Other possibilities include the Girshick-Rubin-Shirayev (GRSh) procedure, which utilizes a more complex approach that combines a step-wise difference indicator with a cumulative difference indicator [10].

5.3.5 Change Point Detection

So, let us assume that we have detected a leak. How do we know when it actually started? One issue with all of the detection approaches described is that the time when the leak is detected does not actually correspond to the time (or specific sample index n_{Leak}) when the leak actually occurred. The detectors always exhibit some level of lag, because the aggregated sample inevitably includes samples prior to the leak and samples from after the start of the leak. This means that passing through the threshold does not provide any estimate of when the leak actually started. One way to solve this problem is to work our way backward from the current time and look for the point of inflection when the leak signal first began to change.

Expressed mathematically, this is equivalent to using the maximum likelihood estimate based on a comparison of summed log-likelihoods assuming various start times n_C:

$$n_{\mathrm{Leak}} = \underset{1 \le n_C \le n_{\mathrm{MAX}}}{\arg\max} \sum_{n=n_C}^{n_{\mathrm{MAX}}} \ln\left(\frac{f_1(n)}{f_0(n)}\right)$$

EQUATION 5.24 Leak Start Time/Index Equation

where n_{MAX} is the number of samples in the aggregate. The *arg max* function takes the largest value of the summed log-sequence working backward from the current time, which should maximize at the point when the leak starts.

5.3.6 Multiple Aggregators

One of the problems encountered using hypothesis significance tests of the types described is that they are designed to be used for either a fixed sample size (and corresponding leak rate) or a fixed hypothesized leak rate (and corresponding estimated sample size). Unfortunately, this makes them both non-optimal if the actual leak rate does not correspond to the hypothesized leak rate. For example, if we assume a fixed sample size approach, then, based on the noise in the data, and assuming the data are white or *iid*, a leak will be alarmed if the rate is larger than the threshold value specified in Eq. (5.11). This threshold is optimized to work for a leak rate that effectively corresponds to the threshold used in the equation. If the leak rate q_{Leak} is larger than the specified threshold, then it will still cause an alarm. However, because the aggregated data will include preleak data averaging to zero, the *minimum* detectable leak size will be inversely proportionate to the number of samples that have occurred since the leak started, or:

$$q_{\text{Leak}} = \frac{n_{\text{AGG}}}{n_{\text{DET}}} VB_T$$

EQUATION 5.25 Off-spec Detectable Leak Rate (Fixed Aggregator Size)

However, if we had actually designed a detector to detect a leak of size q_{Leak} to begin with, then the number of detection samples would, per Eq. (5.11), be:

$$q_{\text{Leak}} = \frac{Z_\alpha \sigma_{VB}}{\sqrt{n_{\text{DET}}}} = \frac{Z_\alpha \sigma_{VB}}{\sqrt{n_{\text{AGG}}}} \frac{\sqrt{n_{\text{AGG}}}}{\sqrt{n_{\text{DET}}}} = \left(\frac{n_{\text{AGG}}}{n_{\text{DET}}}\right)^{1/2} VB_T \tag{5.26}$$

EQUATION 5.26 Optimal Detectable Leak Rate (Fixed Aggregator Size)

The value of q_{Leak} per Eq. (5.26) is always lower than the value calculated per Eq. (5.25). In other words, the minimum detectable leak size for a specified detection time is always lower if we design a detector that is specifically tailored to that size (or detection time) rather than trying to work with a single detector. Regarding the limit, the implication is that if we wish to detect small leaks as rapidly as possible, then we should use a large number of tuned aggregators. This result generally applies to SPRT and other detectors.

5.3.7 False Alarms

If the leak signal is fully decorrelated, and if there is no persistence requirement, then the false alarm rate is easily estimated:

$$r_{\text{FA}} \approx \frac{\alpha}{\Delta t_{\text{AGG}}}$$

EQUATION 5.27 Fixed Aggregator False Alarm Rate (Gaussian White Noise) where Δt_{AGG} is the aggregation time for the detector. If we use multiple detectors, then the number of false alarms must increase because the detectors are independent and each detector has the same probability of emitting a false alarm. Note that if we simply extend the alarm time rather than count a new excursion as a new alarm (by using persistence), then actual alarm rates will be slightly lower than indicated by this equation. However, this equation provides a reasonable estimate of the upper limit as long as $\alpha \ll 1$.

Let us assume that the false-positive probability α is constant over all aggregators, and let us further assume that we will design a system whereby the aggregation time for any aggregator is k times the time for the next smallest aggregator. In such a case, the number of false alarms n_{FA} over any period t is given by the following series:

$$n_{\mathrm{FA}} = \alpha \frac{t}{\Delta t_{\mathrm{AGG}}} \sum_{i=0}^{M} \frac{1}{k^i}$$

EQUATION 5.28 False Alarms with Multiple Aggregators (Gaussian White Noise)

where Δt_{AGG} is now the minimum aggregation time and M is:

$$M = \frac{\ln(t/\Delta t_{\mathrm{AGG}})}{\ln k}$$

EQUATION 5.29 Number of Aggregation Classes

Regarding the limit, the maximum number of false alarms cannot exceed $n_{\mathrm{FA,MAX}}$:

$$n_{\mathrm{FA,MAX}} = \alpha \frac{t}{\Delta t_{\mathrm{AGG}}} \frac{k}{k-1}$$

EQUATION 5.30 Maximum False Alarm Estimator (Gaussian White Noise)

If the signal is autocorrelated and cannot be whitened, then these estimated false alarm event rates are upper limits, because alarms will tend to persist, potentially for a long period. However, in this case, the total alarm time will tend to be the same as long as the underlying probability distributions do not change significantly with processing.

5.3.8 Real-World Adjustments

Real-world leak detection systems may require certain modifications in the equations provided. The following are ways to make these adjustments:

1. Significant modifications to the approaches and equations described are required if the data are not *iid* and cannot be decorrelated.
2. Equations that assume Gaussian noise should be adjusted to use the Student t-test if the number of aggregated points is less than 30 to 50.

3. If the noise is not stationary, then the parameters that describe it, such as the variance, can change over time. Consider using the variance of the aggregated data in combination with the tuning variance based on a separate training set, in accordance with the approach used in the Welch *t*-test [2]. As noted previously, the temporary presence of a high packing rate or other indicator could be a sign of a need to increase the threshold.
4. If the underlying probability distribution is not Gaussian and the number of aggregated points is insufficient to render a normal distribution in accordance with the central limit theorem, then the empirical probability density distribution should be maintained by the leak detection application in tabular or other form based on tuning or analysis of recorded data.

5.3.9 Advanced Signal Detection Approaches

If multiple aggregation-based detectors are used, then the approach described here allows any single detector to declare an alarm as long as the threshold requirements are met. However, as noted previously, a downside is that all detectors have the ability to issue an alarm, which tends to multiply the probability of a false alarm as the number of detectors increases. It also creates a performance issue because a sparse set of detectors could miss a leak or experience a delay in catching it because no one detector is properly attuned to the leak size.

A more advanced form of SPRT is generalized SPRT (GSPRT), in which we allow the parameters of the alternate hypothesis estimator (such as the leak rate μ_1) to be free parameters and then optimize to find the best solution. We then choose the best-fitting model (eg, the one with the highest probability) or:

$$T_{k,n} = \frac{\max\limits_{j}^{m} \left[\prod\limits_{i=0}^{n} f_j(X_i) \right]}{\prod\limits_{i=0}^{n} f_0(X_i)}$$

EQUATION 5.31 Generalized SPRT Best Estimator Definition

where $T_{k,n}$ is the best cumulative sequential test statistic at k of a set of m alternate hypotheses (such as a set of alternate leak rates) at cumulative time index n. If the data are Gaussian, then instead of assuming a fixed leak size μ_1 it is reasonable to assume that the mean of the leak signal sample data over our aggregate or averaging period is the best estimate (the maximum likelihood estimate (MLE)) for μ_1 [10]. Using this assumption, referring back to Equation 5.18, and taking the logarithm of our estimator allows us to develop a new sequential estimator S_n:

$$S_n = \ln T_{k,n} = \frac{1}{2n\sigma_{VB}^2}\left[\sum_{i=0}^{n} X_i\right]^2 = \frac{n\overline{X}_n^2}{2\sigma_{VB}^2}$$

EQUATION 5.32 GSPRT Sequential Estimator (Gaussian White Noise)

In this equation, n is the cumulative number of samples and $\overline{X}_n$ is the average of the aggregate sample values for $i = 0$ to n. This estimator is then compared to our threshold parameters a and b to determine whether one of our hypotheses should be accepted. This calculation is just slightly more complex than the standard SPRT calculation. However, its benefit is that the detection time is always optimal and only one aggregated or averaged data set needs to be monitored. The downside, of course, is that the data must be thoroughly decorrelated. In addition, if the probability distribution is not normal, then the true probability distribution should be programmatically specified based on previously analyzed operating data.

Possible improvements to all of these approaches can be achieved by including the Bayesian prior probability for the leak rate. The Bayesian prior is nothing more than the expected spill or leak probability during the sampling period based on engineering analysis or judgment. See Chapter 13, Leak Detection and Risk-Based Integrity Management for a discussion regarding observed leak incident rates.

Finally, hierarchical Bayesian modeling is a more sophisticated approach that potentially allows error sources (individual instruments, RTTM sources, etc.) to be identified and also allows for more sensitive detection to be achieved in certain cases in which separate leak detection sections are using shared instruments [11,12].

In conclusion, a principled statistical approach to analysis of the leak detection signal will make use of several techniques. These include but are not limited to (1) elimination of outliers in the data stream, (2) signal decorrelation techniques to allow the use of standardized statistical tests that assume randomness in the data, (3) data aggregation to allow the use of statistical tests that assume normality and to reduce the minimum detectable leak rate, (4) use of multiple aggregators or generalized approaches that minimize detectable leak thresholds over a wide range, and (5) selection or calculation of thresholds that minimize both Type I and II errors.

REFERENCES

[1] Zhao B, Chi C, Gao W, Zhu S, Cao G. A chain reaction DoS attack on 3G networks: analysis and defenses. IEEE INFOCOM 2009;2455–63. Available from: http://dx.doi.org/10.1109/INFCOM.2009.5062173.

[2] Welch BL. The generalization of 'student's' problem when several different population variances are involved. Biometrika 1947;34(1–2):28–35. Available from: http://dx.doi.org/10.1093/biomet/34.1-2.28. MR 19277.
[3] Shumway RH, Stoffer. Time series analysis and its applications. Springer; 2011.
[4] American Petroleum Institute Standard. Manual of petroleum measurement standards.
[5] https://en.wikipedia.org/wiki/Hurst_exponent.
[6] Wald A. Sequential tests of statistical hypotheses. Ann Math Stat June 1945;16 (2):117–86.
[7] Zhang XJ. Detecting leakage of fluid from a conduit. U.S. Patent US5355324A, Filing Date: Aug 10, 1992.
[8] Ghosh B, Sen P. Handbook of sequential analysis. Marcel Dekker, Inc.; 1991.
[9] Page ES. Continuous inspection scheme. Biometrika June 1954;41(1/2):100–15. Available from: http://dx.doi.org/10.1093/biomet/41.1-2.100.
[10] Kharitonov E, Vorobev A, Macdonald C, Serdyukov P, Ounis I. Sequential testing for early stopping of online experiments. In: SIGIR'15 proceedings of the 38th international ACM SIGIR conference on research and development in information retrieval; 2015. p. 473–482.
[11] Carpenter P, Henrie M, Nicholas R. Bayesian belief networks for pipeline leak detection. In: Pipeline simulation interest group annual meeting, Williamsburg, Virginia, 11 October–13 October 2006.
[12] Carpenter P, Henrie M, Nicholas R. Automated validation and evaluation of pipeline leak detection system alarms. In: Pipeline simulation interest group annual meeting, Prague, Czech Republic, 16 April–19 April 2013.

Chapter 6

Rarefaction Wave and Deviation Alarm Systems

In this chapter, we examine rarefaction wave (also referred to as negative pressure wave) and deviation alarm systems. This leak detection approach is based on the concept that a sudden-onset leak will create a hydraulic disturbance that will rapidly propagate away from the leak source in both upstream and downstream directions at a predictable velocity. These systems identify the indication of the hydraulic disturbance, specifically pressure and flow changes, as it passes one or more field instrument locations.

Rarefaction wave systems are the more sophisticated and reliable of these two approaches, and we discuss them first. Note that rarefaction wave systems are primarily targeted to liquid commodity pipelines for reasons that are discussed later in the chapter.

6.1 RAREFACTION WAVE PHYSICAL BASIS AND EQUATIONS

Consider a pipeline that is bounded at the upstream end by a pressure source, such as a pump or compressor. If a corrosion pit, rupture, puncture, or other breach of integrity develops abruptly in the pipe at some location, then commodity will suddenly start to flow through the hole. The flow of leaked commodity through the orifice immediately following the onset of the leak will necessarily (as a result of mass conservation) create a flow discontinuity at the leak site: the flow upstream of the hole will be greater than the flow downstream of the leak site. The difference in the two flows must be exactly equal to the leak rate Q_L.

Such a flow discontinuity will immediately create a pair of pressure waves with magnitudes $\Delta p_{WH,Left}$ and $\Delta p_{WH,Right}$ that will originate at the leak location and move upstream and downstream at the speed of sound a_c applicable to the commodity. Each of these waves will be accompanied by corresponding flow deviation waves $\Delta Q_{WH,Left}$ and $\Delta Q_{WH,Right}$ that move in concert with and at exactly the same velocity as the pressure waves. This coupling of the pressure and flow waves occurs via the dynamic behavior of the Navier

Pipeline Leak Detection Handbook. DOI: http://dx.doi.org/10.1016/B978-0-12-802240-5.00006-6

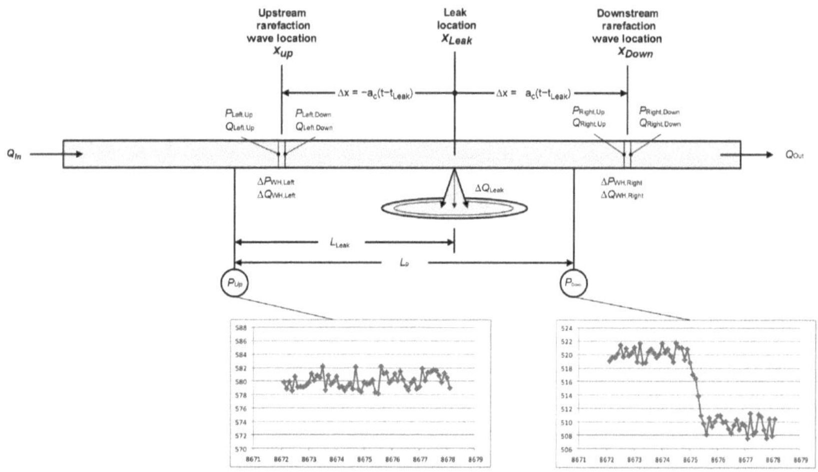

FIGURE 6.1 Negative pressure waves propagating from leak site.

Stokes equations introduced in Chapter 4, Real-Time Transient Model–Based Leak Detection. The behavior of these waves is shown in Fig. 6.1.

Note that there are two wave fronts: one traveling leftward and one moving rightward. Each wave is characterized by a pressure difference across the wave location and a similar flow difference across the location. We also have two stationary pressure-monitoring locations p_{Up} and p_{Down}, which will be used to detect the wave fronts when they pass. The two pressure/time charts at the bottom of the figure indicate that the rarefaction wave has passed the downstream measurement site, but has not yet reached the upstream site.

For the leftward-traveling wave, the pressure difference is therefore:

$$\Delta p_{\text{WH,Left}} = p_{\text{Left,Up}} - p_{\text{Left,Down}}$$

EQUATION 6.1 Leftward Traveling Rarefaction Wave Pressure Difference

Similarly, the flow difference across the leftward-traveling wave is given by:

$$\Delta Q_{\text{WH,Left}} = Q_{\text{Left,Up}} - Q_{\text{Left,Down}}$$

EQUATION 6.2 Leftward Traveling Rarefaction Wave Flow Difference

Comparable equations apply to the rightward-traveling wave front. Pressure magnitudes and flow differences for each wave front are related. The relationship between the instantaneous pressure changes arising from an abrupt flow change in a pipeline is given by the Joukowsky Water Hammer equation:

$$\Delta p_{\text{WH}} = \rho_c a_c \Delta v_c i_{\text{Wave}} = \frac{\rho_c a_c \Delta Q_{\text{WH}}}{A_{\text{Pipe}}} i_{\text{Wave}}$$

EQUATION 6.3 Joukowsky Water Hammer Equation

where Δp_{WH} is the water hammer pressure rise across the propagating flow discontinuity (again moving from left to right), ρ_c is the commodity density, Δv_c is the change in velocity across the wave front at flowing conditions, ΔQ_{WH} is the change in flow across the wave front (the flow on the right side of the wave front minus the flow on the left side of the wave front), and a_c is the speed of sound in the commodity. i_{Wave} represents the one-dimensional wave direction and is $+1$ for a rightward-moving wave front and -1 for a leftward-moving front.

Note that ΔQ_{WH} is expressed at flowing conditions. Because flow rates are generally expressed at some reference temperature and pressure (STP), we can rewrite it as follows for flow at STP:

$$\Delta p_{\mathrm{WH}} = \frac{\rho_{c,stp} a_c \Delta Q_{\mathrm{WH},stp}}{A_{\mathrm{Pipe}}} i_{\mathrm{Wave}}$$

EQUATION 6.4 Joukowsky Water Hammer Equation (Standard Conditions)

where the subscript *stp* has been used to indicate values expressed at standard conditions. Note, however, that throughout this chapter we work with variables at flowing conditions unless stated specifically otherwise.

The commodity speed of sound in the pipe is given by:

$$a_c = \sqrt{\frac{K_c}{\rho_c\left(1 + \dfrac{K_c D_{\mathrm{Pipe}}}{E_{\mathrm{Pipe}} WT_{\mathrm{Pipe}}}\right)}}$$

EQUATION 6.5 Liquid Commodity Speed of Sound Equation

In this equation, K_c is the commodity bulk modulus, D_{Pipe} is the pipe diameter, E_{Pipe} is the pipe material modulus of elasticity (Young's Modulus), and WT_{Pipe} is the pipe wall thickness.

The pressure change across the leftward-moving wave front will be negative because the flow must increase across the wave front (because the leak draws excess flow from upstream). There is always a stationary flow discontinuity at the leak site because the flow must decrease across the leak location by the leak value to conserve mass. Finally, the flow must rise across the wave front moving downstream from the leak location for the same reason: it draws flow from downstream to contribute to the leakage. The region bounded by the waves is therefore always at a lower pressure than would have applied to the original pipeline hydraulic gradient, which is why the waves are referred to as rarefaction or negative pressure waves.

Immediately after the leak starts, the values of both flow waves will be given by:

$$\Delta Q_{\mathrm{WH,Initial}} = \frac{\Delta Q_{\mathrm{Leak}}}{2}$$

EQUATION 6.6 Initial Flow Wave Magnitude

This makes sense because the flow discontinuity across the leak is always equal to the leak size and the two waves are initially at the leak location. Using the Joukowsky equations, the starting value of the upstream pressure wave will therefore be given by:

$$\Delta p_{\text{Left,Initial}} = -\frac{\rho_c a_c \Delta Q_{\text{Leak}}}{2A_{\text{Pipe}}}$$

EQUATION 6.7 Initial Pressure Wave Magnitude (Leftward Traveling)

where A_{Pipe} is the inside cross-sectional pipe area. The value of the downstream pressure wave will have the same absolute magnitude but will be of the opposite sign, per Eq. (6.3).

Note that the sizes of the rapidly propagating pressure and flow wave fronts will *not* remain at these initial values. The wave magnitudes will attenuate due to compressibility and frictional effects over time and with accumulating distance as the wave fronts move away from the leak site.

We can consider the impact qualitatively by examining Fig. 6.2. This figure uses the hydraulic head and local pipeline flow to illustrate the impacts of the leak at a point in time following the start of the leak. We can relate the pressure changes discussed previously to the corresponding head fluctuations by remembering that the head changes will be equal to the pressure deviations divided by the fluid weight density.

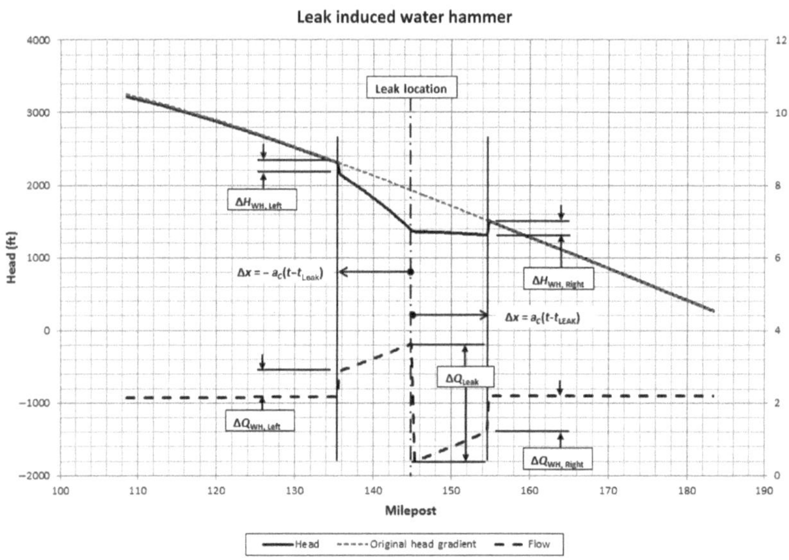

FIGURE 6.2 Transient hydraulic impact from pipeline leak.

The leak is assumed to occur at time t_{Leak}. At a later time, t, the rarefaction wave fronts will have moved a distance $-\Delta x$ (upstream) and $+\Delta x$ (downstream) of the leak site, where:

$$\Delta x = a_c(t - t_{\text{Leak}})$$

EQUATION 6.8

As we noted previously, there will be abrupt pressure and flow changes or discontinuities $\Delta p_{\text{WH,Left}}$, $\Delta p_{\text{WH,Right}}$, $\Delta Q_{\text{WH,Left}}$, and $\Delta Q_{\text{WH,Right}}$ across the left and right boundaries of the wave front. It is important to note that these wave fronts are the absolute limits of the leak-induced disturbance, and that no hydraulic impact will be felt outside of the range between $-\Delta x$ upstream and $+\Delta x$ downstream of the leak site. This figure also shows the impacts of the friction-induced attenuation of the wave fronts. Even though the rarefaction wave front propagates indefinitely in principle, the magnitude of the front will decline as it moves away from the leak site until it eventually becomes undetectable. This decline can be substantial as a function of distance from the leak, and it will ultimately limit the performance of any negative pressure leak detection system, as is discussed later in this chapter.

6.2 PRESSURE SIGNAL AND EVENT PROCESSING

Rarefaction wave detection events are based on the detection of the abrupt negative pressure change associated with a leak as the leading front of the disturbance passes a measurement location. Refer to Fig. 6.3.

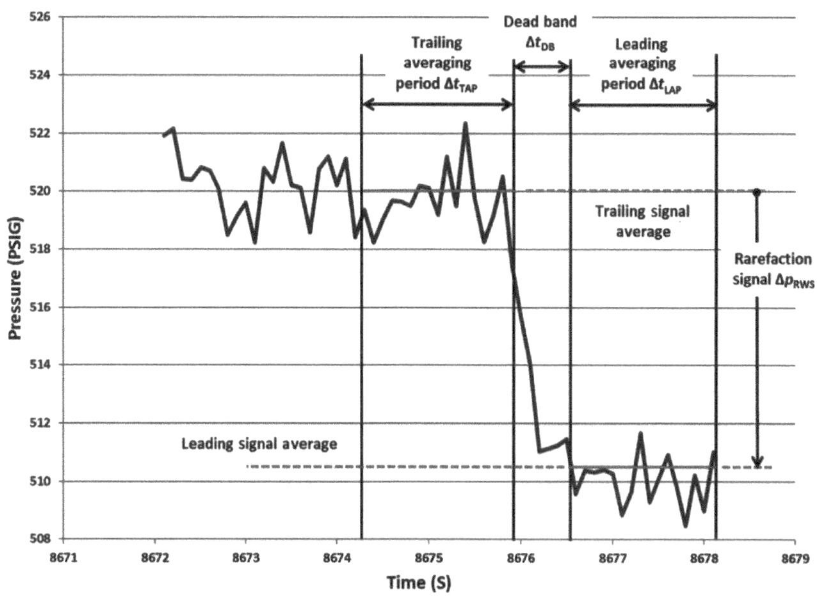

FIGURE 6.3 Rarefaction wave signal processing.

This figure shows a noisy pressure trend at a monitoring site. The noise may be due to instrument error, communications circuit error, or noise from a host of normal hydraulic events propagated from other locations in the pipeline. At some time, there is a distinct drop in pressure associated with the passage of a rarefaction wave front. To detect the wave front, we have two time periods, a trailing averaging period Δt_{TAP} and a leading averaging period Δt_{LAP}. The averaging periods are separated by a dead band Δt_{DB}.

The two averaging periods are implemented to reduce the noise through averaging based on the assumption that the measurement noise is random and independent from scan to scan (see Chapter 5: Statistical Processing and Leak Detection). If this is a good assumption (which is likely over the very short timeframes compatible with the rarefaction wave transient), then the noise for the averaged pressure will be reduced according to the inverse of the square root of the number of samples in the averaging period. We therefore define the rarefaction wave signal $\Delta p_{\mathrm{Rarefaction}}$ as:

$$\Delta p_{\mathrm{Rarefaction}} = \overline{p}_{\mathrm{LAP}} - \overline{p}_{\mathrm{TAP}}$$

EQUATION 6.9 Rarefaction Wave Pressure Signal

where $\overline{p}_{\mathrm{LAP}}$ is the average of the pressure measurements p taken over period leading Δt_{LAP}, and $\overline{p}_{\mathrm{TAP}}$ is the average of the pressure taken over trailing period Δt_{TAP}. The number of measurements n_{LAP} over period Δt_{LAP} that contribute to $\overline{p}_{\mathrm{LAP}}$ is given by:

$$n_{\mathrm{LAP}} = \Delta t_{\mathrm{LAP}} f_{\mathrm{Scan}}$$

EQUATION 6.10 Number of Measurements for Leading Period Δt_{LAP}

where f_{Scan} is the measurement scan frequency in 1/s (Hz). The number of measurements n_{TAP} over period Δt_{TAP} that contribute to $\overline{p}_{\mathrm{TAP}}$ is similarly defined.

If the scan-to-scan pressure measurement noise is random in time and independent (or *iid*, see Chapter 5), and further assuming the noise is normally distributed, then the uncertainty associated with the pressure signal Δp_{RS} is:

$$\varepsilon(\Delta p_{\mathrm{RS}}) = \varepsilon(p_M)\sqrt{\frac{1}{n_{\mathrm{LAP}}} + \frac{1}{n_{\mathrm{TAP}}}}$$

EQUATION 6.11 Rarefaction Pressure Signal Uncertainty

Here, the parameter $\varepsilon(p_M)$ represents the uncertainty in each pressure measurement p_M. If the leading and trailing averaging periods, each with sample count n_{AP} (associated with a corresponding averaging period $\Delta t_{\mathrm{AP}} = \frac{n_{\mathrm{AP}}}{f_{\mathrm{Scan}}}$), are the same length, then:

$$\varepsilon(\Delta p_{\mathrm{RS}}) = \varepsilon(p_{\mathrm{M}})\sqrt{\frac{2}{n_{\mathrm{AP}}}}$$

EQUATION 6.12 Rarefaction Pressure Signal Uncertainty for Equal Sampling Times

We can see that, in this case, averaging reduces the signal uncertainty. It is important to note that this equation ONLY applies to random and independent scan-to-scan noise. Results may differ if the noise has a trend or if the noise values are correlated from scan to scan. In fact, the uncertainty in the pressure measurement signal may not be significantly reduced by averaging in such cases.

The dead band period Δt_{DB} must be long enough to ensure that the leak development completes over the period. Otherwise, the pressure change will be absorbed into the averaged values developed over periods Δt_{TAP} and Δt_{LAP} and will be reduced, thus masking the rarefaction signal. However, the uncertainty regarding when the actual rarefaction event occurs is approximately $\Delta t_{DB}/2$. If the dead band period is too long, then this will increase the uncertainty regarding when the leak occurred and, more importantly, the uncertainty regarding where the leak occurs. This will have important performance ramifications with respect to the performance of the leak detection system, as we shall see later.

Note that in the limit where the dead band and averaging periods are reduced to their minimum values, the rarefaction wave signal $\Delta p_{\text{Rarifaction}}$ becomes:

$$\Delta p_{\text{Rarefaction}} = p_{M,i} - p_{M,i-1}$$

EQUATION 6.13 Rarefaction Signal with No Dead Band or Averaging

In this case, the signal is simply equal to the scan-to-scan change in value. The dead band in this special case is:

$$\Delta t_{DB} = 1/f_{\text{Scan}}$$

EQUATION 6.14 Dead Band for Scan-to-Scan Rarefaction Pressure Signal

We are looking for a rarefaction wave event detection system here. It therefore makes sense to create a site rarefaction wave event RWE only if the signal is less than threshold $\Delta p_{\text{Threshold}}$:

$$\text{RWE} = \begin{cases} 1 & \text{if} \quad \Delta p_{\text{Rarefaction}} < \Delta p_{\text{Threshold}} \\ 0 & \text{if} \quad \Delta p_{\text{Rarefaction}} \geq \Delta p_{\text{Threshold}} \end{cases}$$

EQUATION 6.15 Rarefaction Wave Event Definition

So, how do we set $\Delta p_{\text{Threshold}}$? In general, there is a relationship between the threshold and the normally occurring noise in the data. The number of false alarms will be minimized if we require the pressure deviation signal $\Delta p_{\text{Rarefaction}}$ to be larger than the typical nonleak noise $\varepsilon(\Delta p_{\text{Rarefaction}})$ in the signal. Thus:

$$\Delta p_{\text{Threshold}} \approx -\varepsilon(\Delta p_{\text{Rarefaction}})$$

EQUATION 6.16 Rarefaction Event Threshold Definition

It is worth pausing at this point to consider the meaning of the uncertainty in our evaluation of rarefaction wave systems. As noted previously in Chapter 5, Statistical Processing and Leak Detection, it is not uncommon to equate the measurement uncertainty of a parameter (such as p_M) as being equal to some multiple of the standard deviation of the measurement value near its true value. We define the standard deviation of the measurement noise as:

$$\mathrm{SD}(p_M) = \sqrt{\frac{\sum_{i=0}^{n}(p_{M,i} - p_{\mathrm{True},i})^2}{n}}$$

EQUATION 6.17 Measurement Noise Standard Deviation

where n is the number of samples and $p_{\mathrm{True},i}$ is the true value of the process value at sample i. On the assumption of random and independent noise and only slowly varying hydraulics, a reasonable proxy for the true value would be to calculate the standard deviation by using recorded data from the measurement site and a moving average with the period equal to Δt_{AP}, or:

$$\mathrm{SD}(p_M) \approx \sqrt{\frac{\sum_{i=n_{\mathrm{AP}}}^{n}(p_{M,i} - \overline{p}_{i-n_{\mathrm{AP}},i})^2}{n}}$$

EQUATION 6.18 Measurement Noise Standard Deviation Estimator

Here, $\overline{p}_{i-n_{\mathrm{AP}},i}$ is the moving average of p_M over n_{AP} trailing samples at point i.

For purposes of our analysis, it is important to recognize that uncertainty is always associated with a confidence value Pr_{Conf}, such that the confidence expresses the fraction of sampled values that would be expected to exceed the uncertainty value. If we assume a normally distributed variable, then this confidence is a function of the number of standard deviations desired to minimize the false alarm probability α. Thus:

$$\Delta p_{\mathrm{Threshold}} \approx -Z_\alpha \mathrm{SD}(p_M)\sqrt{\frac{2}{n_{\mathrm{AP}}}}$$

EQUATION 6.19 Rarefaction Signal Threshold Estimator

where $\mathrm{SD}(p_M)$ is estimated by Eq. (6.18) and Z_α is set to a value between 2 and 3 to produce a confidence between 95% and 99.9% (assuming a Gaussian distribution of the error signal). A higher value of Z_α will reduce the sensitivity of the system but will also reduce the number of false alarms.

6.3 LEAK DETECTION AND LOCATION USING RAREFACTION WAVES

We could build a simple system by monitoring the pressure at some specified location and looking for negative step changes in the pressure that might be associated with a leak. However, such a system would be subject to potentially large numbers of false alarms for the following reasons:

- Many normal pipeline operating events can generate negative pressure signals. These include pressures downstream of closing mainline valves, pressures upstream of valves being opened, pump startup or shutdown, pump setpoint changes, opening of side branches, relief events, and others
- There is no way to use a single monitoring location to locate a leak.

A better methodology involves the use of multiple monitoring locations to isolate the leak location. Refer back to Fig. 6.1. At the snapshot in time t for this figure, the leak has initiated at some point in the past, t_{Leak}, and the wave fronts have moved distance Δx away from the leak site. To aid in detecting the leak, there are two pressure transmitters monitoring sites at distance L_P apart, one upstream of the leak and one downstream. The leak is located at distance L_{Leak} downstream of the upstream transmitter.

The leak is closer to the downstream monitoring site than it is to the upstream pressure transmitter. At the current time, the rightward-moving pressure rarefaction wave has already passed the transmitter. Consequently, we see a clear-cut negative pressure step function imposed on top of the random measurement noise. We assume that the step function was observed at time t_{Down}.

The front moving upstream from the leak site has not yet reached the upstream monitoring location. Consequently there is no significant pressure trend at this location outside a smattering of random noise. However, we can assume that a similar impact will be observed at the upstream location when the leftward moving rarefaction wave reaches that location. At some future time t_{Up}, the leftward-moving rarefaction wave will reach this location. Further simplification is achieved by assuming that the speed of sound a_c is a constant everywhere in the pipeline.

Let us move forward in time to the point where the leftward-moving wave has been observed. It should be clear that:

$$t_{\text{Up}} - t_{\text{Leak}} = \frac{L_{\text{Leak}}}{a_c}$$

EQUATION 6.20 Rarefaction Wave Transit Time to Upstream Monitoring Location

Similarly, at the downstream location:

$$t_{\text{Down}} - t_{\text{Leak}} = \frac{L_P - L_{\text{Leak}}}{a_c}$$

EQUATION 6.21 Rarefaction Wave Transit Time to Downstream Monitoring Location

We can combine these two equations to solve for the leak location:

$$L_{\text{Leak}} = \frac{a_c(t_{\text{Up}} - t_{\text{Down}}) + L_P}{2}$$

EQUATION 6.22 Leak Location Equation

The procedure for locating a leak is straightforward. For every pressure-monitoring location, we maintain a queue of detected negative pressure events (defined via Eq. 6.15). Only negative pressure events are added to the queue. Positive pressure deviations are discarded. Many of the events in the queue, of course, may be associated with simple operating changes or instrument problems. How do we eliminate them?

One way to do this is by confirming that the candidate leak is clearly located between the two monitoring sites. This will be satisfied if the delta time between the upstream and downstream events is equal to or greater than the rarefaction wave transient time from one site to the next. If the time difference exceeds this maximum, then the event pair cannot correspond to a leak between the monitoring sites and the oldest event can be discarded.

An example of a real-world rarefaction wave system in action is illustrated by Fig. 6.4. This figure shows the result of a test for simulated leaks with orifice diameters of 1-7/16 inches and of 3/8 inches in a crude oil pipeline with a 48-inch diameter. The tests were performed by implementing a valve in conjunction with the orifice and rapidly opening the valve to simulate the leak and generate a negative pressure wave front. The separation between pressure monitoring stations was approximately 6.5 miles. The location of the monitoring site was approximately 3 miles from the leak site.

The two cases display both the pressure trend at the monitoring site and the state assessed by the leak detection system as the pulse passes. In the first (larger orifice) case, the valve was opened rapidly and then closed after approximately 2 s, as indicated by the immediate decrease in pressure followed by an increase. A true leak would be expected to show a sustained pressure drop. The valve was opened for a longer period in the second case.

It is worth noting the noise in the system that is evident in both of the charts. Field evaluation indicated that this noise in the signal trace was high-frequency (approximately 166 kHz) electrical noise in the analog signal line. This signal originated at the site pressure transmitter and terminated in the site PLC, and it appeared to be generated by the DC power supply in the PLC cabinet. In line with the discussion in Section 6.2, this noise contributed to a minimum detectable leak for the rarefaction wave system. The vendor calculated that this minimum size would correspond to a hole or orifice diameter of 0.4 inches.

Fig. 6.4 indicates that the leak was detected in both cases. However, it should be noted that a range of tests was performed for varying orifice sizes: 6 cases at 3/8 inches, 5 cases at 1/2 inches, 4 cases at 3/4 inches, and 1 case at 1-7/16 inches. For the 15 tests, the event was positively detected as a set of paired events in 11 of the 15 tests but was positively located in only 9 tests (due to technical glitches in the rarefaction wave system). Not surprisingly, the system had the most trouble at the smallest hole sizes: only one of the six tests of the smallest orifice size (3/8 inches) resulted in a positive catch by both detecting the event at both monitoring sites and then locating

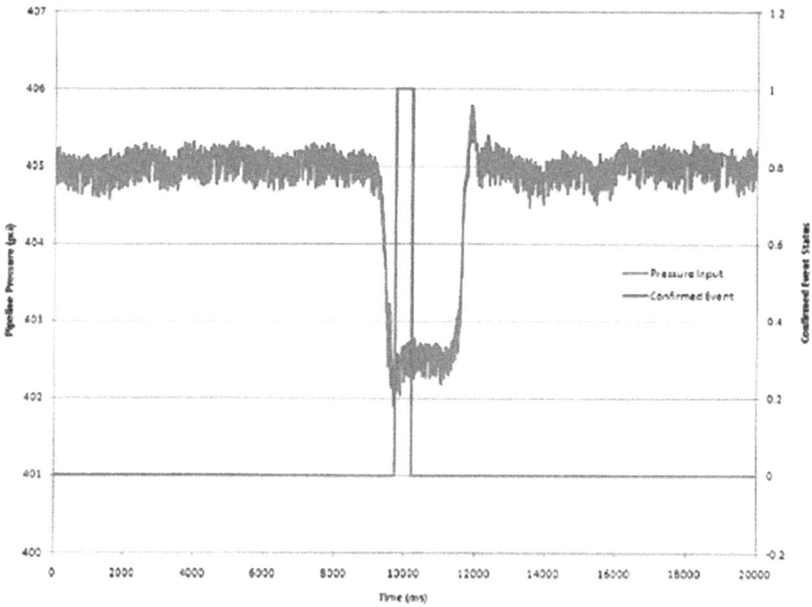

(A)

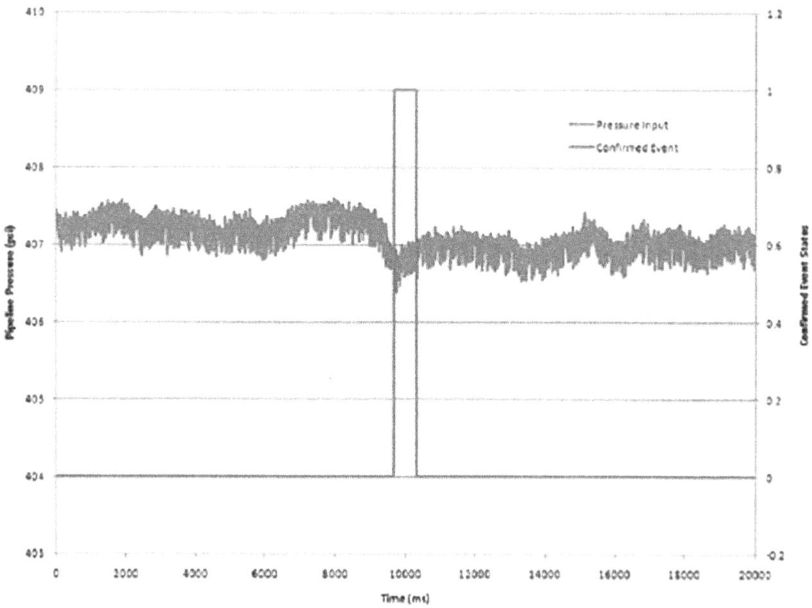

(B)

FIGURE 6.4 Recorded rarefaction wave signals in 48-inch-diameter pipeline for simulated leaks with (Top) 1-7/16-inch and (Bottom) 3/8-inch orifice diameters.

the test leak location, in line with the vendor assessment. The standard deviation of the location error for the successful tests was 486 feet, indicating that this particular system could locate the detected leak with an error of approximately 1460 feet, or one-quarter of a mile. How these limitations come into play is discussed in the next section.

6.4 RAREFACTION WAVE LEAK DETECTION ISSUES, IMPROVEMENTS, AND ASSESSMENT

Negative pressure wave systems have a number of outstanding benefits and also some glaring problems. We discuss these tradeoffs in this section.

A major advantage of a well-designed rarefaction wave system is that it is *fast*. How fast? If we assume that the system will not kick out an alarm unless the leak signature is confirmed at both segment measurement sites, then the average time to detect a leak is fundamentally the sum of the three-quarter segment wave propagation time plus the leading lag time/averaging period plus the dead band time, or:

$$\Delta t_{\text{Detect}} = \frac{3L_P}{4a_c} + \Delta t_{\text{LAP}} + \Delta t_{\text{DB}}$$

EQUATION 6.23 Rarefaction LDS Time to Detect (With Confirmation)

The latter two terms are typically negligible in a reasonably long pipeline segment. If we postulate a 25-mile long segment with a 3500-foot/s wave speed, plus 1 or 2 s to address the other two terms, then the leak detection time will be approximately half a minute (approximately 30 s). Note that if we do not require the signal to be confirmed by both measurement sites, then the detection time is:

$$\Delta t_{\text{Detect}} = \frac{L_P}{4a_c} + \Delta t_{\text{LAP}} + \Delta t_{\text{DB}}$$

EQUATION 6.24 Rarefaction LDS Time to Detect (Without Confirmation)

Another notable advantage of rarefaction wave systems is their ability to accurately *locate* the leak. How accurately? We assume that the event timing error is approximately the sum of the dead band and leading averaging period times. If we use the location error equation previously given in Eq. (6.26), then it is not difficult to show that, for reasonable averaging and dead band times summing to a second or less, the associated location error is much less than a mile. Faster scan times can presumably improve this even further. Similarly, the variation in the wave velocity in a batched liquid pipeline (perhaps approximately $\pm 4-5\%$), and assuming a 25-mile length, will lead to similar location errors. These are very small errors when compared to the typical location performance for an RTTM-based mass balance leak detection system.

We have seen the good. Now, let us examine the bad and the ugly.

One complication involves events arising from outside the pressure measurement site pair. For example, an examination of Eq. (6.22) quickly shows that a problem arises for operationally created negative pressure waves that arise from the left of the leftmost measurement site (ie, a rightward-traveling wave from upstream of P_{Up} in Fig. 6.1) or negative waves associated with normal operations to the right of the rightmost of the two pressure measurement sites (a leftward-traveling wave arriving from downstream of P_{Down} in Fig. 6.1). Such cases will create false alarms or false positives because Eq. (6.22) will indicate a leak at one of the segment endpoints even though the signal was created outside of the measurement section. Some possible ways to handle this include:

1. Utilize a more sophisticated *total hydraulic signal* approach that relies on additional flow measurements to provide more information regarding the direction of propagation for the rarefaction waves. We discuss this in more detail later in this chapter.
2. Identify all operational changes capable of creating such events (such as pump starts and stops, valve position changes, relief events, etc.) and *suppress* the alarms from the rarefaction wave system for a brief period following the event. This is the most commonly utilized approach, and it can work very well. It has the disadvantage of requiring the independent monitoring of other pipeline events, tends to require complex logical rules that can be difficult for system support analysts to maintain, and can result in significant fractions of time in which leak detection is inhibited, particularly in actively operated pipelines.
3. Identify all sources of error $\varepsilon(L_{\mathrm{Leak}})$ for the location of a real leak arising from within the segment per our classification rule (the events must correspond to a source inside the segment) and reduce the effectively monitored segment size accordingly so that events arriving from outside of the monitored segment will not trigger an alarm.

In the case of option 3 here, we modify our classification rule so that the event is a leak only if:

$$\varepsilon(L_{\mathrm{Leak}}) < L_{\mathrm{Leak}} < L_P - \varepsilon(L_{\mathrm{Leak}})$$

EQUATION 6.25 Modified Classification Rule to Prevent False Alarms

where $\varepsilon(L_{\mathrm{Leak}})$ is the uncertainty in the leak location. This approach prevents the LDS from creating false alarms based on signals arising from outside of the segment by inhibiting alarms for leaks that are too close to either end of the segment. Referring to Eq. (6.22), it should be clear that this uncertainty is a function of the uncertainties in the two rarefaction wave detection signal times t_{Up} and t_{Down}, the separation between the measurement sites L_{Leak}, and the wave velocity a_c. Based on the assumption that the uncertainties are all independent, the leak location uncertainty would be:

$$\varepsilon(L_{\text{Leak}}) \approx \frac{1}{2}\sqrt{2\left(\varepsilon(a_c)\left[\frac{L_P}{a_c}\right]\right)^2 + a_c^2\left[\varepsilon(t_{\text{Up}})^2 + \varepsilon(t_{\text{Down}})^2\right] + (\varepsilon(L_P))^2}$$

EQUATION 6.26 Leak Location Uncertainty

such that:

- $\varepsilon(a_c)$ is the uncertainty in the wave speed
- $\varepsilon(t_{\text{Up}})$ is the uncertainty in the upstream pressure signal measurement
- $\varepsilon(t_{\text{Down}})$ is the uncertainty in the downstream pressure signal measurement
- $\varepsilon(L_P)$ is the uncertainty in the distance between the pressure measurements

The approach in option 3 is simple and easy to implement. However, the methodology designed to eliminate the impacts of hydraulic disturbances arising from outside the pressure-to-pressure detection segment as discussed has the disadvantage of producing dead leak detection bands at the segment end points. It also does not protect against normal operating events that arise from inside the segment. With respect to the second issue, a simple extension of this approach can be used to implement similar dead bands at internal sites that are likely to create disturbances, such as relief valve locations.

The approach in option 2 highlights the potential logical/rule-based complexity of rarefaction wave leak detection systems. Such complexity can be difficult to maintain and understand. In addition, if event-based rules are used to inhibit alarms, then this ensures that the system must have a tie-in to the operator's Supervisory Control and Data Acquisition (SCADA) system. Rarefaction wave systems are sometimes sold as stand-alone systems, and such installations will not benefit from this approach.

As we have already seen, the theoretically precise leak location capability of negative pressure wave systems means that the systems must provide data sampling at a relatively high scan frequency. Although the system can still perform competitively with a scan rate of once per second, many SCADA systems and/or communication circuits work with scan cycles of 30 s, 60 s, or even longer. A scan rate of once per minute implies an average leak location error of approximately 15 miles or more for our example pipeline segment, which puts the leak location performance into similar performance territory as a mass balance system. Note that some rarefaction systems operate independently of the SCADA system and, in such cases, the SCADA sampling limitation would not apply. However, the sampling limit will still apply to the communication circuits between the monitoring sites.

In addition, the system (particularly the leak location function) will not work properly if the detection times are not precisely known. This requires time stamping of the data at the field/measurement site combined with precise time synchronization for all sites. Many pipeline data acquisition systems do not time-stamp incoming data at all; if they do, then they only time-stamp the data when they are acquired by the SCADA system. Some systems overcome this issue by providing local global positioning system (GPS) timing and time-stamping capabilities.

Another significant problem for negative pressure wave systems is that the rarefaction wave fronts attenuate rapidly if the hydraulic gradient exhibits strong frictional losses with distance. Refer to Fig. 6.5. This figure shows the calculated rarefaction wave magnitude (numerically calculated using a method of characteristics formulation) as a function of leak rate (expressed as a percentage of the nominal pipeline flow rate) and distance from the leak site. We assume a 10-inch diameter (10.75-inch OD) pipeline with a 15-centistoke crude oil commodity and flow velocities of 3.83 and 11.50 ft/s. Based on the assumption that detection of leak-induced pressure waves less than 0.5 to 1 PSI may be problematic due to noise, these two charts indicate that small leaks approximately 1% or 2% of nominal flow create barely detectable disturbances *even at the leak site* for low nominal flow. Even in the case of high flow, the signals fade into nondetectability when they have propagated a distance of approximately 15 miles from the leak site.

More importantly, virtually all of the high-flow, high-friction leak signals attenuate to undetectable values at a point between 30 to 60 miles away from the leak site. This even applies to extremely high leakage rates approximately 50% of nominal flow.

The implication of this is that rarefaction wave systems require high instrument density to function well. A rule of thumb is that negative pressure wave systems should utilize measurement sites placed no more than 20 to 30 miles apart to achieve high performance expressed in terms of high sensitivity (leaks approximately 1% of flow or larger) and confidence (the disturbance will trigger events at both measurement sites to either side of the leak).

Another problem with rarefaction systems has to do with their operation in conjunction with pump station sites on pressure or flow setpoint control. Consider the case of a pump station with variable speed pumps running on discharge pressure control with a downstream leak and a pressure-monitoring site at the station discharge. When the negative pressure disturbance moving upstream from the leak site reaches the pump station, the discharge pressure will decrease and the pumps will speed up to increase the pressure and maintain the pressure setpoint. Depending on how quickly this control response occurs, it may tend to cancel out the rarefaction signal and obscure the leak trace.

One potential technique that can be used to address this potential issue is the total hydraulic signal approach. This methodology requires the installation of rapid response flow measurement devices (potentially achievable

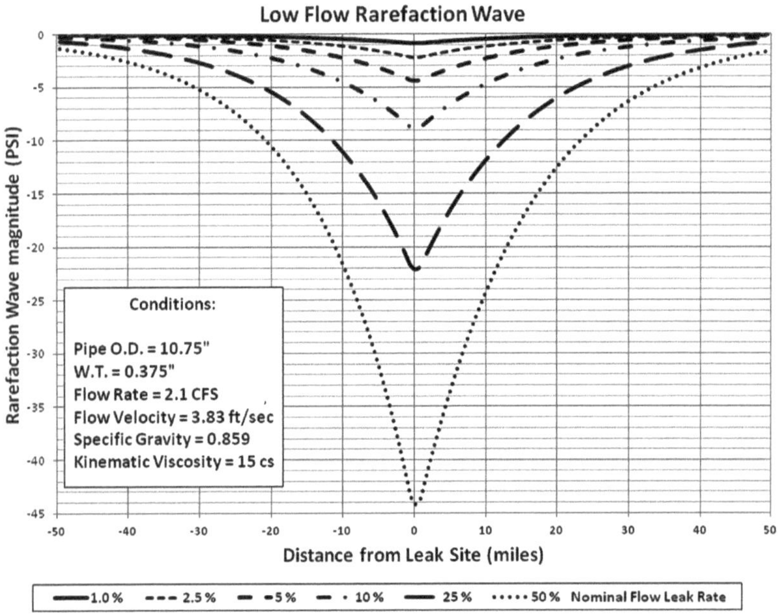

(A)

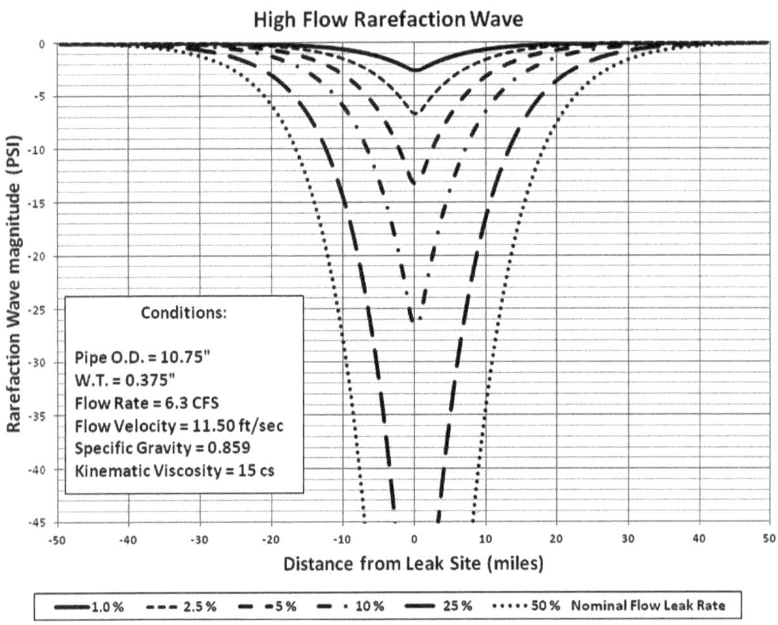

(B)

FIGURE 6.5 Rarefaction wave attenuation for (A) low-flow and (B) high-flow liquid pipeline operations.

through the use of either external/strap-on or internal transmitter-based ultrasonic flow meters) at each pressure measurement site. This achieves two purposes: (1) it allows the direction of propagation for the flow disturbance to be determined via the Joukowsky equation and (2) it allows the calculation of a total hydraulic disturbance signal by adding the flow disturbance signal (suitably corrected via the Joukowsky equation) to the pressure signal. The processed total hydraulic signal at any location is given by:

$$TS_{\text{Rarefaction}} = \Delta p_{\text{Rarefaction}} + \frac{\rho_c a_c \Delta Q_{\text{Rarefaction}}}{A_{\text{Pipe}}} i_{\text{Wave}}$$

EQUATION 6.27 Total Hydraulic/Rarefaction Signal Equation

A total signal is developed for both the upstream and downstream measurement stations. In both cases, the processed transient flow deviation signals $\Delta Q_{\text{Rarefaction,Left}}$ and $\Delta Q_{\text{Rarefaction,Right}}$ are developed using averaging period, dead band, and differencing signal processing similar to that used by the pressure measurement signal processing in Eq. (6.2) regarding the flow meter measurement trends at the measurement sites. The total signal events associated with the total signal are likewise obtained using a thresholding mechanism.

The total hydraulic signal has several interesting properties:

- The hydraulic flow and the pressure components can potentially be used to confirm each other. Thus, if the pressure signal experiences a deviation event due purely to a pressure transmitter problem, then it will not be accompanied by a corresponding flow deviation event, and the total signal event can be immediately discarded as a measurement artifact without any possibility of generating an alarm.
- Strong rarefaction waves that have been triggered by operational events from outside and to the left of the measurement segment and that have not been canceled out due to a system rule can trigger simple pressure rarefaction events at both segment boundaries. This will cause a false alarm because the two events will correlate as an apparent leak located on the left boundary and because a system rule was not violated. However, the same situation will generally trigger a total signal event only at the right boundary. A signal will not be generated at the left boundary because the pressure and flow components will cancel. Consequently, no event pair will be generated by the total signal system and there will be no false alarm. The same reasoning applies for transients generated from outside and to the right of the monitoring segment.
- At measurement locations that are not pressure boundaries, the total signal will generally be twice the size of the rarefaction pressure signal alone, permitting smaller leaks to be detected.
- For measurement segments bounded by control sites, a more reliable signal is likely to be maintained in the face of setpoint changes because any resulting control-related pressure changes will be converted to flow transients and will still be captured in the total signal.

Note the obvious downsides of this approach: (1) it requires more—and probably expensive—instrumentation in the form of flow meters and (2) it is more complicated than a simple pressure-based negative wave approach. We should also note that the existing literature at this time does not provide any indication that this approach has many (if any) real-world installations.

Another important limitation in liquid commodity pipelines is that a basic negative pressure wave–monitoring segment must operate in a tight mode with no slack flow between adjoining pressure measurements to function properly. As discussed previously in Chapter 4, Real-Time Transient Model–Based Leak Detection, slack operation applies to saturated liquid vapor conditions in which the line operates in a two-phase mode whereby the liquid portion of the flow flows in a channel-flow mode along the bottom of the pipe. The slack section is usually attached to a local high point in the line. A basic rarefaction wave system cannot operate using pressure measurements at tight locations on either side of the slack line section because the pressure in the slack section is fixed at the vapor pressure of the fluid. Consequently, a leak-induced rarefaction wave cannot pass through it.

This is not to say that rarefaction wave systems cannot potentially address the leak detection challenge for leaks from tight pipeline segments immediately adjacent to slack sites. Consider the situation shown in Fig. 6.6. This figure shows a leak that has just occurred downstream of a slack section. The leak is some distance upstream of a pressure-monitoring site.

When the leak occurs, the downstream end of the slack section will tend to behave like a constant pressure boundary. The leftward-moving rarefaction wave will therefore reflect off the slack boundary as a coupled, positive, rightward-moving, pressure/flow wave. From the standpoint of a monitoring site that is downstream of both the slack interface and the leak, the pressure will initially drop abruptly (at time t_1) as the rightward-moving rarefaction wave passes. A short time later (time t_2), the reflected leftward-moving rarefaction wave will pass as another rightward-moving pressure wave, but now as a *positive* disturbance (again to be detected using an appropriate variation of the signal-processing mechanisms previously discussed). Refer back to Fig. 6.6. If the distance from the monitoring site to the slack interface L_{ST} is known, then the distance of the leak upstream of the monitoring site L_{LT} is given by:

$$L_{LT} = L_{ST} - \frac{a_c}{2}(t_2 - t_1)$$

EQUATION 6.28 Leak Location When Slack Line is Present

Note that the location of the slack interface must be known. Unfortunately, in real-world pipelines that run under variable flow and pressure conditions, this location is not a constant; it is subject to change. This is an area where an RTTM-based mass balance system and a rarefaction wave

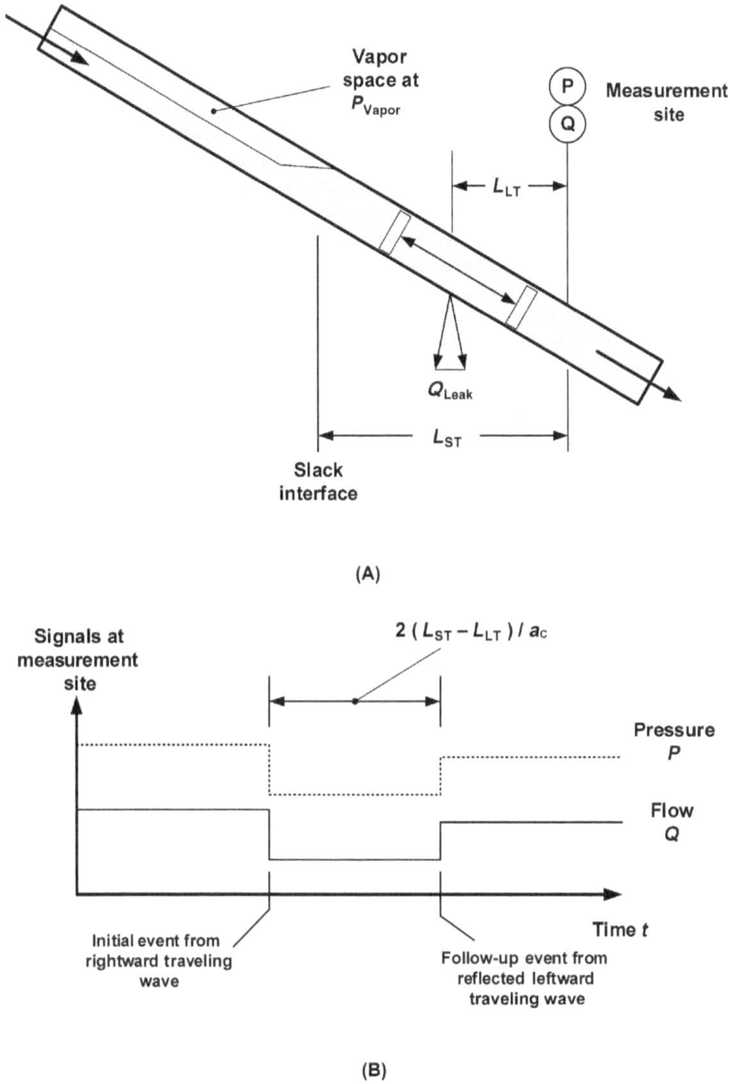

FIGURE 6.6 (A) Rarefaction wave leak detection downstream of pipeline slack site, and (B) reflected rarefaction pressure signal from slack location.

system can complement each other: the transient model can be used to provide the slack interface information and the rarefaction wave system can be used to provide a rapid response mechanism along with a more precise calculation of the leak location. Note that we can again improve performance by implementing flow measurement at the monitoring site and by using a total hydraulic signal approach.

There may also be an issue in that the slack interface is not flat or perpendicular to the pipe front. Thus, any pressure wave impacting the slopped liquid/gas interface may create a wave with a different shape and magnitude that could attenuate faster than the original wave. This is most likely a problem for large-magnitude waves on shallow elevation gradients.

A final limitation regarding the use of negative pressure wave systems is that they are best suited for liquid commodity pipelines. The reason for this is easily understood by referring back to the Joukowsky equation (Eq. 6.3). A key component of these equations is that the size of the pressure disturbance is proportional to the volumetric flow change (effectively the leak size) and the commodity density. At best, the density of a typical pipeline gas commodity under flowing conditions, even under pressure, is usually less than approximately 5% to 10% of a liquid commodity density. The second issue is wave speeds in gases, which are typically approximately 35% to 45% of the wave speeds in liquid commodities. These factors will reduce the initial pressure disturbance to only a few PSI for even enormous leaks. The pressure disturbance will, of course, be further reduced by attenuation. For smaller leaks, the disturbances may be undetectable. For these reasons, rarefaction wave systems are not commonly deployed in gas commodity pipelines.

An assessment of the strengths and weaknesses of these systems is provided in Table 6.1.

TABLE 6.1 Rarefaction Wave Leak Detection Positives and Negatives

Positives	Negatives
• Ultra-fast leak detection • Very precise leak location (if properly supported by fast sampling and time-stamping) • Can provide relatively low false alarm rates if properly supported through signal averaging, appropriate thresholding, and carefully thought-out rules • Can add to or supplement the performance of slower-responding mass balance systems and can benefit from the additional information provided by RTTMs	• Requires very high sampling rates • Signal time-stamping required • Complex logical rules make systems difficult to tune and maintain • Rapid signal attenuation in lines with high flow/friction may require tight spacing on monitoring sites • Will not work well for leaks with slow onset (ie, underground corrosion or slow mechanical failure) • Leak detection near slack lines may be difficult to impossible without special rules or additional instrumentation • Best for liquid commodity pipelines. Will not work well in gas-phase or multiphase pipelines. • High performance may require access to additional information provided by the SCADA system

6.5 DEVIATION ALARM SYSTEMS

Deviation alarm systems are effectively the result of operating a rarefaction wave system without high scan rates and time sampling. These systems look for large flow, pressure, or a combination of flow and pressure deviations from previous norms in terms of flow rates. Consequently, they utilize signal processing similar to the basic approaches discussed previously in Section 6.2. Deviation alarm systems often use logical rules similar to those utilized by rarefaction wave systems, with the exception that, on the assumption that they are not enabled by a high scan frequency, they will not be able to utilize precise timing and location-based rules to locate leaks and minimize false alarms. The systems do offer the pipeline controller the benefit of providing a fairly rapid response to an unexpected or sudden change in pipeline operation and can potentially provide an early indication of a large leak when compared to mass balance systems. That said, from a leak detection point of view, these systems offer fewer of the benefits that are provided by well-designed rarefaction wave systems, with all of the disadvantages.

Refer to Table 6.2 for a summary of the strengths and weaknesses of deviation alarm systems.

TABLE 6.2 Deviation Alarm System Positives and Negatives

Positives	Negatives
• Relatively fast leak detection for large leaks. • Provides an early indication that there may be a problem in the field Potentially useful as a first alert mechanism	• Not particularly useful for small leak rates • Complex logical rules make systems difficult to tune and maintain • Rapid signal attenuation in lines with high flow/friction will slow the response of these systems • Will not work well for leaks with slow onset (ie, underground corrosion or slow mechanical failure) • Leak detection near slack lines may be difficult to impossible without special rules or additional instrumentation • Best for liquid commodity pipelines. Can work in gas-phase or multiphase pipelines, but response will be slower • High performance will require access to additional information provided by the SCADA system • Often subject to high false alarm rates

Chapter 7

External and Intermittent Leak Detection System Types

The previous chapters have primarily presented discussions on internal leak detection technologies such as mass balance, real-time transient model, and rarefaction wave approaches, along with discussions of statistical tools and approaches. These systems are algorithmically based and rely on internal pipeline measurements such as pressures, flow rates, temperatures, and so forth to infer if a leak may be present.

In this chapter, we discuss the other major leak detection classification: external or direct measurement systems. American Petroleum Institute (API) Recommended Practice number 1130 (API RP 1130) [1] describes external or direct measurement systems as devices that operate on a nonalgorithmic principle and rely on physical detection of an escaping commodity. These systems are not reliant on internal pipeline operating measurements such as flow rate, temperature, and so forth.

Although the API 1130 external leak detection classification description is broadly used, in this book we have identified a need to expand this taxonomy. The expanded classification distinguishes between systems that detect the escaping commodity and those that identify changes to the external spill environment. Escaping commodity leak environment detectors are those that identify a change in the pipeline environment when a leak is in process. Examples include acoustic sensors (they detect the sound of a leak) and thermal change detectors (these identify a localized temperature change). The key distinction of this taxonomy group is that the sensors identify a change in the environment created by the escaping commodity, not the presence of the spilled commodity.

Detection systems that identify the presence of the spilled commodity are those we classify as direct detection systems. These leak detection technologies must come into physical contact with the targeted commodity to produce an alarm. Some sensor examples include hydrocarbon sensing tube, cable, and infrared detectors. These systems produce an alarm when the targeted commodity is physically detected.

Pipeline Leak Detection Handbook. DOI: http://dx.doi.org/10.1016/B978-0-12-802240-5.00007-8

The previous taxonomy examples are not inclusive of all external leak detection systems. The remainder of this chapter presents the suite of external leak detection systems, such as direct detection by operating company personnel as well as by third parties, various cable-based sensor systems, acoustic sensors, chemical sensors, and various video/camera sensing systems, as well as methods that are reliant on the use of tracing elements.

Before we discuss the types of technologies in greater detail, we first consider various environmental factors that greatly influence these systems. These external factors determine if and how long it may take to detect an escaping commodity or spill based on the leak's physical location, ground topology, commodity migration path, and other environmental influences, as well as leak detection technology factors.

7.1 SPILL MIGRATION

When a liquid commodity pipeline experiences a leak and the spill begins to accumulate, internal forces inside the pipe combined with environmental forces dictate where the spilled commodity will migrate to and how fast the migration will occur. Internal factors include the orifice size and the pressure inside the pipe. Environmental forces include:

- Gravity
- Soil density
- Water table depth
- Direction of water flow

Orifice size, internal pressure, and gravity are major factors in determining where a liquid spill will migrate. Fig. 7.1 is a simple sketch showing the general relationship between where the leak site is, where the resulting spill may migrate, and the pipe itself.

If the orifice size is large, then the pressure drop across the orifice will be relatively small, and resistance forces in the soil will generally dominate the flow pattern. In this case, the spill pattern will be that of an expanding sphere of commodity in the surrounding earth. If, however, the orifice size is small, or if the fill surrounding the pipe is of very high permeability, as would be the case for well-sorted sand or gravel, then that gravity will dominate (as indicated in Fig. 7.1) and the resulting spill will be pulled downward into the soil. Note that the orifice size for an underground leak could be very small for many corrosion-driven pits. However, it could be large for other breaches due to corrosion or for pressure-driven rupture failures.

The soil permeability or resistance is driven by the material type that surrounds the pipe, such as sand, gravel, or clay. Highly permeable soils have low resistance, and vice versa. In addition, porosity is an indication of how much spilled commodity the soil can potentially absorb. Different pipe trench fill material types have different densities, porosities, and permeabilities

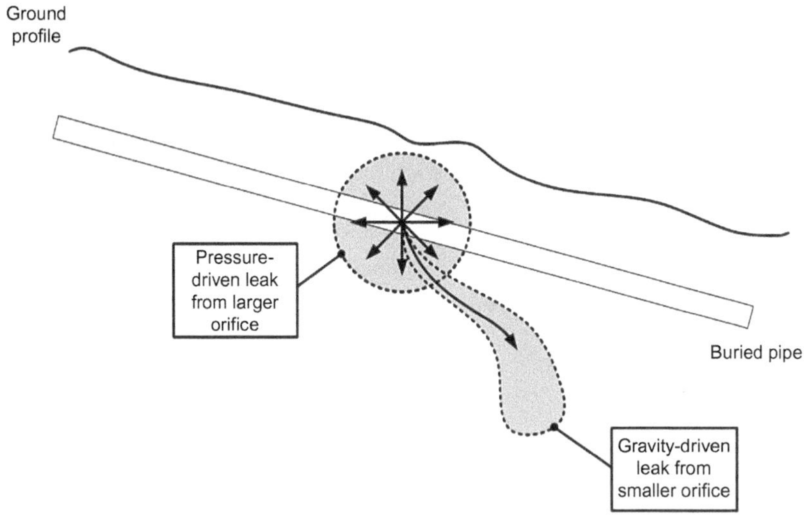

FIGURE 7.1 Pressure and gravity effects on a liquid spill.

depending on organic matter in the soil, texture, particulate size, and packing arrangement. For example, dry clay generally has porosities (again, the fraction of void space if the material is dry) in the range of 40% to 70% and, most importantly, very low permeabilities ranging from 10^{-11} to 10^{-10} cm^2. However, dry, coarse sand is characterized by porosities and permeabilities that can range from 25% to 50% and 10^{-5} to 10^{-7} cm^2, respectively. Sand is somewhat less porous than clay; it packs more efficiently and therefore will hold less spilled commodity. However, the particles in clay are generally finer; therefore, the spaces between the clay particles will be smaller as well. In general, the soil permeability tends to increase in direct proportion to the square of the grain diameter and approximately in proportion to the porosity. Because clay has lower permeability, we can assume that it probably has a finer grain size. The soil resistance is inversely proportional to the porosity, so we can also assume that clay will have a higher hydraulic resistance than sand.

To expand on this, the impact of the material surrounding the pipeline and its associated properties influence where and how fast the spilled commodity will flow. The spill propagation rate is a function of the soil permeability and the soil porosity. The size of pore space either assists movement of the spill traveling through the associated material, such as when the soil has large pore spaces, or hinders it, such as when the surrounding soil has very small pore spaces. High pore space allows easier movement of the commodity compared with low pore space, which contributes to restricting the movement of the commodity through the soil.

Further, soil properties are not static; they can change due to other external influences such as the addition of water. In our example, if we add water to either material, then the density will increase as the water drives out the air. In addition, the available porosity will decrease, indicating that any infiltrating spilled commodity will have to drive out the water. In addition, the addition of either water or spilled commodity will change the resistance of the soil.

A complicating factor of soil density is that it is not constant over time. Depending on the soil type, the density may change not only by the addition or removal of water but also due to compaction, continuing decay of organic materials, and other effects. Compaction occurs during construction and as the pipe settles on the soil, or even as a result of tertiary factors such as earthquakes. It reduces the fill material porosity that contributes to an increase in soil resistance.

Fig. 7.2 shows an ideal case of a spill that has occurred on a buried pipeline. As shown, the pipe is located in an excavated trench that has been backfilled and compacted. In this situation, the encasing soil is of a higher density than the pipe–soil interface. As such, one may expect that the spill will tend to flow along the pipe instead of migrating away from the pipe through the soil, because this is the path of least resistance.

Fig. 7.3 shows a different potential spill migration that could occur as a result of a gravity-driven spill from a tiny orifice into less permeable soil. In this case, the spilled commodity flows downward through the pores of the soil. In this situation, the spill may run to the bottom and migrate along the bottom of the pipe trench.

Another significant spill migration impact is due to the presence of water. Fig. 7.4 provides a potential outcome of a leak from a pipe that is partially submerged by the water table. In this case, assuming the spilled commodity specific gravity is lighter than and immiscible with water, the commodity

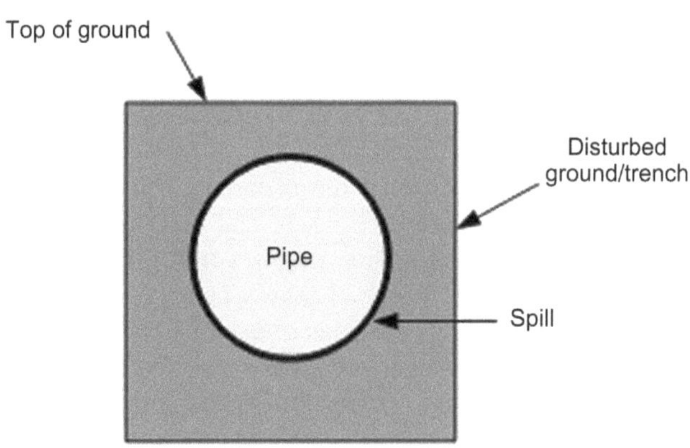

FIGURE 7.2 Spill migration along the pipe.

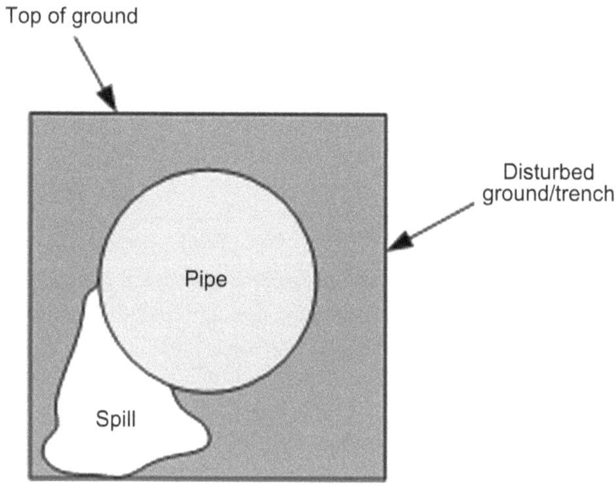

FIGURE 7.3 Spill migration along the pipe with porous soil.

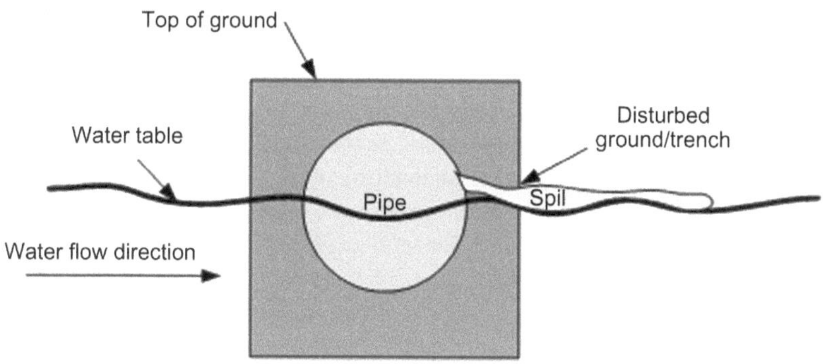

FIGURE 7.4 Water impacts on spill migration.

will float at or on the water table and migrate with the direction of the percolating water flow.

This would also apply to submerged offshore pipelines. In situations in which the pipeline is buried such that water totally surrounds it, any resulting spill will tend to float to the top of the water. This again assumes that the spilled commodity specific gravity is less than that of the associated water, and that the commodity cannot dissolve into the water. At that point, the spill will migrate in the direction of the water flow.

In summary, many environmental factors influence the migration of the spilled commodity. The presented examples are more idealized than what will probably occur in an actual pipeline installation. An actual installation may have a combination of these effects, which could act as a dam or a conduit for blocking the spill or channeling the spill to a different location.

These are just some of the variables that a leak detection engineer must consider when evaluating the potential for leaked commodity to migrate to other locations away from the pipe.

7.2 DIRECT OBSERVATION

This portion of the chapter focuses on direct observation by people. As documented in more detail in Chapter 13, Leak Detection and Risk-Based Integrity Management, this is the number one leak and spill identification method. Fig. 7.5 shows the Pipeline and Hazardous Materials Safety Administration's (PHMSA) count of reported leak identifications from January 1, 2010 to December 31, 2015, inclusively. As shown in this figure, the majority of all spills are identified by people rather than technology. Note that direct observation can be a result of the engagement of any human senses; therefore, the spill might be visually observed (the commodity was seen on the ground or water, or the bright light from a burning commodity was seen, or there was dead vegetation resulting from the toxic effects of the spill), heard (an explosion or the hiss of escaping liquid or gas), smelled (tert-butyl mercaptan odorant in natural gas), or texturally detected (ie, the soil was wet, gummy, or sticky), or the vibration from the leak was sensed.

When we discuss direct observation in the context of pipeline leaks and spills, we must be clear about the channel of observation. Specifically, there are three leak validation observer methods: (1) inadvertent or accidental on-site workers; (2) scheduled, purposeful observers; and (3) inadvertent or accidental third-party observers. Each of these is discussed further in the following sections.

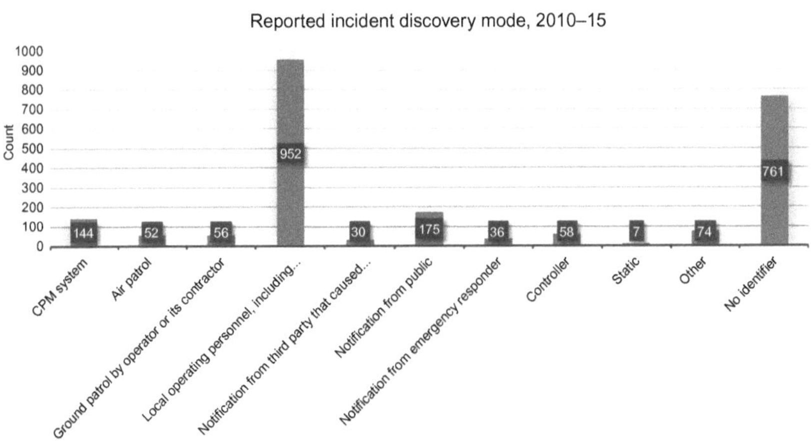

FIGURE 7.5 PHSMA reported spill detection methods.

7.2.1 Site Workers

Regarding direct observation, site workers are specific personnel who work at a facility. These workers include the owner/operator, employees, and/or contractors. We term this site-worker observation (SWO). This occurs when personnel who are working for the company, in one fashion or another, detect a leak or spill as part of their normal duties.

The SWO may also occur when a control room operator sees a change in process that indicates a leak may be in progress. In this situation, local site personnel are dispatched to the area where the leak is suspected to be to validate or invalidate that a leak is present. In this case, the actual detection is performed by the control room operator and the field worker confirms the fact.

The SWO approach is a very successful leak detection method and, from a leak detection point of view, illustrates the value of having workers in the field. As determined by the review of the PHMSA reported spill database, approximately 40% of all leaks are detected by site operator company personnel. This detection rate is fairly high because:

- A significant portion of all PHMSA database leaks (approximately 71%) occurs on operator-controlled property and not on the right-of-way (ROW).
- The probability that personnel are working within operator-controlled facilities is very high. This increases the probability that a person will see the leak.
- Direct observation has a very low false alarm rate

Therefore, because a significant portion of all PHMSA-reported leaks occur within the confines of a facility, because personnel are routinely working within the facility, and given that many of the associated leak rates are from seals and other failed components (and thus low), the probability that a person will see the leak first is very high.

7.2.2 Planned or Scheduled Observer

The second visual observation leak detection method is the planned or scheduled periodic visual observation (PVO). A PVO leak detection method is a process whereby pipeline personnel or contracted parties perform visual inspection of the pipeline on an regular basis.

These scheduled observations may be performed on foot, driving, flying, on a snowmobile, and so forth. The key elements of this observation method are that: (1) it encompasses the full pipeline; (2) it is specifically planned; (3) it occurs on a specific schedule; (4) it is conducted by personnel associated with the pipeline; and (5) it is specifically looking for the presence of a spilled commodity. As identified in Fig. 7.5, approximately 6.82% of all reported liquid pipeline spills (108 in total) in the United States during 2010–15 were detected by PVO.

In addition to being a prudent business practice, PVO is a United States federally mandated requirement. According to Code of Federal Regulation (CFR) 195.412, all hazardous liquid pipeline operators are required to visually inspect the pipeline at an interval not to exceed 3 weeks and at least 26 times per year.

PVO performance, as measured by leak size, time to detect, and false alarms, is highly variable, with no definitive methodology or method of determining a quantitative comparison metric. We do know that the time to detect cannot be any less than the schedule interval. Beyond this, observation lag due physical migration factors and leak rate will generally dictate the time to detect for this method. Detection of a leak/spill occurs when it is of sufficient size and at a location that makes it visible to the observer. The combination of size and location can range from very small, which creates a visible water sheen, to an extremely large underground spill that eventually migrates above ground or causes a significant change in foliage that indicates the presence of a spill.

For example, the leak can be detected immediately after it occurs if scheduled observers happen to be present when the escaping commodity first becomes visible. The Trans Alaska Pipeline 2001 bullet-hole leak is a prime example of very rapid detection time. This leak occurred when an individual fired a rifle at the pipe in an above-ground section of the line. Fortunately, a scheduled aerial observer was almost directly overhead when the individual shot the pipe, and the observer saw the resulting leak. On the other end of the spectrum, several observation cycles could occur before a small underground leak accumulates sufficient spill volume that it surfaces or provides an observable indication that a spill is present. Thus, the PVO leak identification time to detection performance metric can span from seconds to many weeks or even months. Note that there is also no industry standard that specifies a schedule beyond US Federal requirements or that a firm can leverage in their internal analysis of PVO performance.

In line with our discussion, it is clear that PVO performance is established primarily based on controllable factors, such as the inspection schedule, and noncontrollable factors, including the leak size and the physical lags caused by external environmental factors. Regarding the POV leak location performance metric, however, it should be clear that this approach, like all direct observation methods, provides a very precise result. Although the observed spill location may not be exactly where the leak is occurring, the proximities will often (though not always!) be close; therefore, this method does significantly reduce unknown leak locations. This reduces the time required to find the actual leak source as well as to respond to the spill, clean it up, and repair the leak.

In summary, POV leak detection is a key element of the leak detection system that can provide very precise leak location breach.

7.2.3 Third-Party Observation

A review of Fig. 7.5 also indicates that approximately 13.32% (211 total spills) of 2010–15 spill reports originated from the public (175) or emergency responders (36) (collectively known as third-party observers). Third-party observations (TPO) typically occur on the pipeline ROW and not within facilities.

A further review of the PHMSA 2010–15 incident report database identifies that approximately 23% of all spills occur external to the owner/operator-controlled property and on the pipeline ROW. See Fig. 7.6. The ROW is where the cross-country portion of the pipeline is located as it transitions from one owner/operator-controlled property location to the next.

For those leaks on the ROW, TPO are a critical component of the operator's leak detection systems. This is because of the enormous size of the cross-country pipeline and the fact that people are often out and about and in close proximity to the pipeline ROW.

The probability of TPO detecting a leak is a function of the leak location, local population density, time of day, attractiveness of the ROW, and size of the leak/spill, as indicated in Eq. (7.1).

$$P(\text{3rd party detection}) = P(L_l, P_d, t, \text{Size})$$

EQUATION 7.1

where L_l is the leak location, P_d is the population density along the ROW, t is the time of day, and size is the observable size of the leak/spill. The

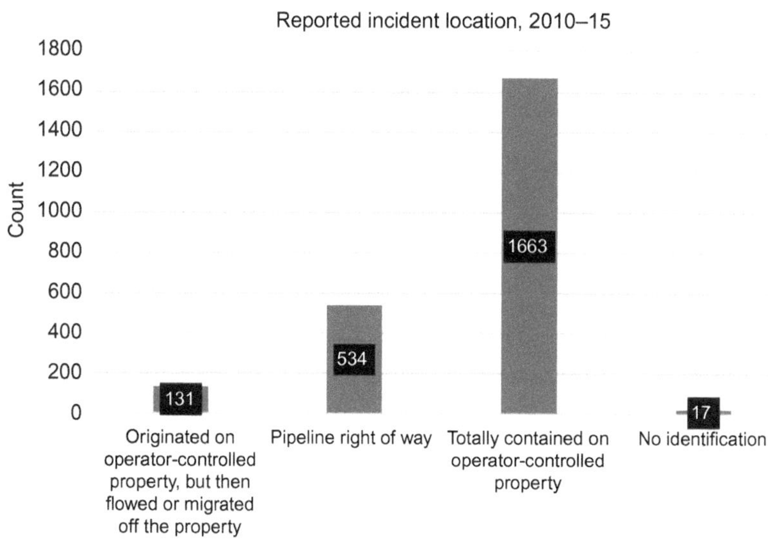

FIGURE 7.6 PHSMA reported spill locations.

probability that a third party will detect the leak or spill first will approach 100% in time, especially if the leak location is within a population-dense area, if the ROW area is attractive (which encourages people to be walking, biking, and so forth in the area), if the incident occurs during the time of day when people are out and about, and if the leak is large and not a small underground seeping or weeping leak. Even if these factors do not apply, the general ubiquity of people in general virtually guarantees that the leak will eventually be detected, although "eventually" might be a very, very long time.

Fig. 7.7 is an example of an actual spill that occurred and that was detected by a third party. This spill occurred in a high-density portion of the town, the spill size was approximately 30 barrels (1260 gallons), and the leak was detected at approximately 2:00 pm local time. As noted in Eq. (7.1), each of these variables contributes to the early detection by an individual over a technology-based system that takes time to derive a potential leak indication.

Certain aspects of incidental third-party leak detection are regulated in the United States. As identified in 49 CFR 195, the owner/operator must include landowner (or, stated another way, third party) awareness training and reporting information.

To maximize the utility of this leak detection method, the owner/operator can:

- Expand or enhance public awareness and coordination through education, information, and communication.
- Enhance the attractiveness of the ROW. This can be achieved by putting in jogging paths, biking paths, walking paths, and lighting to make it safe and attractive.
- Place signage in highly visible locations and at frequent intervals that provide instructions about what to look for and whom to notify.

7.3 DISTRIBUTED CABLE-BASED LEAK DETECTION TECHNOLOGY

Hydrocarbon, water-based, and similar commodity-sensing cable systems, also known as cable-based leak detection systems, comprise a common external leak detection method. These systems use a sensing cable located in close proximity to the pipeline to determine if leak commodity is present outside of the pipe. Cable-based systems must be installed in close proximity to and follow the route of the pipeline. Fig. 7.8 demonstrates one general cable installation design.

Assuming that the sensing cable has been properly located close to the pipeline, it detects the presence of the spilled commodity by change in the cable's physical state. The physical state change could be the introduction of a short circuit, a change in overall cable resistance, or a change in

FIGURE 7.7 High-density population example.

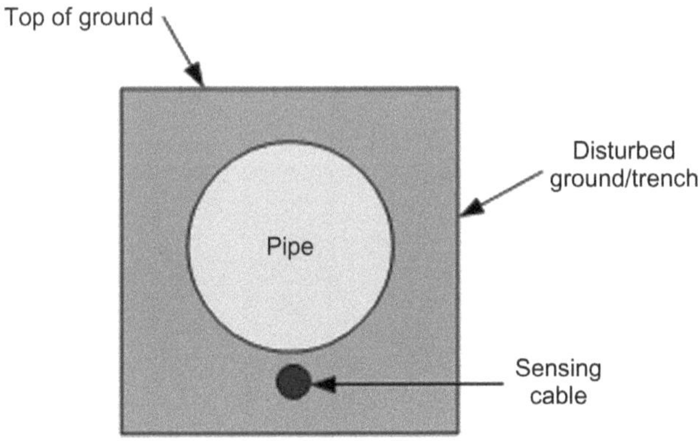

FIGURE 7.8 Sensing cable installation example.

impedance. The cable's normal physical characteristics change due to the presence of the released commodity.

The targeted changed physical state is determined by the cable-sensing equipment. Once the changed state is detected, an alarm is generated. The resulting alarm may be displayed locally and/or transferred to the pipeline supervisory control and data acquisition system for display to the pipeline controller. Because the cable can follow the pipeline over long distances, it is considered to be a distributed (as opposed to a point) sensing system.

An example of one specific type of hydrocarbon-sensing cable leak detection system is a pair of insulated conductors that are located adjacent to the buried pipe. At one end of the cable is a sensing unit, and the other end of the cable may be terminated in some resistance or similar known electrical load. In this situation, the sensing unit provides a specified voltage to the cable and senses the overall current draw during normal, nonleak situations.

When a leak occurs, the resulting spilled commodity comes in contact with the leak detection cable conductors. The action of spilled fluid contacting the cable destroys the insulation between the cable conductors. The resulting loss of insulation causes a short circuit and permits a flow of current from one conductor to the other. This changes the "normal" current measurement of the cable to some new measurement and indicates the presence of the spilled commodity.

The change in current value also provides a means of determining the spill location. This can be achieved through the use of Ohm's law or by using a time domain reflectometer.

With respect to the first approach, let us consider Ohm's law:

$$R = \frac{V}{I}$$

EQUATION 7.2 Basic Ohm's Law

where R is the derived total cable resistance in ohms, V is the applied voltage, and I is the cable measured current in amps. We assume the following conditions (for example):

- constant 24 volts of direct current applied to the cable
- normal cable resistance of 0.0016 ohms per foot
- a measured current of 12 amps

We can use this information to derive the distance of the short. We first apply this information to Eq. (7.3). The key difference between Eqs. (7.2 and 7.3) is that you must divide the total distance by 2 because the current is flowing from the source to the short and back to the source location. This is a doubling of resistance, so the leak location is associated with half of the total cable resistance. Therefore, the calculated overall cable resistance is the round-trip cable length.

$$\text{Distance in feet} = \left[\frac{\left[\frac{24v}{12a}\right]}{0.0016\ \text{ohms/ft}}\right]/2$$

EQUATION 7.3 Derived Distance

The result is 625 feet. In reality, the calculations used for determining the location of the short is more complex than this, but this example demonstrates a fundamental approach.

The other method of determining the location of a change in cable characteristics is through the use of time domain reflectometry (TDR). TDR technology is based on the fact that when a signal is induced into a cable, it takes time for the induced signal to travel through the conductor. The velocity at which this signal travels is defined by Eq. (7.4), where V_p is the velocity of propagation, c is the speed of light, and er is the dielectric constant of the cable.

$$V_p = c/\sqrt{er}$$

EQUATION 7.4 Velocity of Propagation

TDR technology is further based on the fact that if there is a cable impedance mismatch (the cable does not terminate in a load of equal impendence to the cable), then a portion of the induced signal will be reflected to the source. In the case of cable leak detection systems, the location of the leak creates an impendence mismatch that causes a reflected wave to flow back to the source of the induced signal wave.

Using the variables of wave propagation speed and imperfect cable termination impendence provides a means to determine the distance of the impendence mismatch from the transmitting source. This distance is derived from Eq. (7.5).

$$\text{Distance} = V_\rho t/2$$

EQUATION 7.5 Deriving the Distance Equation

where *Distance* is distance to the cable impendence mismatch, V_ρ is the signal transmission velocity, and t is the change in time between when the signal was sent and when the reflected signal was received. To identify where the short is, we must divide the result by 2 because the transmission time is equal to the time from the source to the impendence mismatch and back to the source. TDR provides a very accurate cable length measurement in this situation.

The preceding discussion identifies methods on how leak detection cable sensing technology can provide an accurate leak location estimate. From a leak detection performance consideration, this is a strong positive attribute.

Although identification of the leak location is a positive attribute, the ability to predict how long it will take to generate an alarm is not as precise. To understand the variability in leak or spill time detection, one must understand that a sensing cable detection time is a summation of times associated with how long it takes the spill to reach the cable or spill propagation time (PT) and how long it takes the cable to respond to the change (the cable response time (CRT)) (see Eq. 7.6).

$$\text{Detection time} = \text{PT} + \text{CRT}$$

EQUATION 7.6 Leak Cable Detection Time

Regarding PT, this time can span from seconds to infinity, as noted in our previous discussion of direct detection methods. If the resulting spill can quickly migrate to where the sensing cable is located, then the leak detection PT is minimized. Conversely, if environmental conditions prevent the spill from contacting the sensing cable, then the leak will not be detected and the leak detection propagation detection time becomes infinity.

Fig. 7.9 provides two examples that represent potential PT elements. In Fig. 7.9 (left), the spill has migrated to where the cable is located and will

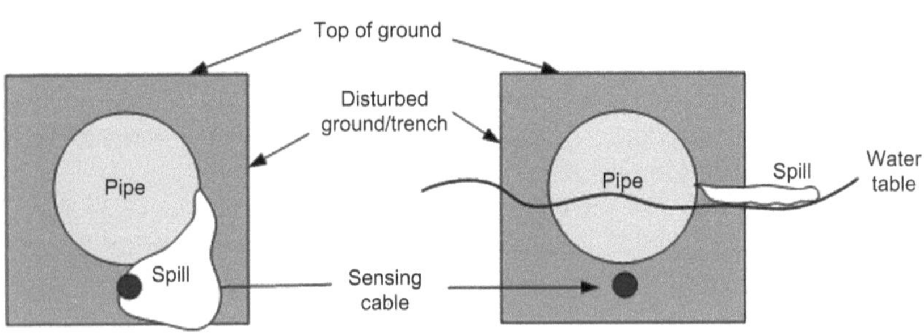

FIGURE 7.9 Sensing cable installation time to detect.

be detected. This could be within seconds or longer. The example on the right indicates that the pipe is partially submerged in a water zone. In this case, the resulting spill is floating on top of the water and migrating away from the pipe according to the water flow. In this situation, the spill will never encounter the sensing cable and the leak will never be detected by this technology.

The second part of the overall detection time is how long it takes for the cable to respond to the presence of the commodity. This time is specific to the function of the cable and the released commodity. Response times can be from a few minutes to several hours. One cable manufacturer specifies that their cable will respond within 12 min to 120 min. The range is driven by the type of commodity that the cable encounters and how fast the cable characteristics change in response to the specific commodity.

Leak detection–sensing cables are not immune to false alarms. In this case, false alarms are defined as an alarm generated when the target commodity release has not occurred. A source of false alarms is third-party commodity encounters. Rather than the cable sensing the spilled commodity, it alarms when there is third-party spill from another source rather than a leak from the operator's pipeline. These types of alarms reduce the value of these systems and increase operating expenses. This could also occur as the result of continued migration of a previous spill from another source.

Another issue with sensing cable leak detection systems is potential recurring operational expenses. Generally, sensing cables are designed such that once they have gone into alarm, the cable must be replaced. Therefore, if the alarm is generated by a pipeline leak or a third-party leak, then it must be replaced. This is a significant issue with this leak detection technology because it not only impacts the ability to perform leak detection until the cable is replaced but also results in additional operating costs and risks. The additional cost includes the cost of the new cable, but of a greater impact are the labor costs of replacing the cable. There is also increased operational risk in replacing a buried cable. This is associated with excavating areas along the pipeline as required to replace the old cable.

Other significant issues with cable-based leak detection applications are associated with retrofit costs and risks. Retrofit costs and risks are associated with installing leak detection cables within an existing buried pipeline ROW. To maximize the potential of identifying a spill in the shortest time possible, the cable must be placed in close proximity to the pipe and generally toward, if not at, the bottom of the pipe. This requires excavating along the length of the pipeline, which is expensive and carries a high risk of accidentally striking the pipe and causing damage to it during the excavation process. Note that although the preceding discussion identified a standard preferred cable location, actual cable placement location is dependent on the physical environment, as was presented previously in the chapter, and could change from this standard.

TABLE 7.1 Sensing Cable Detection Attributes

Classification	Ratings	Notes
Leak detection time	Minutes to infinity	Environmentally driven and system construction–driven
False alarms	Minimal	Generally associated with third-party spills or other external sources
Retrofit costs	High	It is expensive to trench along and in near proximity to an existing pipeline
Retrofit risks	High	Excavating in close proximity to the pipeline carries a significant risk
Distance limitations	Restricted	Cable lengths are generally restricted; multiple cable lengths and sensing sites may be required

Another issue with cable sensing leak detection systems is that the feasible length of the sensing cable is limited. Manufacturers claim maximum cable lengths in the range of 1000 km to 1500 km, but actual distances vary by manufacturer. We have not actually identified a single sensing cable leak detection system that spans a 400-km pipeline. This limitation will require installation of multiple sensing locations to address the need to provide leak detection coverage along longer sections of the pipeline.

Table 7.1 summarizes the various attributes of this technology.

7.4 FIBER OPTIC CABLE–BASED SENSOR SYSTEMS

Fiber optic cable systems are another type of external leak detection system. They are similar to the preceding sensing cable discussion but, rather than using electrical signals, these systems consist of fiber optic cables and the utilization of transmitted light to detect leaks. As with the previously discussed sensing cables, fiber optic leak detection systems are collocated in very near proximity to the pipeline: within 10 cm for gas pipelines and 15 cm for a liquid pipeline. Fig. 7.10 provides a visual reference of where fiber optic leak-sensing cables may be installed for a gas-carrying pipeline and for a liquid pipeline. Note that actual cable installation location is determined by the pipeline commodity and environmental considerations, such as above-ground or underground installation, the presence of water, the soil type, and so forth.

Fiber optic cable systems have been used and are being marketed as pipeline leak detection systems based on their ability to respond to localized remote vibrations or thermal changes. As is discussed more in the next paragraphs, these systems rely on transmitted light Raman and Brillouin scattering phenomena to identify localized changes along the fiber optic cable.

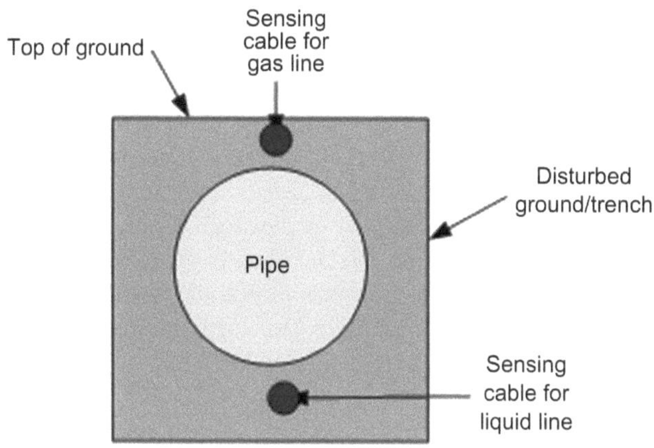

FIGURE 7.10 Fiber optic installation example.

Raman scattering, or the Raman effect, occurs when the transmitted light pulse encounters thermally influenced molecular vibrations. Fiber optic cables consist of one or more doped quartz glass fiber strands. The quartz glass is a form of silicon dioxide with amorphous solid structure. Thermally induced changes to the quartz glass cause lattice oscillations that, in turn, generate an interaction between the transmitted light pulse photons and the electrons of the lattice molecules. This interaction results in light scattering known as Raman scattering. An effect of Raman scattering is a portion of the scattered light is reflected back to the transmitting source, where it is detected.

Brillouin scattering is similar to Raman scattering in that both reflected light pulses are a result of the transmitted light pulse interacting with thermal or vibration-induced changes within the fiber optic cable. The difference between the two is that Raman scattering is the interaction with the lattice molecules, and Brillouin scattering is induced by low-frequency phonons, which are present on localized thermal changes.

As noted, Raman and Brillouin reflected light occurs at specific fiber optic cable locations that experience physical changes as a result of localized sudden temperature changes or induced vibrations. The impact of these sudden temperature or induced vibration changes are often referred to as micro-bends. Micro-bends occur as the cable changes position in response to ground movement, such as would-be triggered vibrations associated with intrusion events, pipe movement, sudden and localized cable movements, or in changes to the optical properties of the cable in response to thermal changes.

A fiber optic leak detection system consists of the fiber optic cable, a light source such as a laser, timing systems, and control logic. The light source as well as the timing and control logic systems (sometimes referred to as the leak detection controller) are located at one end of the fiber optic cable.

As a leak detection system, the fiber optic cable leak detection controller continuously monitors for the occurrence of Raman and Brillouin reflected light. The presence of the reflected light indicates a localized fiber optic cable change. Specifically, the leak detection controller sends out a light pulse and monitors for any reflected signals. This transmit-and-monitor process occurs in a very rapid and continuous sequence.

So, how does the fiber optic cable physical characteristics change in response to a leak? The answer to this question is that, for gas pipelines, it is caused by the commodity Joule-Thomson effect; however, for liquid lines, it occurs as warmer commodity infiltrates the surrounding area. Note that in this second case, this means that the commodity must not be in equilibrium with the soil temperature.

The Joule-Thomson effect describes how the temperature of a gas will change as it is forced through an orifice such as a valve or a small hole (ie, a leak). When gas is forced through an orifice, the escaping vapor will expand, which results in a lowering of the gas temperature. This change in temperature is transferred to the material the vapor comes in contact with. In our case, this will be the pipe wall at the leak site, any surrounding ground, and the fiber optic cable if it is attached or in very near proximity to the pipe and the leak location. The identified items start to cool due to the Joule-Thomson effect and their temperatures will decrease.

If the fiber optic cable is in very close proximity to the pipeline or attached to it, then that specific cable location will also be affected by the change in temperature that the escaping vapor has generated. This lowering in temperature induces physical changes to the fiber optic glass at that location, which generates the scattering of the transmitted laser beam or other light source.

Liquid commodity lines generate changes in the fiber optic cable through the transfer of the spilled commodity heat to the fiber optic cable. This heat transfer assumes that the spilled commodity temperature is sufficiently different than the surrounding pipeline environment and the fiber optic cable steady-state temperature that it will introduce a localized temperature change. The transfer also assumes that sufficient commodity is spilled and that it does alter the temperature of the fiber optic cable as well as the surrounding environment. As with the gas line, the change in fiber optic cable temperature alters the fiber optic glass characteristics, thus generating Raman and Brillouin reflections.

Within the industry, identification of a leak and resulting leak location according to thermal changes is referred to as distributed temperature sensing (DTS). The system is distributed because it senses localized temperature changes along the length of the fiber optic cable. Conversely, if the fiber optic cable, as a whole, slowly changes temperature, then the system will not sense this as an anomaly, and it will not generate an alarm.

To demonstrate this, Fig. 7.11 provides a comparison view of the relationship between the ground temperature and the pipeline commodity

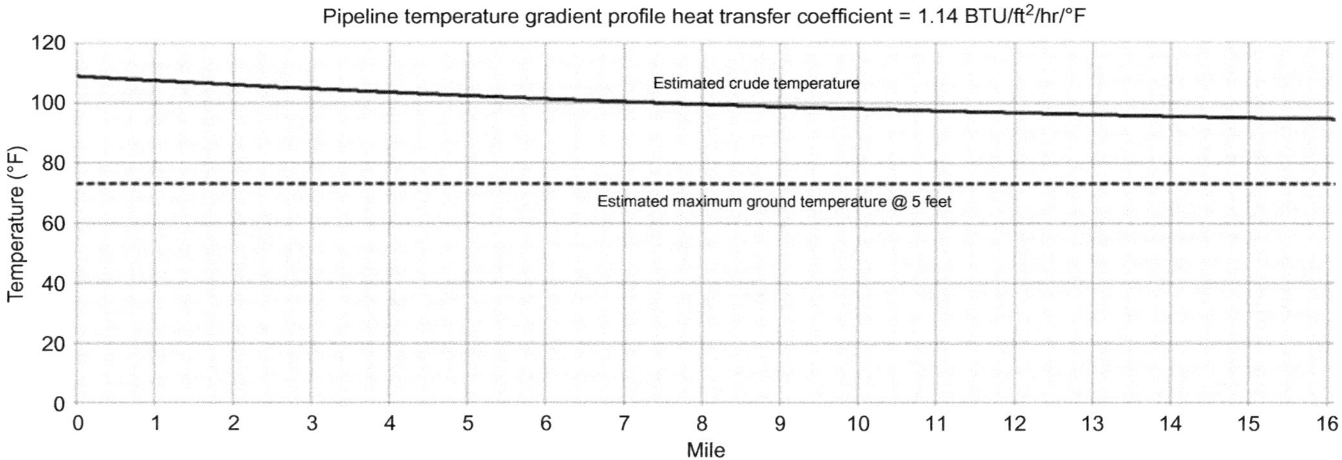

FIGURE 7.11 Thermal delta example.

temperature. As this figure shows, as the commodity moves down the pipeline, it cools down. Thus, the further away from a pump or heat source the commodity moves, the closer to the ground temperature and to the fiber optic cable temperature the commodity becomes. In this example, the commodity's minimum temperature is approximately 95°F (35°C) and the ground temperature is approximately 73°F (22.8°C). This is a sufficient delta temperature to change the fiber optic cable characteristics. However, if the commodity temperature and ground/fiber optic temperatures are equal or almost equal to each other, then there will be insufficient impact on the fiber optic cable to cause an alarm if a spill occurs.

Because the thermal effects on the fiber optic cable are localized and we know very precisely the speed at which the light pulse is traveling, we can determine the leak location very precisely. We rely on Eq. (7.5), but instead of using the V_ρ term, we substitute the speed of light. When we know the time it took for the light wave to travel to the localized thermal changed location and back, we can accurately calculate the distance.

Fiber optic cable leak detection time to detect performance can range from very fast to not at all. The variability in how rapidly the fiber optic system will detect a change in fiber characteristics is a function of how fast the spill-induced temperature change is transferred to the cable and how fast the change in temperature occurs due to the spill. Very fast detection time response is achieved if the commodity spill induces a temperature change in the cable within moments of the release. Conversely, if the cable delta temperature is very small or not present, or if the commodity spill does not transfer sufficient temperature change to the cable, then no detection will occur.

Therefore, how quickly a fiber optic DTS system detects a temperature change is a factor of many environmental variables. These variables are dynamic over the course of the year and the lifetime of the pipeline. As such, the variability in environmental influences creates a situation indicating that the industry has not developed a universally accepted performance mapping methodology or any associated methods to define or calculate the installed system response time.

Although the time to detect a leak is variable, fiber optic DTS systems are generally very resilient to false alarms. If properly installed and calibrated, the DTS monitors for a localized temperature change or ground movement only. It does not respond to temperature changes that impact the full cable, such as the ground warming or cooling. These systems are also fairly immune to slow temperature changes of the pipeline commodity if the system has been properly installed.

Fiber optic leak detection also includes distributed acoustic sensing (DAS) capabilities. DAS is an outcome of research to identify third-party intrusion. Fundamentally, the fiber optic cable generates backscatter light if the fiber optic cable is mechanically "excited" within a localized area. This excitement is a result of very localized vibrations induced into the cable by external forces. Digging and excavating in an area at or very near the pipeline is an example of an event that could introduce vibrations within the fiber optic cable. Another potential cause of vibration could be a leak, particularly if the rapid release of the commodity mechanically disturbs the surroundings.

In summary, fiber optic DTS and DAS systems provide the ability to obtain a very precise location where a spill may be. They are fairly resilient to false alarms and have the potential to detect a spill within seconds. Another benefit of these systems is that you do not have to replace the cables in case of a spill.

Negative attributes of fiber optic leak detection include the potential that the system will take a long time to detect or will not detect a spill due to changing environmental conditions or third-party influences. These systems are also very expensive to install when considering retrofitting a pipeline. Any retrofit effort carries with it a significant risk and cost because this involves trenching and digging within a very close proximity to the full length of the pipeline. In addition, and depending on the length of the pipeline, full coverage may require more than one system. Although fiber optic cables can be extended over relatively long distances, most pipelines may require more than one transceiver and subsequently more than one fiber optic leak detection system.

Table 7.2 provides a summary of fiber optic leak detection systems capabilities.

TABLE 7.2 Fiber Optic Sensing Cable Detection Attributes

Classification	Ratings	Notes
Leak detection time	Seconds to infinity	Environmentally driven
False alarms	DTS systems minimal	DTS fewer. DAS systems may generate more alarms when responding to various environmental acoustic sources such as vehicles
	DAS systems higher	
Retrofit costs	High	Cost associated with trenching and installation along the full pipeline length
Retrofit risks	High	Excavating in close proximity to the pipeline
Distance limitations	Restricted	Cable lengths are generally limited to finite distances

7.5 HYDROCARBON-SENSING TUBES

Hydrocarbon sensing tubes, also known as vapor-sensing tubes (VST), are systems that detect the presence of hydrocarbons within a tube. The tube consists of hydrocarbon-permeable material that allows migration of commodity through it but prevents entrance of water and other vapors.

Fig. 7.12 provides a simple overview of a VST system. It consists of an air source at the inlet, the vapor-sensing tube, and a vapor sensor at the outlet.

VST fundamental operation assumes that hydrocarbon vapors will enter the tube if spilled commodity encounters it. These hydrocarbon vapors are then transported down the tube by movement of the air inside it to where they are detected at the output. The time to detect is a function of the air velocity, the location where the vapors enter the tube, and whether or not the system is continuous or intermittent, as shown in Eq. (7.7).

$$t_D = V_a \Delta D + K$$

EQUATION 7.7 Basic VST Time

where t_D is time to detect, V_a is the velocity of the air within the tube, ΔD is the distance between the end of the tube sensing unit and the vapor entry point, and K is the intermittent operating time period, which for continuous operation is zero.

As indicated in Eq. (7.7), VST systems may operate in a continuous or intermittent mode. Continuous operation occurs as the inlet air source is operating nonstop, which keeps the air flow moving all the time. Under continuous operation, the system time to detect may be the shorter of the two operating modes. The time to detect is a function of the velocity of the air moving through the tube and where the vapor enters the tube. Continuous operation does not provide a means to locate where the vapor entered the system, but it eliminates the time delay associated with the intermittent relaxation time.

Intermittent operation occurs by running the air inlet source at defined periodic times and only long enough to purge the tube of the current air volume. During quiescent periods, the air within the tube is allowed to relax. Intermittent operation extends the time to detect as a function of the air

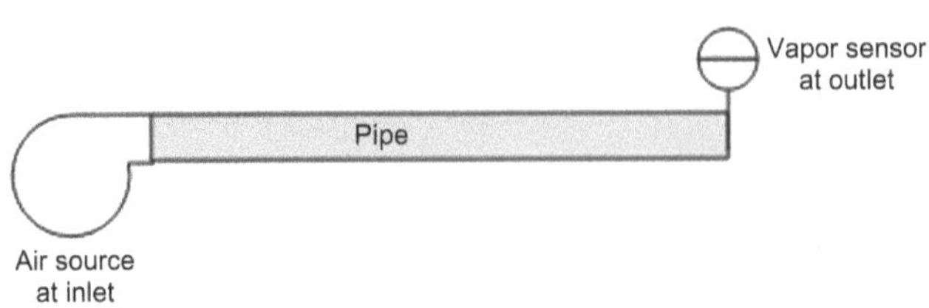

FIGURE 7.12 Basic VST system layout.

purge periodicity cycle. Although detection of a leak may take longer with this operating mode, it provides a reliable means of identifying where the vapor entered the system. Eq. (7.8) is the fundamental equation for detecting the location where the vapor entered the system.

$$D = V_a t$$

EQUATION 7.8 Deriving VST Distance Equation

where D is the distance between the vapor-sensing device and where the vapor entered the tube (the leak location), V_a is the air velocity within the tube, and t is the time between when the air movement started and when the vapor was sensed.

VST systems are very sensitive to the targeted vapors. This enables the system to detect very small spills. Unfortunately, this also creates a situation in which they may generate false alarms if other sources of vapors enter the system and are also detected. VST systems must also be installed in very close proximity to the pipeline being monitored. The sensing tube must be installed in a location that maximizes the potential that the spilled commodity will contact the sensing tube. The preferred installation location and method are functions of where and how the pipeline is constructed, the targeted commodity, and overall environmental considerations such as whether the monitored pipeline is encased with a double wall pipe system or is buried next to the pipe.

VST systems also have a distance limitation of approximately 50 km if the air supplies are located between two sensing tube sections that are each 25 km long. Table 7.3 provides a summary of this technology.

TABLE 7.3 VST Detection Attributes

Classification	Ratings	Notes
Leak detection time	Minutes to infinity	Environmentally driven and a factor of the air source velocity
False alarms	Medium	VST systems are very sensitive to vapor sources; third-party sources can generate alarms
Retrofit costs	High	Must place the sensing tube in very close proximity to the targeted system
Retrofit risks	High	Excavating or working in very close proximity to the pipeline
Distance limitations	Restricted	Maximum distance of each system is approximately 50 km

7.6 FIXED/DISCRETE SENSOR LEAK DETECTION SYSTEMS

Fixed or discrete leak detection systems use sets of individual or discrete sensors to detect a targeted spill-specific physical attribute. These discrete sensors are placed singly or as a set of sensors located at intervals along the pipeline and are linked by a communications cable or wireless networks.

7.6.1 Fixed Infrared and Spectrographic Detectors

This section addresses fixed systems that use electromagnetic (EM) radiation to detect leak signatures.

One EM radiation type is the fixed infrared detector. The infrared spectrum encompasses the frequency range of $0.003-4 \times 10^4$ Hz with a corresponding wavelength range of 1 μm−750 mm. When infrared radiation with wavelengths of 3.3−3.5 mm encounters hydrocarbons, the hydrocarbons will absorb portions of the photon energy. Infrared leak detection systems utilize this absorption rate to identify whether hydrocarbons are present within the area that the infrared beam passes through.

One method of infrared leak detection deployment is through an open path or line-of-sight detectors. Open path systems consist of a transmitter and receiver unit separated by some distance but in line of sight of each other. The infrared signal is transmitted between the two units to determine the potential presence of hydrocarbons. In operation, the system actually transmits two infrared beams. The detection beam is set for the 3.3-μm wavelength. A second beam, called the reference beam, is transmitted at the same time as the detection beam. The reference beam wavelength is selected so that it is slightly different than the 3.3-μm wavelength used by the detection beam. The reference beam wavelength is also selected so that hydrocarbon energy absorption does not occur. The different wavelengths allow the receiving unit to compare the energy level received for each transmitted beam. Comparing the received reference beam magnitude to the detection beam level allows the system to cancel out environmental influences and identify the presence of the targeted hydrocarbon vapor, if present. This results in higher confidence that the system can detect when hydrocarbons are absorbing a portion of the detection beam infrared wavelength energy.

Open-path EM infrared detectors are very effective. At the same time, these systems have several limitations. First, the open channel system must have a clear line of sight between the transmitter and the receiver unit. If the infrared beam is blocked, then it will not function. The second major limitation is length. These systems have a finite operating distance. One vendor specifies that its system will operate up to 150 meters (approximately 492 feet).

Open channel systems are applicable for very specific and localized leak detection. The operating length limitation prohibits this technology's application over several hundreds of miles of buried pipeline.

Another type of EM infrared (IR) leak detector is known as the point IR system. This type of system is fully contained in an instrument case. Rather than test for the presence of hydrocarbons across a long distance, it tests for the presence of hydrocarbons at a fixed point or location. These systems rely on the hydrocarbon vapor entering the fixed point device and altering the infrared signal.

Fixed EM infrared detectors can be very effective. However, they also have limitations. One limitation is that the hydrocarbon vapor must enter the device. The probability that this occurs is a function of where the hydrocarbon vapor source is, wind direction, and device location. If the quantity and location of detectors are insufficient to provide full area coverage, then detection of the spill could take a long time or might never occur. These systems can generate false alarms because hydrocarbon sources other than a spill can trigger the device. However, their installation is relatively low-cost and there is minimal installation risk.

7.6.2 Infrared Imaging

IR imaging leak detection systems operate very differently than EM infrared detection systems discussed in the previous section. IR imaging sensors work on the fundamental principal that all objects emit thermal energy. The emission frequency is below the visual energy spectrum but higher than microwave frequencies. Infrared energy is found within the 0.7-μm to 300-μm wavelength band.

IR imaging devices are generally—but incorrectly—called cameras. Rather than develop an image due to reflected light, as cameras do, they detect the range of thermal energy within their field of view, which results in a thermal image. Using this principal, IR leak detection "cameras" can identify if a spill is present due to the different thermal energies emitted by a liquid spill or vapor release as compared to the normal background thermal energy radiation.

IR imaging devices are constructed and used as handheld units or mounted on aircrafts or watercrafts. They can have a fixed base, with the imaging device mounted in a location and the area continuously monitored. They are also transportable because they can be mounted on an aircraft or other vehicle so that images of the covered terrain are produced. In our case, the area of interest would be on the pipeline ROW. They are also used as handheld devices, with personnel carrying the imaging device and viewing localized areas.

Current technology for this approach provides a very robust system that can be used to identify the presence of gas external to the pipeline or active liquid spills. The systems are not designed to continuously monitor the full pipeline, which is a limitation to this technology.

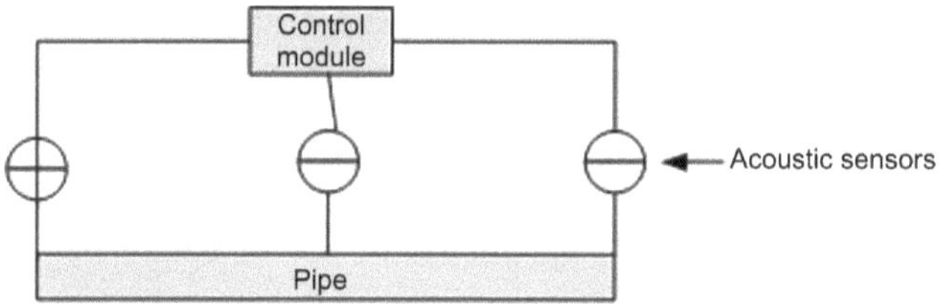

FIGURE 7.13 Basic acoustic system layout.

7.6.3 Fixed Acoustic Sensing

Fixed, external, acoustic leak detection sensors are based on the premise that when a leak occurs, it can be detected using acoustic methods. One method "listens" to the low-frequency acoustic sound that accompanies a leak-induced wave front as it passes proprietary sensors. Fig. 7.13 shows a basic system layout using three acoustic sensors connected to a central processing system. The communication link between the field sensors and the processing module may be through dedicated "wired" communication channels or through wireless connections.

By using multiple sensors, the system can find the general location of the leak. This is possible because the sound associated with the leak is moving upstream and downstream of the leak location at the speed of sound for that commodity. By detecting the precise time when the sound wave passes the upstream and downstream sensors, a location can be derived.

The positive attribute of this approach is that the owner/operator is not required to trench or excavate the full pipeline length. This reduces installation cost and risk. Another positive attribute is that by installing multiple sensors, leak location is possible. Vendors also claim that the systems are subject to very low false alarm rates.

Negative attributes are that the sensors must be placed on the pipeline and that background noise or attenuation of the signal due to distance from the source may block or mask the sound of a leak.

7.6.4 Fixed Hydrocarbon-Sensing Probes

While generally not applicable to the full pipeline, there are situations when the owner/operator wants to monitor specific locations such as a buried valve pit, basement of a building, a sump, and so forth. An external leak detection technology applicable to these types of locations is the leak detection probe.

Leak detection probes are devices that sense the presence of the targeted commodity and generate an alarm. Hydrocarbon-sensing probes are generally designed to not alarm in the presence of water, but they can be specifically designed to identify the presence of the hydrocarbon floating on top of the water.

Installation of these devices can be fixed or floating. Fixed installations would target those locations where it is anticipated that any accumulated water would not cover the sensor. When the water level rises and falls, the tube can be attached to a float so it rises and falls with the water as well.

Hydrocarbon sensing time can be as short as a few seconds but may take longer depending on the hydrocarbon involved, relationship of where the hydrocarbons are entering the area, and the sensor location, as well as other factors. At least one manufacturer of these devices indicates that its unit can be cleaned and reused if a hydrocarbon is sensed.

These devices generally do not generate false alarms. However, the device may issue an alarm based on the presence of hydrocarbons from third-party sources or from the surrounding environment. If the owner/operator has no control of the third-party sources or other environmental sources, then these types of alarms may result in deactivation of the alarm system.

Retrofit of these devices to an existing pipeline produces minimal risk, unlike retrofitting a pipeline with hydrocarbon sensing cables. Reaction time is typically fast and the location is precisely defined. At the same time, they are not designed for application along long stretches of pipeline ROWs.

7.6.5 Fixed Vapor or Tracer Element Sensors

Fixed vapor-sensing technology involves the process of detecting the presence of a targeted vapor or tracer element. The targeted vapor could be one generated by a gas pipeline release, a vapor emitted from a liquid commodity spill, or tracer elements that have been injected in the pipeline commodity.

In each situation, the leak detection system consists of a tracer element sensor and associated monitoring and alarming equipment. The system can be installed as a fixed unit or can be transportable. An example of fixed-site installation would be a system in a tank farm, valve vault, and so forth. These are left in place to continuously monitor the installed location.

Another deployment method is the use of a transportable system. In this application, the vapor-sensing system is contained within a portable device. The user then moves with the portable device through the area of interest, such as along a pipeline ROW, to determine if the targeted vapor is present.

Another approach is to place sensing probes or cables along the ROW. After the probes or cables have been in place for a period of time, they are analyzed for the presence of the tracer compound.

These systems can detect very low levels of targeted vapors or tracing elements, which improves the detectability of small spills. Negative attributes are that they are not continuously monitoring systems and are labor-intensive to deploy and utilize. They are also not designed to monitor long pipelines.

7.7 OTHER EXTERNAL METHODS

This section presents an overview of some other methods of external leak detection that may be applicable within specific pipeline environments.

7.7.1 Ultrasonic Meter External Leak Detection

Ultrasonic leak detection can be described as a hybrid internal/external leak detection system. The internal aspect of this leak detection method relies on the flow rate of the commodity as it moves through the pipeline as well as the commodity temperature. The external portion of the system generally relies on clamp-on ultrasonic flow meters.

The fundamental approach of this method is effectively a flow balance approach, as described in Chapter 6, Rarefaction Wave and Deviation Alarm Systems. The pipeline is segmented by installation of clamp-on ultrasonic flow meters across the targeted leak detection area, as shown in Fig. 7.14. Each flow meter measures the flow rate of the commodity at its location as well as the commodity temperature. This information, for each measurement site, is transferred to a central master station. This master station contains an algorithm that calculates a volume balance for each flow segment.

$$Vb_{seg} = \beta V_o \Delta T_{in} - \beta V_o \Delta T_{out}$$

EQUATION 7.9

As shown in Eq. (7.9), for each segment, the master station then compares the temperature-adjusted volume entering a segment to the volume leaving. If inlet and outlet volumes are different by a location-specific volume, then this is an indication that a leak may be present between the associated measurement sites.

Implementation of this type of leak detection system requires the following supporting infrastructure:

- Electrical power
- Telecommunication connecting the field devices to the master station

FIGURE 7.14 Ultrasonic leak detection example.

- Ultrasonic flow meters
- Temperature sensors

Limitations to this type of system include:

- Commodity inlet between sensor locations is not allowed because the additional inflow is not measured by the two flow meters
- Commodity outlet between sensor locations is not allowed because the off-take flow volume will appear as a leak
- Intermittent slack line conditions within a liquid pipeline will require higher leak thresholds

7.7.2 Intermittent Leak Detection Systems and Methods

Other external-based leak detections systems are described as intermittent systems. Other methods within this classification are typically associated with smart pigging applications.

One such scraper device incorporates an acoustic data acquisition device within a spherical, free-floating, instrumented pig device scraper. As the device travels through the pipeline, it "listens" for the sound associated with a leak.

Another type of intermittent scraper-based leak detection system utilizes technologies such as magnetic flux and ultrasound. These devices measure pipe wall thickness to determine if a pressure containment breach has occurred.

Both systems are intermittent in that the scraper is passed through the pipeline on a periodic or intermittent basis. The systems are also very sensitive to very small leaks. This provides the owner/operator an opportunity to discover small leaks. The systems also provide a very precise leak location capability.

Although these systems can identify small leaks precisely, they do not provide a continuous leak detection monitoring system. As such, the time to detect a leak is a function of how often the device is passed through the pipeline, how long it takes to transition the pipeline, and how long it takes to analyze the data (see Eq. 7.10).

$$t_D = t_P + t_T + t_A$$

EQUATION 7.10 Intermittent Device Detection Time

where t_D is leak time-to-detect, t_P is the time between runs or periodicity time, t_T is the pipeline transit time, and t_A is the analysis time. These systems are good at verifying pipe wall integrity as part of a broader pipeline integrity program.

7.7.3 Unmanned Aerial Vehicle Leak Detection Technology

Unmanned aerial vehicle (UAV)—commonly referred to as "drone" leak detection technology—is based on an aircraft piloted by remote control or onboard computers. For leak detection, this technology is still in the experimental stage. Drones do not have onboard visual observers or pilots to visually see the leak. Rather, they rely on various onboard leak detection technology sensing systems (many of which have been described previously in this chapter) to determine if a commodity release may have occurred.

Drone technologies include rotary-based vertical takeoff and fixed-wing forward-movement takeoff systems. Vertical takeoff and landing drones provide the capability to launch and recover the systems in smaller areas than fixed-wing aircrafts.

Each type of drone can be configured with different leak detection technologies as part of the payload. Such technologies could include:

- Forward-looking infrared (FLIR) cameras
- High-resolution visual cameras
- Laser-based methane gas detectors
- Multi-spectral imaging
- Short-wave infrared (SWIR)
- Synthetic aperture radar (SAR)

Strengths of drone leak detection technology include:

- Lower operating costs than fixed-wing or helicopter aerial observation systems
- Operate at lower speeds than fixed-wing aircrafts, which should provide improved detection capabilities
- Operate at lower continuous altitudes for improved detection capabilities
- Operate when cloud levels are lower, which would prevent the use of other aircrafts
- Reduced risk to personnel because the use of a drone eliminates the need for a pilot and observer on the aircraft

Limitations to the deployment and use of drones include:

- Limited payload capabilities that restrict the size of the detection technology that can be deployed. This restricts the use of some higher-resolution sensors.
- For drone stability, it is often critical that sensor to be very stable. The smaller drones have much higher susceptibility to motion as a result of wind and thermal turbulence.
- Current operating restrictions require visual contact. This limits the distance and area in which the drone can be operated. Surveying a long pipeline would require multiple flights, which could be accomplished using multiple drones or repetitive launch and recover cycles of a single drone.

- Flight time, depending on the size of the drone and payload, and the operating time may be limited, such as no more than an hour.
- Highly dynamic legal state. Although the legal requirements of operating drones will stabilize, current requirements are highly dynamic and subject to change.

As noted, the application and use of drones as a leak detection technology is experimental but expanding. The pipeline operator must clearly define the leak detection mission with respect to logistical area to cover, terrain, and payload. These will help define the type of drone that should be used and potential legal requirements that must be met.

7.8 GENERAL ASSESSMENT

External leak detections systems are applicable to a range of pipeline infrastructures. We conclude this section by delving into a general assessment of how these systems may be deployed and the positive and potential negative issues with the various applications.

As discussed, there are many environmental conditions that affect the effectiveness of external leak detection systems. Some of these external conditions include surrounding soil conditions, underground water table, and ambient conditions. Selection of any external system must take into consideration the environmental variables to maximize selecting the most appropriate external system and installation method.

A particular concern associated with fixed external systems is the amount of coverage required to detect leaks from underground pipe. Leaks in underground sections of the pipeline are likely to have significantly different characteristics than leaks that occur in above-ground sections. External detection of above-ground leaks is directly linked to atmospheric conditions, with little interference due to the surrounding medium other than wind and weather.

Migration of underground leaks will be impacted by the surrounding soil conditions and properties, and potentially by the associated water if the pipe is below or near the water table. Unless the leak is caused by third-party damage, in which case we can assume some exposure at the leak site to the surface and/or atmosphere, any leak flow path must make its way through the soil to other locations. However, the underground leak could be a large rupture. In that case, the surrounding soil resistance may be overcome and the leak will erupt or emerge above the ground. Alternately, the leak may be very small, such as a pinhole corrosion leak; in this case, the soil will present considerable additional resistance to flow, and leak rates for a given orifice size may be much smaller than for an above-ground leak. Also, if the pipe is above the water table, then the leaked commodity may tend to flow downward, but it will still be influenced by the pipe pressure at the leak source and surrounding soil density.

It is important to recognize that the surrounding soil is governed by its own constitutive relations. The soil strength and permeability will also be different depending on whether it is saturated with water or whether it comes to be saturated with the released commodity. In addition, the presence of the released commodity within the soil will potentially cause the soil to expand and may change the constitutive relationships defining the soil. In particular, a highly viscous commodity may cool significantly and strengthen the soil, bottling up the leak, or may increase the soil resistance to the point where the leak will be contained.

Finally, leaks from pipelines that are buried beneath the water table or otherwise submerged will flow preferentially upward until the water table is reached. Once the released commodity reaches the top of the water, it will flow along the interface in accordance with the water flow and the impact of gravitation. Note that any released liquid commodity that is in contact with saturated soil below the water table or free water in a streambed will tend to diffuse light ends into the water. Once dissolved, the light ends will flow with the water and will have a much different flow pattern than the original petroleum vapor plume. Hydrocarbon leaks in offshore lines will create oil slicks that will rapidly expand and be transported by ocean currents to remote locations.

Thus, the spill flow path will be complex and subject to the influence of: the circumferential position where the leak occurs; the size of the hole; the local terrain/elevation gradient; the permeability of the soil to released commodity flow; the pre-existing moisture content of the soil; the strength of the soil; the depth of the leak above or below the water table; the impact of the spill on the strength of the soil; the pipeline gradient/route; internal pipe pressures; whether it is onshore or offshore; and many other factors. As a result of this, it is difficult to estimate how well an externally placed detector will detect a leak. Even a detector that is placed directly below the leak may take a long time to respond to the released commodity.

In general, it is probably safe to say that underground leak detection is improved by:

- More detectors
- Detectors close to the pipe
- More circumferential coverage
- Detectors in the soil
- Detectors located at lower elevations when the pipe is above the water table
- Detectors co-located with the water table if the pipe is below the water

However, these generalizations will tend to translate to significant cost, particularly if the installation is a retrofit to a pre-existing buried pipeline. Systems that use terrain management to reduce the number of detectors by locating them at strategically selected locations based on the terrain will have to deal with potentially long detection times because it will take time for the spilled commodity to diffuse or flow to the monitoring site. Furthermore, such detectors will still have to resolve issues regarding the preferential flow path.

REFERENCE

[1] American Petroleum Institute Standard 1130. Computational pipeline monitoring for liquid pipelines. September 2007.

Chapter 8

Leak Detection System Infrastructure

Leak detection systems do not exist in a vacuum. These systems require input data as well as a means to present leak alarm information to the appropriate operator staff. To fulfill these needs, the leak detection technology must reside within the broader pipeline infrastructure. The following sections provide a review of this infrastructure from the leak detection application viewpoint and needs.

8.1 FIELD INSTRUMENTATION

All internal as well as some external leak detection systems require instrumentation in the field, and the performance of these leak detection systems is directly dependent on the quality and availability of appropriate field instrumentation. Individual chapters elsewhere in this book discuss the instrumentation requirements of different types of systems. For example, Chapter 4, Real-Time Transient Model–Based Leak Detection, Section 4.3 discusses the instrumentation needs of an RTTM-based system. A key instrument attribute which must be analyzed is instrumentation errors.

While a complete analysis of instrument error is beyond the scope of this book the reader can obtain more information from API Technical Report 1149 [1], which addresses this topic in much more detail. However, it is useful to discuss the various types of instrument errors and to describe how they affect a leak detection system with a focus on computational pipeline monitoring (CPM) systems.

8.1.1 Measurement Uncertainty

In this section, we provide a general overview of instrumentation uncertainty. Measurement uncertainties result from several contributions, including:

1. Reference accuracy
2. Ambient temperature and effects of other field variables
3. Time skew

Pipeline Leak Detection Handbook. DOI: http://dx.doi.org/10.1016/B978-0-12-802240-5.00008-X

Reference Accuracy

Reference accuracy is reported by instrument manufacturers and consists of at least three different effects: hysteresis, nonlinearity, and nonrepeatability, which are measured at laboratory reference conditions.

Hysteresis is the difference between readings of the same value (such as pressure) reported when the value is increasing as compared to decreasing. Nonlinearity is the deviation of the observed value as compared to a straight line between the zero-point and maximum value. Nonrepeatability is the maximum deviation between readings of the two identical values. The combination of these factors is often referred to as accuracy and is reported by the manufacturer at reference conditions (a specific ambient temperature, process temperature and atmospheric pressure).

Influence of Ambient and Process Conditions

Instrument manufacturers also report, as a separate error source, the effect of ambient temperature and process temperature. For example, the effect of a 50°F change in ambient temperature may be approximately the same as the reference accuracy of the instrument. Pipeline instruments could experience a range of ambient temperatures more than 50°F over a given period. In this case, the ambient temperature effect can be significant. Therefore, for an instrument that has a reference accuracy range of 0.1%, this impact may double the measurement inaccuracy, or worse. Note that we must stress that manufacturers' specifications are based on laboratory tests, not actual field conditions. Therefore, one can expect that as-installed results will not meet the laboratory specifications.

Furthermore, it is critical to note that there is a difference between resolution and accuracy. Leak detection systems are more sensitive to changes in instrument readings and to the differences between two different instruments than to the absolute accuracy of the instrument. Being able to detect changes is critical and warrants high-resolution data, even when the instrumentation accuracy is not precise.

Bias

Sometimes instruments exhibit specific bias due to calibration, incorrect installation, or other constant errors. This results in the instrument output exhibiting fixed offset, or bias, from its process value.

Note that none of the instrument accuracy components discussed (hysteresis, nonlinearity, repeatability, and ambient temperature) would result in constant bias in an instrument reading.

8.1.2 Time Skew

Time skew refers to the difference between the true sample times of each instrument and the time stamp applied to the value as it is input to the leak

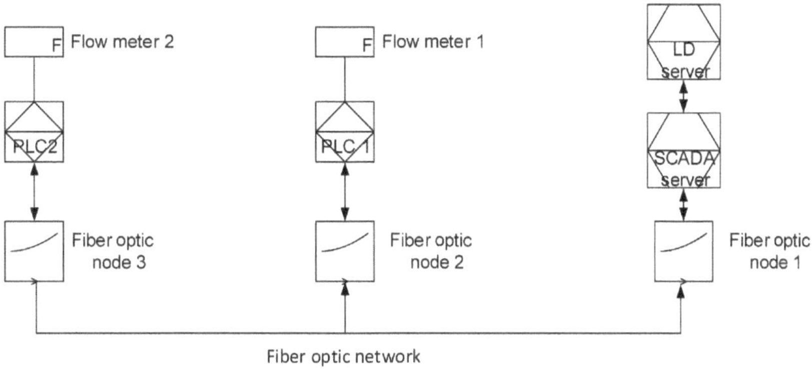

FIGURE 8.1 Simplified flow meter pipeline network example.

detection system. Ideally, each instrument reading would be time-stamped in the field and the true sample time would be acquired along with the data value. In this case, one might design the leak detection system to compensate for the time differences (time skew) between the field measurements. However, such field data time-stamping is rare or nonexistent.

Fig. 8.1 is a simplified pipeline that includes the control center supervisory control and data acquisition (SCADA) system, the leak detection system, and two flow meter measurement sites. A fiber optic telecommunication network connects the control center and flow meter measurement sites.

Although this is a simplified example, it provides sufficient detail to demonstrate the areas where time skew arises. One source of time skew is propagation delay associated with the telecommunication network. One cannot avoid physical realities when dealing with telecommunication systems. Many SCADA systems, on a scheduled basis, sequentially scan data from each of the remote programmable logic controllers (PLCs). In our example, we assume a fiber optic network. The first reality is that it takes time for the fiber light to travel from the SCADA server location to the field site and then return. The fiber optic cable refractive index identifies how fast the light travels down the fiber optic cable. As such, if the first site is, for example, 50 miles from the SCADA server and the second site is 100 miles from the SCADA server, then it will take twice as long for the signal to travel to the second site and back to the first.

The second physics rule we cannot avoid is that every "box" that the fiber optic signal passes through or terminates at increases the telecommunication system time delay uncertainty. It is also reasonable to assume that there will be more devices to pass through to reach remote site two than remote site one. Consequently, remote site two delay is greater than twice the optical transit time. We refer to the telecommunication time lag as t_{Lag}.

Another source of time skew occurs when SCADA systems sequentially obtain field data by polling sites. In this case, the SCADA host requests data

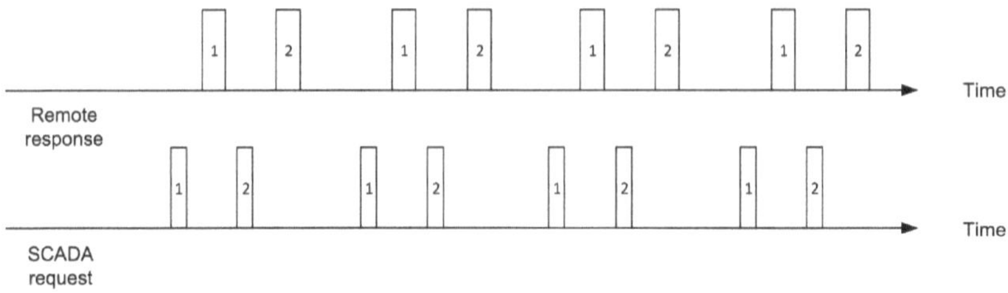

FIGURE 8.2 Simple poll response timing example.

from field site one and waits to receive it, then it requests data from the next site, and so forth, until all field data has been received. To complicate things even further, many SCADA configurations include the process of retrying to obtain a field site's data multiple times if the original request fails. Fig. 8.2 is an example of this poll and response sequence.

SCADA receives data skewed in time from one site to the next. As the SCADA system scans more and more sites, this will increase the time skew between different sites. Time skew is also possible even if the SCADA system is capable of transmitting parallel polling requests. While initiating all field data requests simultaneously, the system will still encounter various equipment delays, different communication network lags, communication retries, and so forth. Although parallel data requests can assist in reducing skew, they do not eliminate skew. The variable telecommunication time lags and time differences resulting from the way that the SCADA system acquires the field points result in total time skew between data points.

Time skew between data points can be particularly troublesome for flow rates. This is because flow rates are often obtained from metering accumulators rather than from direct flow rate measurements. In this case, the flow rates are derivatives computed from the accumulator changes and may fluctuate wildly on a scan-to-scan basis.

Time skew is also very troublesome during fast transients. When values are changing rapidly, time skew can and will result in significant uncertainties in the values sent to the leak detection system. The impact of time skew creates measurement uncertainty solely because of the fact that the time at which the data are acquired is not precisely known. It will delay the ability of a leak detection system to respond to a leak and (in an RTTM-based system) cause modeling errors. In a rarefaction wave leak detection system, as previously discussed in Chapter 6, Rarefaction Wave and Deviation Alarm Systems, this uncertainty can have a significant negative impact on leak identification and location calculations.

One can also anticipate that time skew will delay the ability of a leak detection system to respond to a leak. That delay is likely to be at least several times the time skew.

8.1.3 Data Sampling Processing Best Practices

We now discuss issues related to sampling of the field data and provide recommendations for best practices for sampling and conditioning of data used by a leak detection system. Most of the data required by leak detection systems are time-varying data, and the pattern of the variation in time is important. Unfortunately, most SCADA systems are designed to report the state of the pipeline system rather than the rate of change of the pipeline system state. For leak detection applications, greater focus needs to be given to digital signal processing principals than is usually applied in SCADA data acquisition, particularly when one wants to be able to achieve the greatest level of sensitivity and rejection of false alarms.

Consider a measurement that will be used as an input to a real-time transient model. The measurement (eg, pressure, temperature, or flow rate) responds to process variations as well as to a leak that the system needs to detect. One would like the sampled data to represent the process variations as accurately as possible.

Respect the Nyquist-Shannon Sampling Theorem

Pressure, temperature, flow, and other values are continuous time signals that are sampled at discrete points in time by the SCADA system. The Nyquist-Shannon sampling theorem [2] states that the highest frequency that can be represented in the sampled data is one-half the sampling frequency. If data are sampled every 10 s (0.1 Hz), then the highest frequency that can be represented in the sampled data corresponds to a periodicity of 20 s (0.05 Hz).

Best practices require the data to be filtered prior to sampling to remove frequencies that are higher than one-half the sampling frequency. Failure to do so can result in aliasing of the underlying signal, which is when a sampling artifact results in frequencies observed in the sampled data that are shifted from the true data signal.

Fig. 8.3 illustrates aliasing of a 0.1-Hz signal that is sampled every 11 s. The sampling "aliases" the true 0.1 Hz to approximately 0.01 Hz. Although the original signal had a periodicity of 10 s, the sampled signal has an observed periodicity that is slightly greater than 100 s.

An alternative to applying a filter to the data prior to sampling is to filter after sampling. This cannot prevent the aliasing described. However, process variations that are faster than can be properly represented in the sampled data will appear as "noise" in the sampled data. In this case, a heuristic rule of thumb we use is to apply a digital filter to data with a cutoff frequency of at least one-seventh the sampling frequency. Therefore, if data were sampled every 10 s (0.1 Hz), then the data would be passed through a digital filter with a cutoff frequency of 1/70 Hz.

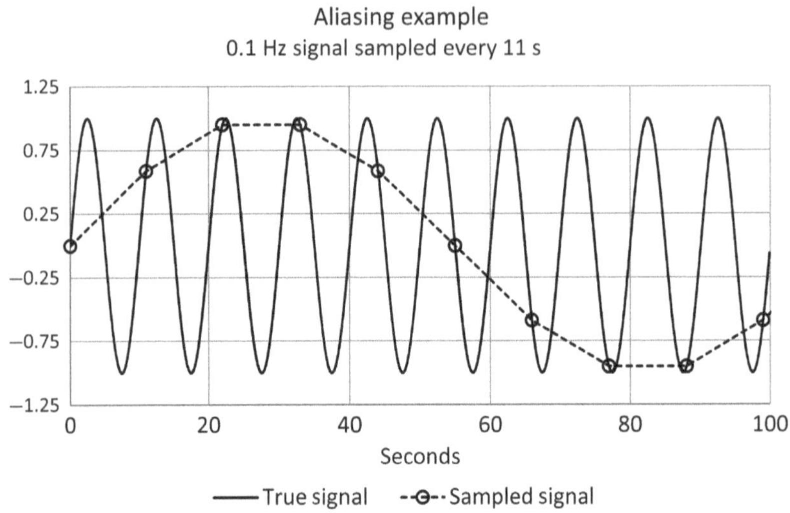

FIGURE 8.3 Aliasing example.

Use Report by Exception With Care

Reporting by exception in data acquisition provides data updates only when values have changed by more than a specified amount. We refer to this as the instrument dead-band. This can be very useful in reducing communication overhead, but it is usually detrimental to leak detection because it reduces the observed sensitivity of the field instrument. If possible, eliminate report by exception on all continuous value field measurements such as pressures, temperatures, differential pressures, and valve open fraction. For analog measurements, set the dead-band smaller than the minimum resolution of the analog-to-digital (A/D) converter.

Prefer 16-Bit A/D Converters or Digital Interfaces

Analog instrument measurements rely on one of many resolution A/D (analog to digital) convertors. A 16-bit A/D converter has 1 part in 65536 resolution. In contrast, a 12-bit A/D converter provides only 1 part in 4096 resolution. For a pressure transmitter ranging from 0 to 1500 psi, a 12-bit converter gives only 0.37-psi resolution. This is limiting, particularly for gas pipeline leak detection and for leak location estimation. Current smart transmitters provide digital interfaces that eliminate A/D issues and are preferred when feasible.

Note that there is a difference between resolution and accuracy. Leak detection systems are more sensitive to changes in instrument readings and to the differences between two different instruments than to the absolute accuracy of the instrument. Being able to detect changes is critical and warrants high-resolution data, even when the instrumentation accuracy may not.

Filter Data Obtained Through A/D Conversion

Even when data have been properly filtered at the source prior to sampling by the SCADA system according to the Nyquist-Shannon sampling theorem, data that are then converted by an A/D converter will likely have fluctuations of the order of the sensitivity of the A/D converter (variations in the least significant bit). Leak detection algorithms will generally benefit by filtering out this noise.

Apply Input Data Filtering Consistently and Appropriately

When low-pass filtering of the sampled data is necessary to remove noise (a term we use loosely) in the data, we suggest the following:

1. Apply the low-pass filter uniformly. Because a leak detection system typically relies on multiple data inputs, and generally on the comparison of one input with another, it is usually best to apply the same filter to all inputs.
2. Prefer a Bessel digital filter. A Bessel filter preserves the wave shape of the filtered signal by having a maximally flat phase response. The phase delays of the frequencies passed by the filter are minimally shifted with respect to each other. A second characteristic that is important to leak detection is a time-domain step response of the Bessel filter with no overshoot [3].

Prefer Continuous Over Discrete Signals

Pipeline process changes are generally continuous rather than discrete. For example, a valve is not open or closed. Instead, it can be fully open, fully closed, or anywhere in between. Opening or closing a valve may cause a high-amplitude process change. However, the value of the process change is a function of the valve position history and the characteristics of the valve. Consequently, it is far better to have an accurate representation of the valve open fraction than the more common open, closed, or in-transit status.

Sample at an Appropriate Rate

It is difficult to provide a clear guideline on sampling frequency requirements other than to say that the fastest that a leak detection system will be able to detect a leak, even a very large one, is likely to be several times the data sampling periodicity. The smaller the leak, the greater the number of data samples that will be required for detection.

Note that for rarefaction wave leak detection systems, the instrument sampling rate should be much higher than the rate based on the hydraulic wave transit time. This sample rate is calculated from the pipeline segment length and the commodity speed of sound.

8.1.4 Dealing With Calibration and Other Instrument Maintenance

The leak detection system is critically dependent on the availability and reliability of its instrumentation inputs. Knowing when an instrument is unreliable is critical.

Calibration of instrumentation inputs should be regularly performed. During the calibration process, however, the instrument ceases to provide valid data to the leak detection system and instead will likely provide values at the extremes of its range. Furthermore, it may be useful to show the instrument readings in SCADA during this period, but it is *never* useful to provide unrepresentative data to the leak detection system.

Other maintenance activities such as instrument replacement or pipeline repairs may also temporarily interfere with the usefulness of instruments for leak detection.

Leak Detection System Should Have Input Override Capability

To accommodate temporary instrument data loss, it is important for the leak detection system to provide a user interface that allows one to manually override any instrument input or to flag it as invalid. This capability is distinct from capabilities provided by the SCADA system because there may be times when the actual instrument reading is useful in SCADA but not useful for the leak detection system.

The input override capability should include the following:

1. Allow one to override any leak detection input measurement
2. Allow one to identify any input measurement as invalid
3. Provide a text comment field to record the reason for the action
4. Permanently log all actions in the historical data archive (see Section 8.4)

Clearly Communicate the Status of Instrument Overrides

The leak detection system display should prominently display all points that have been overridden or taken offline, whether the action was performed through the leak detection system or in SCADA. Procedures should be in place so that during shift turnovers, pipeline controllers are informed of the status of all instrumentation overrides that affect the leak detection system.

Calibrate Field Instruments Frequently

Instruments that are inputs to leak detection systems should be calibrated at least every 6 months. However, the maintenance frequency should be increased for instruments that have known drifts that occur more rapidly.

8.2 SUPPORTING TELECOMMUNICATION AND NETWORK INFRASTRUCTURE

Telecommunication and network infrastructure requirements are different for different types of leak detection technologies. Consider that some rarefaction wave and external leak detection technologies require a telecommunication infrastructure that links two or more remote field locations. Conversely, CPM systems tend to require a telecommunication and network infrastructure that links a number of remote sites to a central location. In the rarefaction wave/external leak detection example, the telecommunication infrastructure may encompass tens of miles but the CPM system may extend hundreds of miles (see Fig. 8.4).

8.2.1 Telecommunication Infrastructures

In this section, we discuss the types of telecommunication infrastructures generally used, telecommunication issues that impact leak detection systems, redundancy requirements, and data transfer rate best practices.

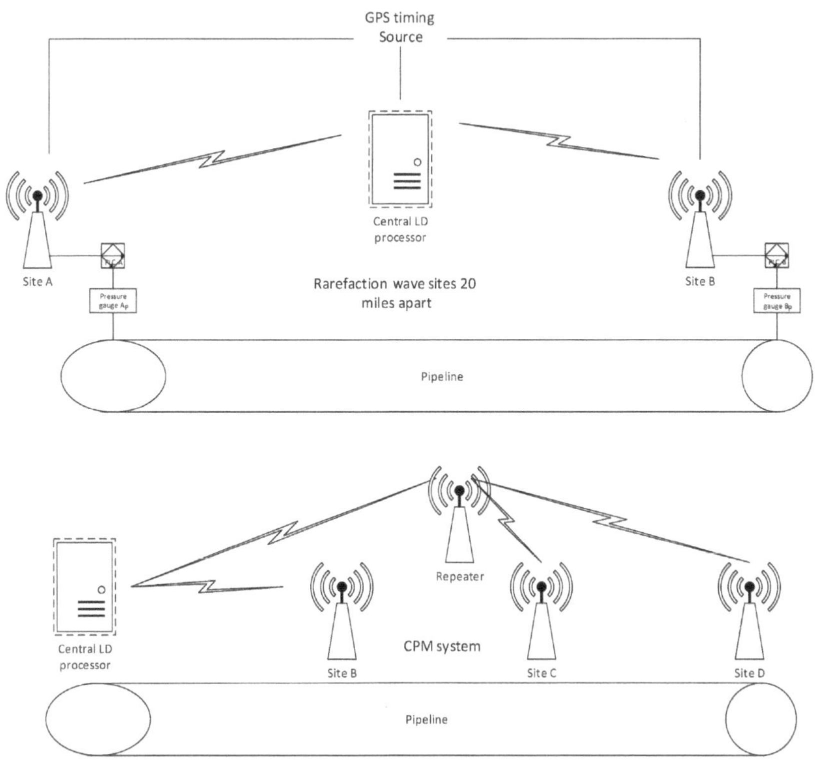

FIGURE 8.4 Telecommunication infrastructure example.

When it comes to leak detection systems and telecommunication infrastructures, two things are constant. Firstly, over time, the means and methods of transferring data from remote to local sites have utilized virtually every technology-based telecommunication infrastructure, with the exception of flag semaphore and other similar methods. In addition, multiple technologies may be used to link sites together in the same system. The second constant is that there will always be a data transfer bottleneck somewhere.

Fiber Optics Telecommunication Systems

Let's start by examining the most reliable telecommunications infrastructure: fiber optic telecommunication systems. These systems are capable of transferring a large amount of data rapidly. This rapid transfer of data can easily exceed the speed at which the leak detection system can process it. Thus, the leak detection system processing speed becomes the principal telecommunication bottleneck.

Fiber optic infrastructures are also immune to external noise sources that could disrupt data transfers. External interference such as solar flares, radio transmitters, high-power electrical transmission lines, and such will not negatively affect fiber optic data transfer. These systems are also able to send signals over great distances without the need for repeaters.

Fiber optic systems also have a cost advantage if installed when building the pipeline. The fiber optic cable and associated electronics cost per mile is low. The installed fiber optic cable would also include additional fibers, which are available for other operational requirements such as phone lines or even another leak detection technology system such as a fiber optic thermal detection system.

Although fiber optic telecommunication system installation costs are generally lower if installed during pipeline construction, retrofitting an existing pipeline is much costlier and has some risk. Considerations for a retrofit system include right-of-way agreements, risk in potentially burying the fiber in close proximity to the pipeline, network access points, and so forth.

Microwave Systems

Microwave systems are a very common telecommunication infrastructure. Fig. 8.4 shows an example of a microwave telecommunication installation. These systems predate fiber optic and satellite telecommunication technologies, and, for many years, were the primary means of transferring data over long distances.

The data transfer rates of these systems can easily meet and exceed leak detection system requirements. The pipeline industry has used such systems extensively for decades.

Some downsides to this technology include data interference caused by solar flares and other radio transmissions. Rain and snow can also affect data transfer rates. Although system designers consider these environmental impacts during the design phase, extreme weather can still affect their functionality. The systems may introduce increased data latency and time skew depending on the number of microwave repeater sites and the network's configuration.

Microwave systems are line-of-sight systems, which means that one site must have a clear view of its neighbor to work. This may require an increase in the number of repeater sites for a complete network. Many commercial providers maximize the distance they can cover by installing microwave infrastructures on top of mountains. However, mountain top locations are very expensive to build and service.

Satellite Communications

Satellite communication is another telecommunication infrastructure. Almost universally, operators who rely on satellite communications use a commercial vendor who has access to a geostationary satellite. Geostationary satellites appear to remain at one point in the sky, because the orbital period is the same as the earth's rotational velocity.

Because the satellites are geostationary, they have a constant footprint (area of coverage) on the earth's surface. A positive attribute of satellite communications is that they overcome the line of sight issue that microwaves have. As long as the Earth-based transmit/receive locations can "see" the satellite, communications will occur. Procurement of satellite data rates that will meet most leak detection system requirements is possible.

A challenge to satellite communications is that solar flares, as well as the sun in and of itself, affect the ability to transfer data. Sun data transfer outages occur twice every year when the Earth-based station is in direct alignment with the satellite and the sun. Although generally short, these outages occur every spring and fall.

Impairment of satellite communications also occurs through other signal blocking events. More than one operator has experienced a communication failure only to discover that someone parked a semi-truck in front of the satellite dish, thereby disrupting the telecommunication link.

Satellite communications are also subject to signal fade caused by rain, snow, and sleet. During significant adverse weather events, disruption of service can occur.

Other Telecommunication Systems

Other less commonly used but still available communication infrastructures include very high-frequency (VHF) radio systems and telephone landlines. Generally, these communication mediums are "last mile" solutions rather than full system deployments. That is, the telecommunication provider uses one of these technologies to link the remote site to a "broadband" telecommunication hub such as a fiber optic or microwave system. Generally, these systems lack the same broad bandwidth capabilities of the other methods. They also are limited in distance and subject to more data outages.

In summary, selection and deployment of a telecommunication system are dependent on the physical environment, available telecommunication technology options, and system requirements. Typically the operator has more than one option, including different telecommunication providers as well as a fully owned and operated systems. Each system has specific positive attributes as well as negative ones. Matching the telecommunication infrastructure to the leak detection system needs is the final objective.

8.2.2 Telecommunication Redundancy

When it comes to leak detection, the associated telecommunication system should be redundant. As discussed in Chapter 10, Human Factor Considerations in Leak Detection, leak detection systems are safety systems and need to have virtually 100% system availability. This requirement necessitates the use of redundant telecommunication infrastructures.

Telecommunication system redundancy, in its simplest form, consists of two separate telecommunication channels that are physically distinct from each other. An ideal telecommunication redundancy network could use a separate fiber optic network and microwave systems. In this case, the telecommunication system is immune to simultaneous failure, barring a catastrophic event such as an earthquake or flood destroying both the fiber optic cable and a microwave tower.

We do acknowledge that self-healing fiber optic rings and some highly redundant microwave systems are available. These systems can also provide an enhanced level of telecommunications reliability as well, because they allow rerouting of communications in the case of a single fiber optic break.

As a word of advice, if you contract your telecommunication service, ensure that the contract wording specifically requires completely independent telecommunication networks. Even with that, do not be surprised if both telecommunication systems fail at once; the authors have seen this occur on more than one occasion. Typically, the vendor agrees to provide requested service. Then, over time, they make changes to their network and the independence is lost. This situation exists until that single point of failure occurs. It happens, so *caveat emptor*.

8.2.3 Telecommunication Issues

Telecommunication systems, like every other form of information exchange, can experience issues in transferring data. As an example, in this chapter, we discussed the cause and consequence of data time skews.

For some leak detection systems, it is very important to know almost precisely when the acquisition of the data occurred and to know that the acquired field data contain minimal to no skew. Chapter 6, Rarefaction Wave and Deviation Alarm Systems presents details on the rarefaction wave system, which functions best with precisely timed acquired data. As shown in Fig. 8.4, we demonstrate the use of a GPS timing source. GPS timing is a method of synchronizing time at all locations. In this configuration, having the data time-stamped at the source allows computation of a potential leak and leak location with greater precision.

Other telecommunication issues include total outages, garbled messages, dropped messages, and so forth. A well-designed redundant telecommunication system provides resilience, which allows the SCADA system to obtain data from either channel when a failure occurs in one. This resilience will help to maintain system operations in less than optimal circumstances.

8.2.4 Telecommunication Best Practices

Although some operators design, implement, and maintain their own telecommunication infrastructure, most lease these services from commercial suppliers. For these lease arrangements, the operator relies on the commercial vendor's expertise and capabilities to provide the required level of service. *It is the responsibility of the pipeline operator to define the required minimum level of service.* Key service contract elements include reliability, availability, redundancy, and bandwidth requirements.

The definition of reliability is the system's ability to perform all required functions for a defined period without failure at a defined confidence level. One such specification could state that each telecommunication circuit will have a reliability of 95% at a 99% confidence level. This indicates that we are 99% confident that that the system will be operational at least 95% of the time. If we have two completely independent circuits with a 95% reliability factor, then this would prove an overall telecommunication circuit reliability of 99.75% with 99% confidence. This seems like a fairly stringent specification, but it still allows for 21.9 outage hours per year. The question to the operator is: are you willing to have your leak detection system out of service for nearly a full day every year?

Reliability is just one part of the equation. The other part is availability. The definition of availability is the probability that the system is operating properly when a data transmission event occurs. Remember that we calculated that a fully independent and redundant telecommunication system could still have 21.9 outage hours if we assume each circuit has a 95% reliability. During those 21.9 hours, the system is not available. Therefore, the system is not 100% available.

Eq. (8.1) provides two approaches to calculate availability. As shown, availability is a function of up time and down time. Another way to look at it is the relationship between mean time between failures (MTBF) and mean time to repair (MTTR). The authors feel that mission-critical circuits should have an availability of 99.5% or more. This will provide an overall redundant circuit availability of more than 99%.

$$A = \frac{t_{\mathrm{up}}}{t_{\mathrm{up}} + t_{\mathrm{down}}}$$

or

$$A = \frac{\mathrm{MTBF}}{\mathrm{MTBF} + \mathrm{MTTR}}$$

EQUATION 8.1 Mean Time Between Failures/Mean Time to Repair Equations

In Eq. (8.1), A is circuit availability, t_{up} is up time while t_{down} is down time.

When discussing telecommunications, an essential design requirement is overall bandwidth requirements. Often you will hear the following: the more bandwidth, the better. Our experience is that it does not matter how much bandwidth you contract for, because it will always be consumed. It is also our experience that SCADA requirements drive bandwidth requirements, which, in turn, establishes the periodicity of data transfer to the leak detection system.

To determine the minimum bandwidth, one must understand that the total SCADA system needs usually dwarf that of the leak detection system. As such, bandwidth needs are set by the SCADA system requirements. Therefore, one must work with the SCADA vendor to identify maximum (worse case) data transfer levels. A worse case may be when a major upset occurs that impacts all field data. This results in a "data flood" coming across the network. Once you know the amount of data flood, the transmitted data increase that by at least 10% and then round up to the next "standard" bandwidth value. This should ensure that the SCADA system requirements are met and, subsequently, that the leak detection system needs are met as well.

In summary, best practices for mission-critical circuits require highly reliable and available circuits with sufficient bandwidth. These circuits should be fully independent with no common mode failure potentials.

8.3 SCADA SYSTEM CONSIDERATIONS

In Chapter 10, Human Factor Considerations in Leak Detection, we discuss various approaches used to interface leak detection and SCADA systems.

The amount of data, as well as the direction of data exchange, that are shared between these systems is specific to the leak detection technology. The following paragraphs outline the various interactions and conceptual methods to achieve the data exchange and considerations that the leak detection and SCADA engineers/analysts should apply to new installations or system upgrades.

The simplest leak detection system and SCADA data exchange occurs when the only data transfer is that of the LDS alarms and statuses to SCADA. This can be the case when the leak detection system is not reliant on SCADA data. Still, even with such limited data exchange, the following must be considered: (1) communication redundancy; (2) alarm transfer approach; and (3) how to ensure that critical alarms are not "lost" if a communication failure occurs.

Local area network (LAN) communication redundancy is another factor in the system's mission-critical data transfer reliability and availability requirements. This is distinct and separate from telecommunication circuit redundancy. It should be a requirement that all LAN communications between the LDS technology and SCADA be fully redundant and as independent as possible. Applying this design requirement increases the probability that when data transfer is needed, a LAN communication link will be in service.

Other communication design considerations include: (1) defining which system will establish or designate the active communication channel; (2) specifying how the system determines that a loss of a communication channel has occurred; and (3) defining how the system switches back to the primary communication circuit once it has been restored.

With redundant communication circuits, transfer of data can occur in three potential ways. One data exchange method (not a recommended one) is having the communications from the leak detection system to the SCADA system occur on one circuit while the SCADA communications to the leak detection system occurs on the other. In this configuration, data transfer is nondirectional. This configuration loses the benefits of the redundant circuits. Other configurations provide better communication capabilities.

A second method involves sending and receiving data simultaneously across both communication circuits. In this configuration, identical data transmission occurs on each circuit. The positive aspect of this approach is that if one circuit fails, the data will arrive over the other circuit. There is no delay resulting from the system detecting a communication circuit outage and having to resend the data over the other circuit.

A concern for simultaneous data transfer is that the receiving application must determine which data set it will accept. This could be a simple heuristic that says all data on port one will be used unless that circuit fails. As long as the communication circuits are contained within a LAN without bandwidth limitations, this approach works.

The third redundant communication utilization approach involves identifying one circuit as primary and the other as backup. In this configuration, all data exchanges occur on the primary circuit unless an outage occurs. Once an outage has occurred, the systems switch over to the backup circuit and continue to operate.

This method is commonly used, but the engineer/analysts must take into consideration the rules or logic required to select which circuit is primary, which is backup, what logic is used to determine when to switch, and which system determines which circuit is active or primary. These rules and requirements are defined in the interface control document (ICD), as detailed further in Chapter 11, Implementation and Installation of Pipeline Leak Detection Systems.

Once the primary communication circuit rules are developed, other rules must be established regarding when a circuit will be declared out and how the system switches back to the primary circuit once it has failed to the backup. Loss of communication rules usually include a few timing components. Signal loss timing tests detect when no data are received within some predetermined time (3 s, for example). A second loss of communication timing test involves the other system failing to acknowledge a data transfer within some time window (within 2 s, for example). If either of these events occurs, then the system in charge switches communications to the backup circuit and flags the primary as out of service.

Once a switch from primary to backup occurs, this action initiates the third consideration: how and when does the system return communications to the primary circuit? Best practices require controller or analyst interactions to force this switch rather than having the systems make the determination to perform the switch.

The reason for this is the issue of intermittent circuits. There are situations when a communication circuit switches back and forth from good to bad and back to good. If the controlling system was automatically switching every time it detected a good circuit, then an endless series of communication circuit switches could ensue. Therefore, by requiring controller or analyst involvement, they can determine that the primary communication circuit is good and stable. At that time, they can force the transfer to the primary circuit.

Another design consideration is the schedule of data exchange between the leak detection system and SCADA. Generally, one of two approaches is used. The first approach is that SCADA continuously requests alarm and status updates from the leak detection system. In this configuration, because during the majority of time no changes have occurred within the leak detection system, a normal system status is all that is sent back to SCADA.

A downside of this approach is repetitive communications. On the positive side, the repetitive requests and subsequent responses validate that the two systems are still operating and can communicate. The update periodicity establishes how soon a leak alarm will be sent to the SCADA system.

The other method of transmitting data is called quiescent or report-by-exception transmissions. In this approach, data are transferred from the leak detection system to the SCADA system only when it changes. Normally operating leak detection systems generally have no leak alarms or system state changes present. In this situation, no data may be transferred for long periods. A risk to this communication method is that a loss of one or both communication circuits could occur without any warning or indication. To avoid this, many report-by-exception systems are configured such that the SCADA system requests a "heartbeat" or "I'm alive" type of scan request message at a predefined interval.

Another essential communication consideration is to ensure that all critical leak detection alarms are successfully transmitted to SCADA. Circumstances can occur that prevent a successful transfer of critical alarm messages from the leak detection to SCADA. Because leak alarms are critical safety alarms they must not be lost due to a communication outage or other problem. To prevent this requires that the leak detection system inform SCADA of how many alarms it is sending, after which SCADA must tell the leak detection system how many alarms are received. If the number of transmitted alarms is different from the number of received alarms, then all alarms must be sent again until the SCADA system confirms a successful receipt of all alarms.

Each of the considerations discussed are applicable to all leak detection systems that interface with the SCADA system. When the leak detection system must receive as well as send data to the SCADA system, other considerations must be taken into account. Chapter 11, Implementation and Installation of Pipeline Leak Detection Systems, discusses those issues.

8.3.1 SCADA HMI Considerations

How SCADA displays leak detection alarms, system status, and diagnostic data is an area of major consideration. As discussed in Chapter 10, Human Factor Considerations in Leak Detection, control room management policies, procedures, and ergonomic design must take these data components into account. The objective is to provide the controllers with essential information that provides firm situational awareness without overloading them with a flood of less essential data.

Another task is to determine the range of data that will be displayed. At the minimalist level, the SCADA system may display just the leak alarms as part of the normal SCADA system alarm page. At the other extreme is a full replication of the leak detection screens in the SCADA

HMI. There is no clear "best" design principal for this consideration. The operator's control room procedures, capabilities of the dedicated leak detection system displays, the capabilities of the SCADA HMI, and leak detection to SCADA data communications load are some of the factors that must be considered.

8.4 HISTORICAL ARCHIVING OF DATA

It is important to archive the input data to the leak detection system (especially for mass balance/RTTM systems), selected leak detection system outputs (particularly alarms), and control affecting the leak detection system.

8.4.1 Archiving Measurement Data

As a practical matter, it is important to maintain a continuous multi-year (preferably permanent) archive of all of the measurement data that are input to the leak detection system, particularly for an RTTM-based system or, for that matter, any mass balance–based system. Ideally, this archive would be saved at the periodicity that it is available for leak detection processing. Assuming that the leak detection system processes "snapshots" of data every *T* seconds, the data should be archived every *T* seconds, preferably by a resilient application specifically designed for the task so that the data are archived as continuously as possible. The historical archive should be designed such that the leak detection system can process the archived data offline and in "fast time." The definition of fast time is the processing of archived input data significantly faster than real time. This facilitates the following:

1. Testing and validation of the leak detection system
2. Evaluation and tuning of leak detection parameters to improve sensitivity and/or reduce false alarms
3. Evaluation and tuning of modeling parameters to improve the fidelity of an RTTM
4. Identification, evaluation, and mitigation of instrumentation issue
5. Incident analysis
6. Through auxiliary utilities, leak events might be superimposed on the archived data to evaluate the ability of the system to detect leaks as part of leak detection performance analysis (see Chapter 9: Leak Detection Performance, Testing, and Tuning)

When beginning the process of implementing a new leak detection system, development of the historical data archive should be one of the first tasks. The archived data will facilitate offline tuning and testing of the leak detection system.

The following considerations should be used in the design of the data archive.

1. The data archive should be portable and selectable for an arbitrary period of time. It should be easy to deliver the data to those working on the leak detection system. Note that by portable we mean that the archive should be able to be transferred from computer to computer, not that its structure is in some industry-standard portable format.
2. Updates to the data archive should be deliverable to its users (eg, leak detection support personnel) in an incremental fashion. In other words, if one already has the data for 12 months, then delivering the 13th month should only require delivering the additional month's data.
3. Because measurement points are likely to be added or deleted over time, it is important for the archive itself to contain its own data dictionary. For example, one should be able to work with the data archive without having to resort to external configuration files.

A relational database such as Microsoft SQL Server or Oracle's RDBMS may not be the best choice for the historical data archive. Furthermore, any program that is configured to compress data by smoothing over or rejecting small changes (eg, OSIsoft PI) is rarely appropriate for this task. The requirements for portability and incremental delivery make these options problematic because a multi-year database of data archived every few seconds will likely be very large.

One file-based way to organize portable and incremental data archives that has proved useful is to save (or export from the primary storage) the measurement data snapshot by snapshot. The data should be stored in hourly files that are named by date and time such as 2015.03.23_22EST, where the name's first 10 characters identify the date and the final five identify the hour and time zone.

All of this presupposes that the leak detection system has the ability to read and process archived data in an offline fast-time mode. This is a capability that should be required of any RTTM-based leak detection system because of significant capabilities for postevent analysis.

8.4.2 Archiving Leak Detection Results and Control Actions

For incident analysis, and for regular maintenance of the leak detection system, an archive of key outputs of the leak detection system (such as mass or volume balances for an RTTM and all leak alarms) should be maintained in a database.

In addition, manual control actions taken by users, such as clearing an alarm and resetting leak detection calculations, should be archived, preferably with sufficient information to identify why the action was taken.

Pipeline activity that impacts the leak detection system should also be logged and archived to facilitate maintenance of the leak detection system. Examples of activity that should be archived are:

- Pipeline maintenance or construction activities that affect the availability or reliability of the field measurement data or the flow paths in the pipeline.
- Instrumentation calibration
- Instrumentation overrides or other actions that disable the real-time validity of the instrument

8.5 RESILIENT SYSTEM DESIGN

Leak detection systems are safety systems. The intent and purpose of these systems are to provide the operator with reliable notification that a breach in pipeline integrity has occurred. As such, the system design must be resilient so it can graciously handle abnormal states while continuing to meet the design intent.

That said, what is a resilient system design? We can define resilience as the ability to gracefully handle both abnormal events and recovery. However, to expand on this further, we look at resilience as the ability of the system to handle variations, disruptions, and abnormal events without catastrophic failure.

We can demonstrate this with a balloon metaphor. Once inflated, we can push our finger into the balloon, but it does not pop. It changes shape and adapts to our prodding it. Once we remove our finger, it returns to a normal state. You can even have multiple people push on the balloon and it just changes shape until removal of these external forces occurs. However, you can exceed the balloon limits and stress it beyond its resilient level, and then it will pop. Implementing a resilient design does not guarantee that the system will not fail, but it does result in a system that can survive a broader set of variations, disruptions, and abnormal events.

Resilient system design includes all of the following:

1. Telecommunications system redundancy between the field and the central leak detection location.
2. Redundant communications between SCADA and the leak detection system.
3. Leak detection system redundancy. The leak detection system should be installed on at least two independent hardware or virtual server platforms with automatic synchronization between the two on redundant communications circuits.
4. Robustness of the leak detection system's algorithms and software. Garbage data should not cause the leak detection system to fail, although the results may cease to be meaningful until the data quality is restored.

Redundant field devices also contribute to a resilient system design. Unfortunately, it is rare that the field data itself is redundant. Therefore, even if all other aspects of the system are resilient, the system may be disabled by field data failures. For this reason, field maintenance personnel must treat instrumentation problems that impact the leak detection system as high-priority incidents.

REFERENCES

[1] API Tech Report 1149. Pipeline variable uncertainties and their effects on leak detectability, 2nd ed.; September 2015.

[2] Wikipedia Contributors. Nyquist–Shannon sampling theorem. Wikipedia, The Free Encyclopedia; 1 Apr. 2016. Web. 2 Apr. 2016.

[3] Wikipedia Contributors. Bessel filter. Wikipedia, The Free Encyclopedia; 14 March 2016, 09:20 UTC, https://en.wikipedia.org/w/index.php?title=Bessel_filter&oldid=709992326 [accessed 23.03.16].

Chapter 9

Leak Detection Performance, Testing, and Tuning

It goes without saying that in the absence of regulatory fiat, no rational pipeline operator is likely to go through the hassle and expense of implementing and supporting a pipeline leak detection system unless there is some measurable benefit. Measurable benefit implies that value is provided or assessed in useful and *quantifiable* terms. The ability to use software-enabled methods to provide such quantification is starting to become available.

To that end, this chapter provides discussion and background needed to describe the performance of a pipeline leak detection system (LDS). We start at the beginning, with a discussion of the metrics that are likely to be most useful in terms of evaluating the system performance. After that, we move on to the challenges of predicting leak detection system performance based on pipeline and other variables that apply independently of the leak detection system. Then, we continue by discussing the methods used to assess the performance of installed systems. We conclude with a discussion of how the tools described in this chapter can be used to aid in tuning the LDS.

9.1 PERFORMANCE METRICS

Consider the questions that you as an operator might be asked by a journalist, regulator, or elected official interested in the performance of your leak detection system:

- How big of a leak can your system detect?
- What period of time will it take to detect the leak?
- How much of the time is your system operational?
- How accurately can you locate the leak?

And some real kickers:

- What fraction of leaks occurring in your pipeline will be detected by the LDS?
- How will you know whether a leak alarm corresponds to a real leak?

Pipeline Leak Detection Handbook. DOI: http://dx.doi.org/10.1016/B978-0-12-802240-5.00009-1

Performance metrics allow different leak detection systems and approaches to be compared to each other and against absolute corporate, regulatory, and other legal requirements. A list of metrics used to define leak detection system performance is shown in Table 9.1. Let's start by looking at easily evaluated primary metrics that are based on direct measurements of the LDS in operation. This will include a discussion of performance mapping, which shows how the various metrics are related to each other, and how these relationships can be usefully visualized. We conclude by considering methods and approaches by which these primary metrics can be combined with other information to derive high-level metrics that can assist in determining how efficiently the LDS performs its primary task.

9.1.1 Primary Performance Metrics and Leak Detection Performance Mapping

Primary LDS performance metrics are defined as measurement standards that can be obtained via straightforward configuration (such as the fraction of the pipeline being monitored), observation (ie, the false alarm rate), or in situ testing (the detectable leak rate, for example) of the installed system. These metrics fall into the following categories:

- *Leak Detection Scope*: These are metrics that either parameterize or describe the performance of the LDS at the system level. These include global parameters such as the proportion or fraction, f_{M}, of the pipeline being monitored by the LDS and the evaluation period, t_{EVAL}, but they also include system performance indicators such as the system availability/service factor, f_{SF}, and the baseline (or false) alarm rate, R_{ALM}.
- *Leak Detection Sensitivity*: Metrics that define the ability of the LDS to detect leaks. These include the minimum detectable leak rate Q_{DET}, the leak time to detect t_{DET}, and the leak detection confidence (or probability of detection) P_{DET}. The reader should note that these three variables are not independent of each other. The detectable leak rate can be raised or lowered by appropriately varying the time to detect or the required detection confidence. The linkage of these variables is addressed via leak detection sensitivity performance mapping, which is discussed in more detail later.
- *Leak Location Performance*: This category includes metrics that define the ability of the LDS to locate leaks. The most important of these are the leak rate Q_{DET}, the leak location error (either the absolute error LE_{ABS} or the relative or normalized error LE_{REL}), and the leak location confidence P_{LOC}. As with the leak detection sensitivity metrics, these variables are dependent on each other via the LDS leak location performance map.

TABLE 9.1 List of Leak Detection System Performance Metrics

Symbol	Name	Metric Type	Description
f_M	Monitored fraction	System	The fraction of the pipeline that is monitored by the LDS. Typically expressed without units, or as a percentage. Calculated using: $f_M = \frac{L_M}{L_{TOT}}$ where: • L_{TOT} = total length of pipeline in the operation • L_M = total length of pipeline being monitored by the LDS
t_{EVAL}	Evaluation period	System	The period of time into the past during which the LDS monitors for signs of a leak. Typical units are seconds, minutes, or hours. This parameter is an indicator of the maximum lag time that the LDS may take to detect a leak
f_{SF}	LDS service factor	System	The fraction of time that the LDS is fully operational. Typically expressed without units or as a percentage. The value of f_{SF} should be as close to 1.0 (or 100%) as possible
R_{ALM}	Baseline alarm rate	System	The rate at which alarms are generated by the LDS in the absence of an actual leak. Also referred to as the false alarm rate, for obvious reasons. A high value may indicate that the LDS is tuned too aggressively. This metric is a component of $P_{(Leak\|ALM)}$: the likelihood that any particular alarm was caused by an actual leak. Units are typically alarms/year
f_{ALM}	Baseline alarm fraction	System	The fraction of time during which the LDS is in an alarm state due to the presence of false positives/baseline alarm rate. Typically expressed without units or as a percentage. The value of *f*SF should be as close to zero as possible

(*Continued*)

TABLE 9.1 (Continued)

Symbol	Name	Metric Type	Description
Q_{DET}	Detectable leak rate	Leak detection sensitivity	The leak rate that can be detected in time t_{DET} or less with confidence P_{DET}. The function that relates these three parameters defines the *leak detection sensitivity performance map* of the LDS. Units are typically consistent with units used to express nominal flow rates for the pipeline (BPH, MMCFPH, etc.)
			Note that a special case for this parameter evaluated for $t_{DET} = t_{EVAL}$ is the minimum detectable leak rate $Q_{DET,MIN}$. This defines the smallest leak that can be detected by the system with desired confidence P_{DET}
t_{DET}	Leak detection time	Leak detection sensitivity	The period after onset required to detect a leak of size Q_{DET} with confidence P_{DET}. Units are seconds, minutes, or hours. As noted previously, this parameter is dependent on the detectable leak rate Q_{DET} and the leak detection confidence P_{DET} (the probability or confidence that the LDS can detect and alarm a leak equal to or greater than Q_{DET} within a period of time t_{DET}). Normally expressed without units or as a percentage. Often standardized to be 95% or 99%
P_{DET}	Conditional leak detection probability/ confidence	Leak detection Sensitivity	The probability or confidence that the LDS can detect and alarm a leak equal to or greater than Q_{DET} within a period of time t_{DET}. Normally expressed without units or as a percentage. For fixed time t_{DET}, we can expect that P_{DET} will increase if Q_{DET} is likewise increased. Often standardized to be 95% or 99%. An element of the *leak detection sensitivity performance map*

(Continued)

TABLE 9.1 (Continued)

Symbol	Name	Metric Type	Description
LE_{ABS}	Absolute leak location error	Leak location	The absolute distance error that the LDS can be expected to exhibit when calculating the location of a leak for leak rate Q_{DET} with confidence P_{LOC}. The function that relates these three parameters defines the *leak location performance map* of the LDS. Units are typically miles or kilometers for internal leak detection systems, but may be feet or less for some external systems
LE_{REL}	Relative leak location error	Leak location	The relative distance error that the LDS can be expected to exhibit when calculating the location of a leak, expressed as a fraction or percentage of the total pipeline length. Calculated using: $LE_{REL} = \frac{LE_{ABS}}{L_{TOT}}$ where L_{TOT} = Total length of pipeline
P_{LOC}	Conditional leak location confidence	Leak location	The probability or confidence that the LDS can locate a leak with an error of LE_{ABS} or LE_{REL} for a discharge rate equal to Q_{DET}. Normally expressed without units or as a percentage. Often standardized to be 95% or 99%. An element of the *leak location performance map*
$\eta_{LED,NET}$	Net leak event detection efficiency	Derived	The net probability $P_{(ALM\|Leak)}$ that the installed LDS will detect and alarm a randomly generated leak event based on the expected conditional probability density function $f_L(q)$ of the leak incident rate for the pipeline, including the impacts of LDS service factor and false alarms. See Eq. (9.7). Normally expressed without units or as a percentage

(*Continued*)

TABLE 9.1 (Continued)

Symbol	Name	Metric Type	Description
$\eta_{LRD,NET}$	Net leak rate detection efficiency	Derived	The net efficiency with which the installed LDS measures all leakage flows (as opposed to leak events) based on the expected conditional probability density function $f_L(q)$ of the leak incident rate for the pipeline and including the impacts of LDS service factor and false alarms. See Eq. (9.8). Normally expressed without units or as a percentage
η_{ALM}	LDS alarm efficiency	Derived	The posterior probability or *likelihood* that the LDS alarm corresponds to a real leak, that is, the probability $P_{(Leak\|ALM)}$ of a leak given the presence of an alarm. See Eq. (9.9)
η_{EVT}	LDS event efficiency	Derived	The fraction of events that the LDS properly captures, including all real leaks properly detected (true positives), all real leaks not detected (false negatives), and all alarms not corresponding to actual leaks (false positives). This metric describes how well the LDS creates alarms in the presence of pipeline leaks without creating extraneous alarms. This parameter is equal to 1.0 (or 100%) if the LDS always alarms in the presence of a leak and never creates false positives. See Eq. (9.10). Normally expressed without units or as a percentage

The goal for many of the parameters in Table 9.1 is fairly obvious. We generally want the service fraction, f_{SF}, to be as close to 100% as possible and the baseline leak rate, R_{ALM}, to be as close to zero as possible, although just how close is a topic we return to later. For other metrics, the data supplied by the metric may be ambiguous or incomplete. An example of this is the LDS evaluation time t_{EVAL}, which is the maximum amount of time the LDS is given to catch a release following leak onset. Expressed another way, it identifies the amount of past data that the system incorporates into its statistical processing. t_{EVAL} is the maximum value for the time-to-detect metric t_{DET}.

Note that t_{EVAL} generally only addresses the configured data processing time used by the LDS, and it does not normally address the additional physical lag, t_{PL}, required for leak/spill information to reach the field instruments. This lag may be relatively unimportant because it is very short for many systems (mass balance systems for pipelines carrying liquids operating in tight mode or negative pressure wave systems). However, t_{PL} can be fairly large for mass balance systems liquid pipelines operating in slack or gas pipelines (because the hydraulic transients will change only slowly for sensors at segment endpoints as the line unpacks due to the leak) or for certain external detection systems (because the transient thermal pulse due to the leakage, for example, may have to diffuse slowly through some distance of soil to reach the detector). Operators should take care to ensure that they know whether or not their evaluations of t_{DET} include the physical lag time.

Furthermore, because the evaluation period is clearly an indicator of leak detection lag, it is tempting to conclude that this parameter should be minimized to minimize the time to detect t_{DET}. This is a valid assumption for some, but not all, systems. As an example, the value of t_{EVAL} is automatically minimized for rarefaction wave and other systems that generally process only current data or cumulative data over a brief period of time (one or two sonic waves pass over the pipeline segment in the case of the rarefaction wave system, generally amounting to no more than a few minutes). However, the majority of leak detection installations are mass balance computational pipeline monitoring (CPM) systems, and short t_{EVAL} are not optimal. In reality, these systems benefit from having a large value of t_{EVAL} in that the LDS can catch leaks over a range of time periods depending on the detectable leak rate, Q_{DET}, and the desired confidence of detection, P_{DET}. In such a sense, increasing t_{EVAL}, if feasible, extends the performance of the LDS by reducing the minimum detectable leak size at the cost of a longer detection time, t_{DET}. Thus, the system should catch large leaks quickly and small leaks slowly. The relationship between Q_{DET}, t_{DET}, and P_{DET} is provided via the LDS leak detection performance map.

A leak detection probability sensitivity performance map (instead of a time leak detection sensitivity performance map) is usually calculated or evaluated for a range of leak rates. An example of such a chart is shown for a CPM system applied to a large gas pipeline system in Fig. 9.1A. The value of t_{EVAL} for the system is 5 hours. Note that the detectable leak rate is zero for 10 to 15 minutes following the start of the leak because it takes that long for the system to acquire enough evidence to determine that the deviations being observed in the data are significantly different from the normal operation of the pipeline. After this, the probability that a leak has been observed starts to increase. However, it increases *much* more slowly for the smallest leak rate. Thus, a leakage of 50,000,000 SCFPH (equivalent to a full pipeline rupture in this system) will be detected within less than 15 min in the control

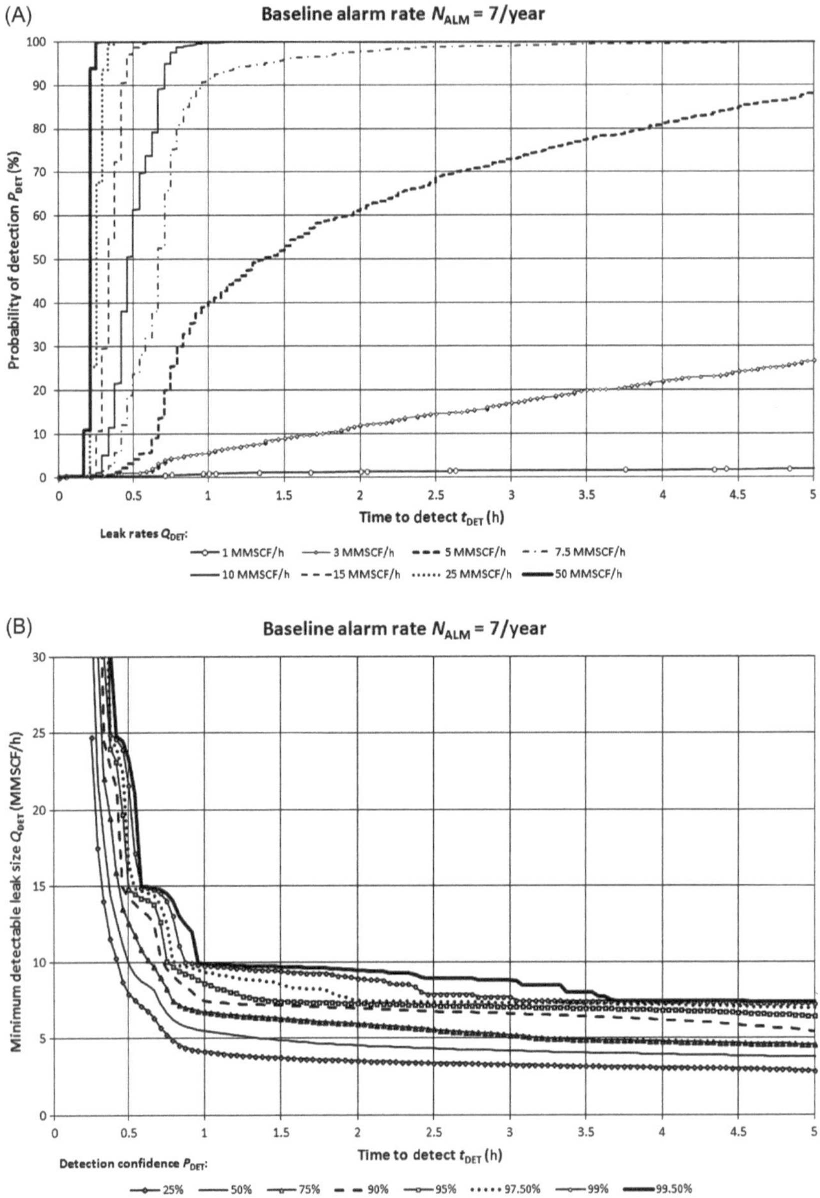

FIGURE 9.1 LDS leak detection sensitivity performance maps: (A) detection probability compared with time and (B) leak rate compared with time.

room. However, the smallest leak rate shown (approximately 2% of nominal pipeline flow) is effectively undetectable over the 5 hour evaluation period.

An alternate way of displaying the same information is via the detectable leak rate (compared with the detection time) sensitivity performance map, which is usually performed for a range of detection probability values. See Fig. 9.1B. In line with the previous chart, the detectable leak size is off the map for periods of less than 15 min and then drops rapidly as the detection period increases further. However, there is a minimum detectable leak size in this map. This is very characteristic of mass balance leak detection systems. It is an inevitable consequence of the behavior of the speed of decline in the aggregated error of the leak detection signal, the value of which can never go down faster than the inverse square root of the elapsed time following leak onset. This is discussed in Chapter 5, Statistical Processing and Leak Detection.

The detectable leak rate will always be greater as we increase the confidence that the signal represents a real leak. As before, leaks below 3–4 MMSCFPH (approximately 6–8% of the nominal pipeline flow) are effectively undetectable over the evaluation period for this particular LDS applied to this particular pipeline.

Some things to note with respect to these two leak detection sensitivity maps are:

- For this LDS installation, increasing the evaluation period would produce some benefits in terms of increasing the leak detection confidence (or reducing the detectable leak size) at the cost of increased leak detection time. However, note that based on the rates of change at the right side of both charts, the benefits of a doubling of t_{EVAL} appear to be marginal at best. As noted previously, this is not unusual behavior for mass balance systems.
- The baseline alarm rate for the examples shown is 7 false alarms/year. Whether or not this is an acceptable rate will be revisited in the next section of this chapter, and again in Chapter 10, Human Factor Considerations in Leak Detection.
- The maps are for illustrative purposes only and should not be construed to indicate leak detection sensitivity for any particular pipeline or LDS!

The second important kind of leak detection performance map is the leak location performance map. These maps typically show either the absolute or the relative leak location error as a function of leak rate for various leak location confidence values. An example of one such map is shown in Fig. 9.2 for an real-time transient model (RTTM) mass balance CPM system being used on a large crude oil pipeline.

Not surprisingly, the performance indicates that the leak location error, LE_{ABS}, declines as the leak rate, Q_{DET}, increases in size. It also shows that the error increases as the required leak location confidence, P_{LOC} (ie, the confidence that the leak is actually inside the bounds $X_{LDS} \pm LE_{ABS}$, where X_{LDS} is the leak location predicted by the LDS), is increased. Note the following:

- The maximum error in a leak location performance map cannot exceed the segment length.
- An equivalent normalized map can be built by showing the relative error as a function of Q_{DET} or normalized leak rate, Q_{DET}/Q_{NOM}.

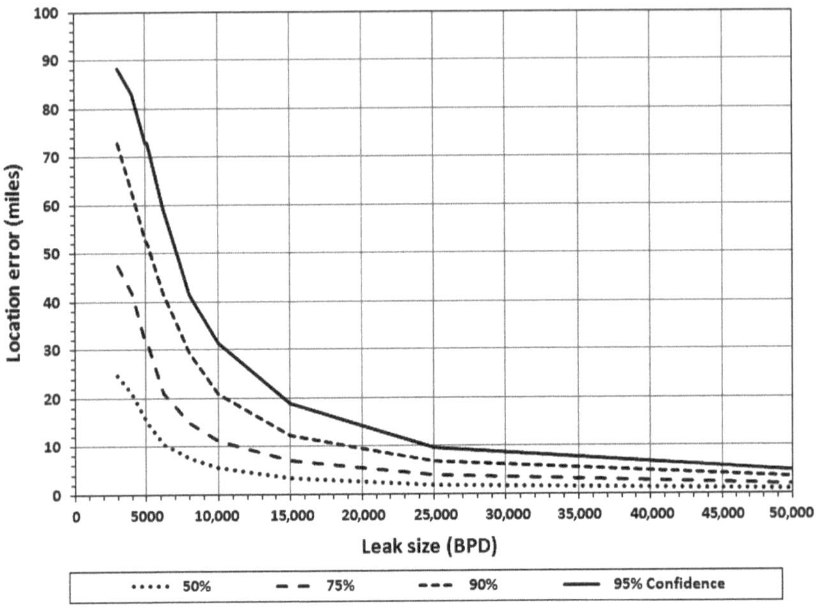

FIGURE 9.2 LDS leak location performance map.

- The estimate of leak location error in a mass balance system tends to improve asymptotically with time. The leak location map is therefore generally evaluated at the current time, *t*, based on data collected over the previous period, t_{EVAL}.

As before, the leak location map shown in Fig. 9.2 is specific to just one implementation. Note that most efforts to implement formal performance mapping of pipeline leak detection systems have focused on mass balance CPM systems.

9.1.2 Derived Metrics and LDS System Efficiency

The parameters discussed in the last section are referred to as primary metrics because they can be obtained via testing and performance mapping of the LDS as installed to the pipeline without reference to other parameters, such as the degree to which the target pipeline is actually at risk. (This topic is discussed in some detail in Chapter 13, Leak Detection and Risk-Based Integrity Management.) Unfortunately, these metrics do not allow us to ascertain the immediate value of the system in terms that a risk manager or operator would find useful. These include fundamental questions such as: *How likely is it that an alarm issued by the LDS corresponds to an actual leak?* or *How reliable is the LDS as a leak detector?*, or even *What should I do when I receive an alarm from the LDS?*

The last question is an issue discussed in Chapter 10, Human Factor Considerations in Leak Detection. As for the other questions, there is another class of metrics (also included in Table 9.1) useful for answering these questions: *derived metrics*.

Let's start by considering the degree of pipeline risk. We formalize or quantify the degree of pipeline risk by specifying two parameters: the pipeline system leak incident rate, R_{Leak}, and the conditional leak probability density $f(q_{\text{Leak}})$. The leak incident rate for a pipeline simply defines the rate at which the system may expect to experience spill or rupture events and can be defined as:

$$R_{\text{Leak}} = \int_0^{L_{\text{TOT}}} \kappa_{\text{Leak}}(x) \cdot dx$$

EQUATION 9.1 Leak Incident Rate Equation

where κ_{Leak} is the leak rate per unit distance (ie, leak incidents per mile per year) and L_{TOT} has been previously defined as the total segment or pipeline length. If the incident rate per mile is constant over the pipeline and not a function of location, then we have:

$$R_{\text{Leak}} = \kappa_{\text{Leak}} \cdot L_{\text{TOT}}$$

EQUATION 9.2 Pipeline Leak Incident Rate for Constant Incident Rate Per Unit Distance

The leak incident rate can be calculated on an average basis for all of the pipelines in the United States based on analysis of Department of Transportation (DOT) Pipeline and Hazardous Materials Safety Administration (PHMSA) data. It is also possible that, based on maintenance and accessibility of long-term records, or even on engineering judgment, an estimate can be provided for R_{Leak} on the basis of an individual pipeline or system. We discuss these issues further in Chapter 13, Leak Detection and Risk-Based Integrity Management.

Beyond the overall leak rate, we have already seen from the previous section that virtually all leak detection systems will experience a soft or probabilistic decline in their ability to detect leaks as the leak rate declines. Consequently, it should be clear that if most of the expected leaks are beneath this floor, then the utility of the leak detection technology will be limited. To get a handle on this, we define the conditional cumulative distribution function of the leak rate, $P_C(q_{\text{Leak}})$. This is the cumulative probability, P_C, that a randomly generated leak rate, Q_{Leak}, is less than or equal to a specified leak rate, q_{Leak}, given specified location x and the presence of a real leak at location x defined via a Boolean or binary leak state, s_{Leak}, or:

$$P_C(q_{\text{Leak}}) = P(Q_{\text{Leak}} \leq q_{\text{Leak}} | X = x, s_{\text{Leak}} = \text{True})$$

EQUATION 9.3 Leak Rate Conditional Cumulative Distribution Function Definition

$P_C(q_{Leak})$ can be calculated as the integral of the conditional probability density function $f_L(q)$ of the incident rate with respect to the leak rate, q, again given location x and the presence of a real leak state, $s_{Leak} = \text{True}$, or:

$$P_C(q_{Leak}) = \int_{-\infty}^{q_{Leak}} f_L(q) \cdot dq$$

EQUATION 9.4 Leak Rate Conditional Cumulative Distribution Integral

Alternately:

$$f_L(q_{Leak}) = \frac{\partial}{\partial q}[P_C(q)]_{q=q_{Leak}}$$

EQUATION 9.5 Leak Rate Conditional Probability Density Function

Note that in both of these equations, the assumption that a leak is already present at location x is implicit. It would be rigorous to assume that $f_L(q)$ can vary in a known fashion at every location in the pipeline, but, as seen in Chapter 13, Leak Detection and Risk-Based Integrity Management, this is probably pretty optimistic because determination of this function is difficult due to the sparseness of the data. Consequently, we assume for the remainder of this section that this function is the same everywhere in the pipeline.

Let's assume that a leak occurs at some location x. Further assuming that the LDS is functional and not already in a false alarm state, we can define the operating leak event detection efficiency $\eta_{LE,OP}$ as the probability $P_{(ALM|Leak)}$ that the operational LDS will detect and alarm a randomly generated leak. We can calculate this as the integration of the product of the leak conditional probability density and the LDS leak detection confidence (evaluated at the end of the LDS evaluation period, t_{EVAL}, because this choice of detection time maximizes the probability of detection):

$$\eta_{LE,OP} = P_{(ALM|Leak)} = \int_{0}^{Q_{MAX}} [f_L(q) \cdot P_{DET}(q, t = t_{EVAL})] \cdot dq$$

EQUATION 9.6 Operating Leak Event Detection Efficiency Equation

In summary, if there is a leak within some particular range between q and $q + \Delta q$, then the probability of detecting it is the rate increment Δq times the probability density f_L evaluated at q times the probability that the LDS can actually detect a leak of size q. To get the probability of an alarm, we then integrate over the range of possible leak rates from zero to some maximum expected rate Q_{MAX}, where Q_{MAX} is approximately the pipeline nominal flow rate.

It is important to note that the leak detection confidence $P_{DET}(q,t)$ is NOT the same as the operating leak detection efficiency $\eta_{LE,OP}$. The first parameter is the probability or confidence that the LDS will generate an alarm given a leak of a specific size q within some time t, and it is a strong function of the LDS threshold. The second is the probability that the fully functional LDS will generate an alarm given both the expected distribution of pipeline leaks and the LDS leak detection confidence, each expressed as a function of leak rate.

We can further address the impacts of service factor and periods when the leak detection system is in a false alarm state by defining the net leak event detection efficiency $\eta_{LE,NET}$:

$$\eta_{LE,NET} = f_{SF} \cdot (1 - f_{ALM}) \cdot \eta_{LE,OP}$$

EQUATION 9.7 Net Leak Event Detection Efficiency

The leak event detection efficiency is not the only way of grading the ability of the LDS to detect leakage. If we consider that the damage done by a leak is a function of the spill volume, then we can also view the leak detection system as a kind of global flow meter that measures leak rates. By extension, we would care most about detecting the leaks that have the highest flow rate. To that end, we can also define the net leak rate detection efficiency $\eta_{LR,NET}$ as:

$$\eta_{LR,NET} = \frac{f_{SF} \cdot (1 - f_{ALM}) \int_0^{Q_{MAX}} \left[f_L(q) \cdot P_{DET}(q, t = t_{EVAL}) \right] \cdot dq}{\int_0^{Q_{MAX}} \left[q \cdot f_L(q) \right] \cdot dq}$$

EQUATION 9.8 Net Leak Rate Detection Efficiency Equation

The rate efficiency effectively weights each detected leak event by its associated leak rate and then divides by the flow-weighted integration of all of the leak events. This efficiency is often much higher than the leak event detection efficiency, which tends to be significantly reduced if the leak distribution density function is centered below the effective threshold for the system. Note that the value of a leak detection technology is multifaceted. As such, the rate efficiency, like the event efficiency, does not by itself provide a complete picture of the system performance based on the downsides or consequences of experiencing a leak. Other factors beyond the leak flow rate enter into this issue, as discussed in Chapter 13, Leak Detection and Risk-Based Integrity Management. A complete understanding of the leak technology performance requires a judicial weighting of all of these parameters.

If we rely more on the net leak rate detection efficiency, $\eta_{LR,NET}$, than on the net leak event detection efficiency, $\eta_{LE,NET}$, to assess the leak detection capabilities of our system, then would we care about events and event efficiencies at all? The answer is *yes*. Alarms are discrete. Insofar as each one might signal a potential spill but that might *not* be a spill because it is a false alarm. Evaluation of alarms consumes some time, effort, and expense. Also, additional overhead to perform a field callout and/or shut down the pipeline may be required. This is simply a consequence of the uncertainty surrounding the alarm: is it an actual leak with immediate action required to protect people, property, and the environment from the harmful effects of any resulting spill...or is it bogus? Think about the consequences if you fail to respond and it is real!

To provide some degree of rigor around this issue, let's specify a new system metric called the LDS alarm efficiency, η_{ALM}. This is the posterior probability or *likelihood* that the LDS alarm corresponds to a real leak. Recalling that R_{ALM} is the false alarm rate, this is defined as:

$$\eta_{\mathrm{ALM}} = L_{(\mathrm{Leak|ALM})} = \frac{\#\,\text{True positives}}{\#\,\text{False positives} + \#\,\text{True positives}} = \frac{\eta_{\mathrm{LE,NET}} \cdot R_{\mathrm{LK}}}{R_{\mathrm{ALM}} + \eta_{\mathrm{LE,NET}} \cdot R_{\mathrm{LK}}}$$

EQUATION 9.9 LDS Alarm Efficiency

This is a key parameter. This efficiency is a measurement of the value of LDS alarms as leak event indicators. In principle, one could calculate this directly for any installed LDS/pipeline combination using the history of false and true positives. However, for most pipelines, the actual rate leak events would be too low to obtain a good estimate of the true positive rate. Therefore, an alternate approach would be to refer to nationally compiled incident databases to develop a reasonable incident rate, as discussed in Chapter 13, Leak Detection and Risk-Based Integrity Management. The leak event efficiency could then be determined based on formal testing of the LDS, which we will be discussing shortly.

One last LDS performance metric and we are done. This one is called the LDS overall event efficiency, η_{EVT}, and it is defined as the fraction of events that the LDS properly captures, including all real leaks properly detected (true positives), all real leaks not detected (false negatives), and all alarms not corresponding to actual leaks (false positives):

$$\eta_{\mathrm{EVT}} = \frac{\#\,\text{True positives}}{\#\,\text{False positives} + \#\,\text{False negatives} + \#\,\text{True positives}} = \frac{\eta_{\mathrm{LE,NET}} \cdot R_{\mathrm{LK}}}{R_{\mathrm{ALM}} + R_{\mathrm{LK}}}$$

EQUATION 9.10 LDS Global/Overall Event Efficiency

This efficiency measures how often the LDS is correct on a leak detection event basis. It goes without saying that the overall event efficiency is always lower than the alarm efficiency.

9.2 TUNING AND TRADEOFFS

It should be clear by now that there is no single metric that will fully define LDS performance. Tradeoffs are implied because the most useful derived metrics are not independent of each other in the sense that changing any one will also alter the others. In this section, we investigate the general relationships between the various performance metrics. To do this, we are also going to assume that we have a good understanding of our leak detection signal (based, say, on an analysis of recorded LDS output signal data) so that we can show how the signal behavior affects the performance of the LDS.

Let's utilize the simplest possible combination of leak signal error and LDS models to illustrate these tradeoffs. Our illustrative LDS is a mass balance CPM system utilizing a standard volume balance signal for a single pipeline segment and a very simple instrumentation error model. Our volume balance error model is pre-whitened/decorrelated, fully random Gaussian noise with a periodicity or update period that averages 5 minutes. The periodicity is important: it implies that the errors of the volume balance for periods less than five minutes are fully correlated with each other. For periods more than 5 minutes, however, errors are generally uncorrelated and independent of each other. The mean error of the volume balance is zero, but the illustrative standard deviation of the sample-to-sample error is 2.5% of nominal flow. This addresses all contributors, including flow balance errors, pressure and other instrument errors, and modeling errors.

The LDS operates by sampling every 5 minutes. The leak detection methodology is equally simple: we use continuous sampling of the data, and a leak will be declared if the signal exceeds a user-specified threshold (see Chapter 5: Statistical Processing and Leak Detection). Let's further assume that for regulatory reasons, company policy, or other reasons, we want to detect the smallest possible leak in no more than 12 hours. To accomplish this (and to simplify our analysis), we will pool all of our current volume balance data in a single aggregator using the period $t_{\mathrm{EVAL}} = 12$ hours. We will update our data set every 5 minutes; every time we add a new sample point, the impact of the oldest point in the queue is removed.

What we would like to do is calculate a threshold value that will be high enough to minimize the type I probability that we will have a false alarm, but not so high that we will experience a type II error and fail to catch a leak. To analyze this problem, we return to a number of equations we reviewed in Chapter 5, Statistical Processing and Leak Detection. Let's start by assuming that our $P_{\mathrm{Type\ I}}$ and $P_{\mathrm{Type\ II}}$ errors are the same, or:

$$P_{\mathrm{Type\ I}} = P_{\mathrm{Type\ II}} = \alpha$$

EQUATION 9.11 Illustrative Case Type I/II Allowed Error Probabilities

We are fundamentally performing a hypothesis test on two Gaussian distributions H_0 and H_1:

$$H_0{:}q_{VB} \leq 0; \quad H_1{:}q_{VB} > 0$$

EQUATION 9.12 Illustrative Case Leak Rate Hypotheses

where H_0 is our null hypothesis (there is no leak and the leak rate q_{VB} is zero or less) and H_1 is our alternate hypothesis (there is a leak and the leak rate q_{VB} is greater than zero). We can assume that the null and alternate hypothesis volume balance standard deviations σ_{VB} are the same. From Chapter 5, Statistical Processing and Leak Detection, we recall that if we are comparing two Gaussian or normal distributions, then:

$$VB_T = \frac{q_{VB}}{2} = \frac{Z_\alpha \sigma_{VB}}{\sqrt{n}} = \frac{Z_\alpha \sigma_{VB}}{\sqrt{t_{\mathrm{EVAL}}/t_S}}$$

EQUATION 9.13 Illustrative Case Leak Rate Threshold Calculation

where n is the number of independent samples, VB_T is the implied leak detection threshold, and Z_α is the number of standard deviations required to achieve a one-tailed confidence of α. At every update of our pooled data, we will issue an alarm if the volume balance exceeds VB_T.

Based on this analysis and the signal parameters specified, the behavior of the implied threshold for varying type I/type II probability values is shown as a function of t_{EVAL} in Fig. 9.3. Not surprisingly, the value declines with increasing evaluation time. Because we are trying to maximize the system sensitivity, we select a value of 12 hours over which to pool our volume balance data. Note that for the 12-hour evaluation period, the error probability in this chart changes dramatically (by approximately five orders of magnitude) over a relatively short threshold span that is less than 1% of nominal flow.

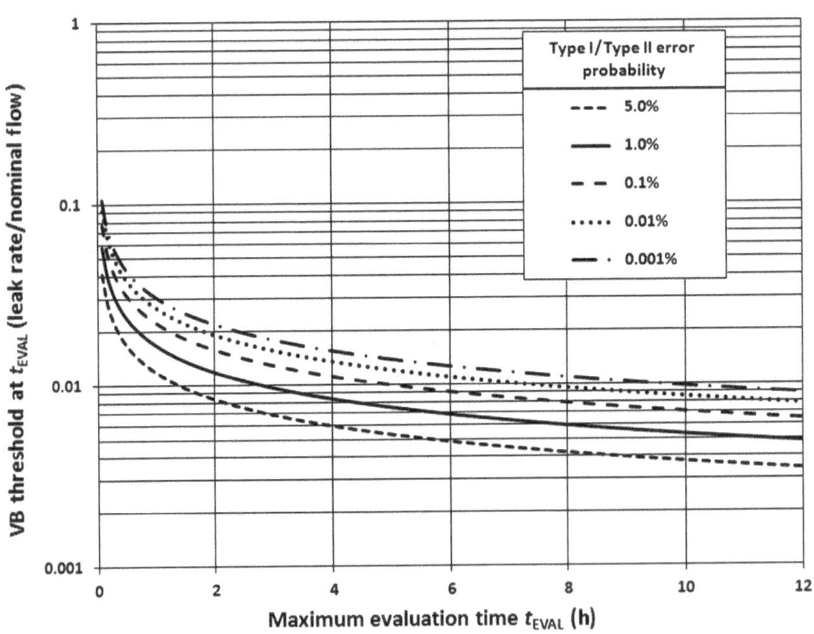

FIGURE 9.3 Simplified/illustrative LDS maximum evaluation time thresholds.

To see how this error behavior influences the selection of our 12-hour threshold, and how the threshold impacts the LDS performance, let's select an error probability $\alpha = 0.1\%$. By averaging our white noise stream over 12 hours, we have reduced the effective number of independent periods to only 2 per day, or 730 periods per year. Per our false alarm discussion in Chapter 5, Statistical Processing and Leak Detection, the error probability value, when combined with the number of independent evaluation periods in one year, results in a false alarm rate of 0.73 alarms/year. Referring to Fig. 9.3, we see that this corresponds to a threshold of 0.64% of nominal pipeline flow, if we assume that the LDS has 12 hours to detect the leak.

Note that although we use 12 hours of data to check for leaks, our simplified LDS can still detect leaks over shorter periods, again in line with Chapter 5. This is because an alarm will be created as long as the *average* signal over the 12-hour period exceeds threshold:

$$Q_{\text{DET}} \frac{t_{\text{DET}}}{t_{\text{EVAL}}} > VB_T$$

EQUATION 9.14 LDS Aggregator Off-Design Leak Detection Equation

Thus, the detectable leak rate for this simplified LDS is proportional to $1/t_{\text{DET}}$, as we can see via the corresponding leak detection sensitivity performance map shown in Fig. 9.4A. Note that Figs. 9.3 and 9.4A are NOT the same. The first figure provides the optimal set of thresholds based on various averaging periods, whereas the second is the detectable flow rate that will apply if we use a single 12-hour period and select the thresholds for that period from Fig. 9.3.

Let's assume for the sake of argument that the leak rate function $f_L(q_{\text{Leak}})$ is a log-normal probability density function with a mean rate of 5% of nominal pipeline flow and a standard deviation of 15% of flow. This distribution is shown as a function of leak rate in Fig. 9.4B. Although the selection of parameters for the illustrative curve is rather arbitrary, it is conceptually consistent with the authors' observation that most pipeline leaks are small (often well below a few percent of nominal flow), although there will always be a few very large leaks at a much lower probability. The distribution of leak sizes is discussed further in Chapter 13, Leak Detection and Risk-Based Integrity Management.

If we walk vertically up the performance map in Fig. 9.4A for $t_{\text{DET}} = t_{\text{EVAL}}$, we can also plot the leak detection confidence for the LDS as a function of leak rate. This is shown plotted alongside the leak distribution function in Fig. 9.4B. This figure indicates that we will eventually detect almost any leak with a discharge rate more than one percent of nominal pipeline flow with our LDS (though it is likely to take 12 hours to do so). Good news! Let's also use Eqs. (9.6 and 9.7) to calculate our net leak event detection efficiency as a function of leak rate. Not so good news! When integrated to a maximum leak rate equal to the pipeline nominal flow, our event detection efficiency (again referring to Fig. 9.4B) maximizes at only 65%, leaving 35% of all leaks unaccounted for by our LDS. This is obviously

(A)

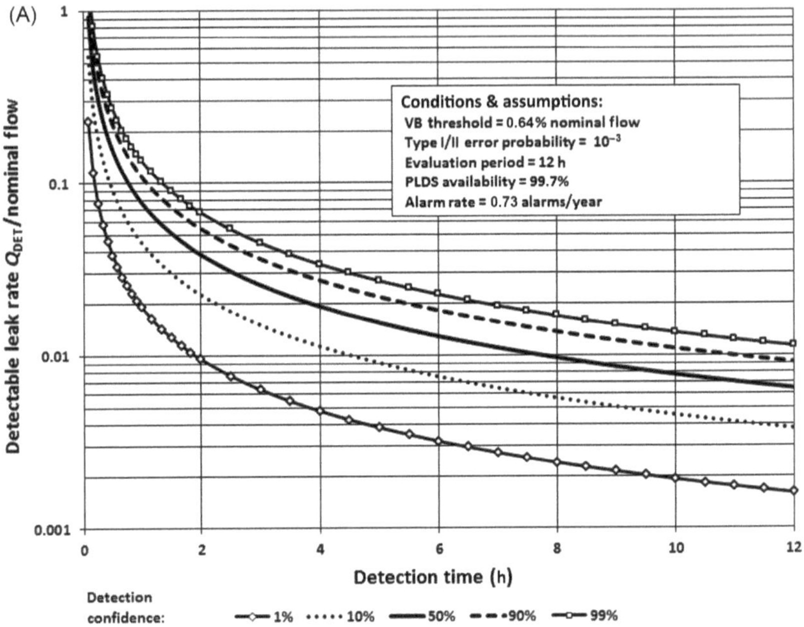

(B)

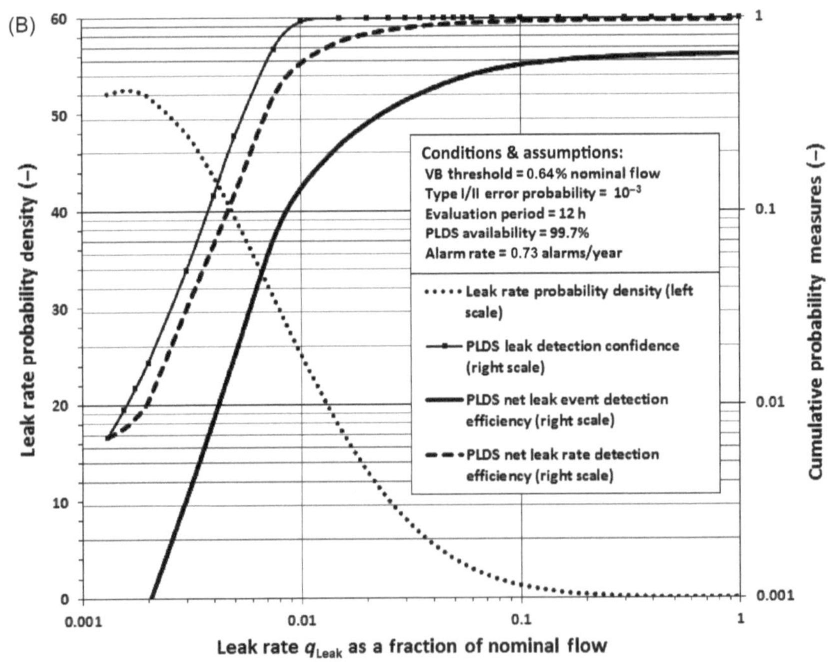

FIGURE 9.4 Simplified/illustrative LDS performance map: (A) overall performance map for $VB_T = 0.78\%$ nominal flow and (B) leak distribution and illustrative LDS cumulative leak probability functions evaluated at $t_{EVAL} = 12$ h.

because these leaks are effectively below the threshold of our system. We do much better if we use the leak rate detection efficiency (Eq. 9.8) as a performance guideline. This parameter maximizes at 98%, indicating that although the LDS is not terribly impressive on the basis of leak event detection, it is very good at estimating all of the leakage flows from the pipeline.

So based on an error probability of 0.1%, and as long as we are willing to wait as long as 12 hours, we can detect 65% of leak events (and 98% of the leakage flow) from our pipeline at the cost of less than one false alarm/year. It is tempting to ask ourselves if we can do better by lowering the threshold. All we need to do is repeat this process for a range of threshold values and observe the impacts on the baseline leak rate, leak event detection efficiency, and leak rate detection efficiency. We can also calculate the leak alarm and event efficiencies by way of Eqs. (9.9 and 9.10).

Results are shown in Fig. 9.5. This figure shows that lowering the 12-hour threshold from 0.64% of nominal pipeline throughput to 0.4% has the benefit of raising the leak event detection efficiency from 65% to 71%. It also produces a significant increase in false alarms, which go from less than one to nearly twenty alarms/year. Raising the threshold to a value of 1% of nominal flow, on the other hand, reduces the leak detection event efficiency to approximately 56% but also reduces the false alarm rate to a minuscule level. Interestingly, the leak rate detection efficiency is relatively insensitive to these threshold changes, declining only slightly from 98% to 96% as the threshold is raised from 0.64% to 1.0% of nominal pipeline flow.

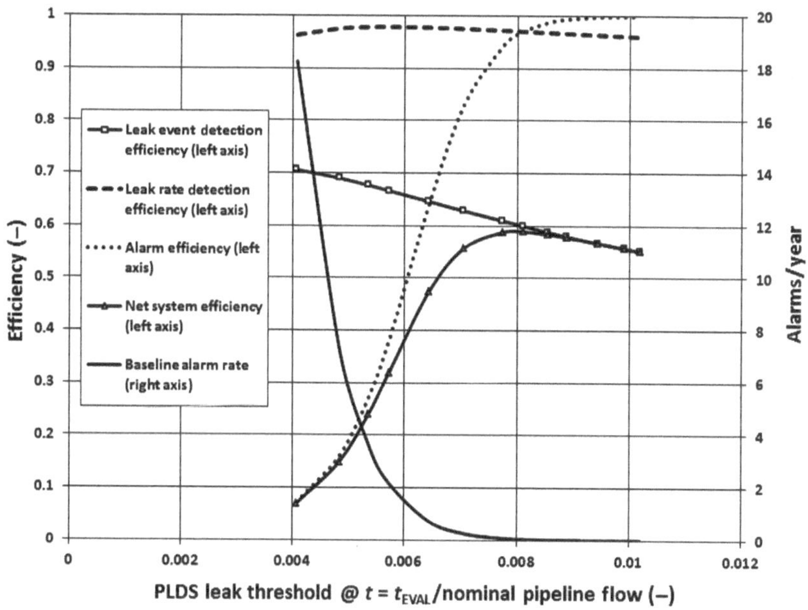

FIGURE 9.5 Impact of threshold on alarm rates and LDS efficiencies.

The LDS alarm and system efficiencies follow this trend. For our baseline threshold of 0.64%, the LDS alarm efficiency indicates that the likelihood that any particular alarm actually corresponds to a real leak is only approximately 63%. Lowering the threshold further causes the value of this efficiency to collapse. Raising the threshold has the opposite effect: for a threshold of one percent of flow, the chance that any particular alarm actually corresponds to a leak increases dramatically to nearly 100%. The improvement of the system efficiency is also significant, if not quite as impressive, increasing from 47% to 59% as the threshold increases from 0.64% to 0.80% of nominal flow, and then declining for higher threshold.

In short, the LDS system efficiency is a *maximum* at approximately 0.80% of pipeline flow. With the increasing threshold, this efficiency will initially increase and eventially maximize as it approaches the declining leak event detection efficiency curve. After the system efficiency maximizes, both curves merge, eventually become indistinguishable from each other and decline together with increasing threshold. This phenomenon indicates that if we value false alarms and failures to detect as equivalent negatives (which is, of course, not necessarily the case, and we return to this issue in Chapters 10 and 13: Human Factor Considerations in Leak Detection and Leak Detection and Risk-Based Integrity Management), then the point where the system event efficiency maximizes is the most *optimal* threshold for the LDS.

Some comments regarding our illustration of LDS tradeoffs are in order:

- Our simplified error model is Gaussian. However, in line with Chapter 5, Statistical Processing and Leak Detection, many real-world signal errors have fat or heavy tails. This can potentially cause the detectable leak rates to increase dramatically over shorter detection periods. For longer detection times, the performance will generally approach the behavior of the leak detection sensitivity map of this model as pooled errors utilize more data and become more Gaussian, in line with the central limit theorem.
- The periodicity of the volume balance error stream for this model was simple and arbitrarily fixed at 5 min. In reality, periodicity of your pipeline's volume balance signal is best determined based on analysis of the LDS leak detection signal that uses real-world data as an input.
- More importantly, the time series behaviors of most volume balance signals are a composite of many summed errors with differing periodicities. If the periodicity of any component is extremely large, then it will often tend to look like an error *bias* or *autocorrelation*. In all cases in which there are components of the signal with periodicity longer than the sampling frequency, the error behavior is likely to decline more slowly than shown in Fig. 9.3. This is often the result of a failure to properly decorrelate the signal, since, as we discussed previously in Chapter 5 (Statistical Processing and Leak Detection), pooled random Gaussian errors that are independent and identically distributed tend to decline according to the square root of the number of samples (our proxy for

the passage of time), while pooled errors for correlated signals decline more slowly. This will degrade LDS efficiency.

- To address this, recognize that random error components of the signal that have periodicities longer than the LDS sampling period cause signal values at different times to be correlated with each other. An error predictor can therefore decorrelate the signal by eliminating the time-correlated portion of the current signal based on previous signal values. A simple way of doing this is to subtract a moving average of previous values, but more sophisticated methods may also be used. Refer again to Chapter 5, Statistical Processing and Leak Detection for a discussion of decorrelation.
- Our conceptual LDS used a single pooled sampling period. Increased sensitivity over the entire evaluation period can be achieved by using multiple evaluation (or averaging) periods t_{Pi} such that $i = 1, NP$, where t_{Pi} is period i, NP is the number of periods, and $t_{Pi} \leq t_{\mathrm{EVAL}}$. (This assumes that each averaging period is independent.) However, this is likely to be accompanied by increased false positives. For example, if we use an additional 5-minute evaluation period to rapidly detect large leaks, then the number of potential false alarms for this aggregator is proportional to the number of 5-minute periods in 1 year. A value of $\alpha = 0.001$ would expose us to an additional 105 false alarms per year. The message here is not that we should not pursue increased performance by using shorter time frame aggregators, but that the permissible Type I/II error value may have to be reduced to control false positives.
- Our simple LDS effectively uses a fixed sample size Z-test, where the sample size is based on the signal periodicity and the maximum time to detect. Sequential methods such as SPRT do not use a fixed sample size and can respond significantly faster than a fixed sample size test.
- Finally, the error model is *stationary* in that the parameters defining the leak signal randomness, such as the mean, standard deviation, and form of the error distribution (in this case normal), do not change over time. In a real-world LDS, the errors may be larger in transient scenarios, or if instrument error behaviors change abruptly. Consequently, a real-world LDS may change its threshold when a pump starts or stops, or if the pipeline flow mode changes (ie, the flow becomes slack at some location).

These issues are significant with respect to real-world systems. However, they do not alter the fundamental findings provided by our simplified LDS and error models. These are as follows:

1. There is no single number that can be used to evaluate LDS performance. Using a range of useful metrics provides greater insight into the value of the system.
2. Leak detection sensitivity and false alarms are not independent of each other. Changing one parameter will generally change the other, and sometimes in dramatic fashion. It is up to the user to determine desired trade-offs and develop an optimum solution.

3. More principled metrics such as leak detection event efficiency, leak detection rate efficiency, leak detection alarm efficiency, and LDS system efficiency provide greater insight and make it easier to find an optimum solution that will address operator needs.

9.3 LDS PERFORMANCE TESTING AND EVALUATION

The previous section demonstrated that pipeline leak detection system performance evaluation is multifaceted. Insofar as the various performance metrics are related to each other, minor changes in selected tuning parameters that are targeted in achieving system gains for one metric may have dramatic and undesirable impacts on other metrics. Beyond the obvious utility to be gained in optimizing performance to appropriately balance false positives and negatives, understanding LDS performance can be an important regulatory issue for the pipeline operator. As we discuss in Chapter 12, Regulatory Requirements, operators of hazardous liquid pipelines in the United States that implement one commonly utilized category of internal pipeline leak detection systems (CPM systems) are required to perform periodic performance testing of those systems (see API 1130) [1].

In this section, we investigate practical methods that can be used to evaluate the performance of CPM systems and other LDS categories. The primary methods that we discuss are commodity withdrawal testing, field point modifications, and software-based testing. After that, we turn our attention to the important topic of LDS tuning.

9.3.1 Commodity Withdrawal Testing for CPM Systems

CPM system commodity withdrawal testing is performed by removing commodity from the target pipeline system in an unmetered fashion (in the sense that SCADA and the LDS are blind to the withdrawal) and confirming that the leak detection system appropriately alarms the apparent leak. This methodology is identified [1] as one possible method for testing CPM systems (the largest category of internal systems as discussed in Chapter 12: Regulatory Requirements). Many operators and regulators tend to view this approach as providing the highest degree of confidence and confirmation that the LDS performs in line with expectations, and it cannot be argued that it is certainly gratifying to withdraw some quantity of oil or product from the pipeline and see the leak detection alarm actually generated in the pipeline control room.

It should be noted, however, that the commodity withdrawal method can come with some considerable baggage. Problems and issues that accompany this approach are the following:

1. This method requires the removed commodity to be temporarily stored during the test process and then reinjected into the pipeline at the

conclusion of testing. If permanent pipeline tankage is not available, then this means that vacuum trucks, tanker trucks, temporary skids, or other methods must be provided to enable the process.

2. Implementation of temporary storage may require hot-tapping or other modification of the pipeline. The number of locations where this can be easily implemented is necessarily limited by cost and accessibility issues, and this will limit the degree to which leak location capabilities can be assessed.
3. If permanent storage is used for the withdrawal, and if levels or pressures in the tanks are normally tracked by the LDS to perform any mass balance calculations, then these field measurements must somehow be disabled or modified during the test process.
4. In addition to complicating the test, any supporting field point modifications tend to contaminate data trends stored in historical data repository systems.
5. Storage volumes for withdrawn commodity are generally limited. This may limit the scope of testing in terms of testable leak rates or detection times.
6. If commodity is permanently removed from the pipeline, then meters and supporting custody transfer systems have to be adjusted after the test.
7. The test process is generally disruptive to normal operations. If performed too often, then it will also get in the way of the normal processing of the LDS, which is to look for leaks.
8. Because the number of withdrawal tests that can be practically performed is therefore necessarily limited, it is impossible to develop detailed performance maps of the type discussed earlier in this chapter.
9. Many modern leak detection systems utilize a large number of variables to determine the presence of a leak. In other words, they are far more sophisticated than the simple thresholding system we visited in Section 9.2. This means that even more SCADA points may have to be adjusted to enable the test than just tankage measurements.
10. If the number of field points that require modification to successfully execute the test are excessively high, then this may increase the chances that the withdrawal circumstances are not sufficiently similar to a real leak to have confidence in the outcome of the test.
11. Commodity withdrawal may be more difficult to implement for gas phase pipelines due to the need to accommodate higher pressures and because of the likelihood that larger volumes will have to be withdrawn to accommodate potentially longer detection times and higher effective thresholds required for leak detection in compressible commodity systems.
12. This approach is generally not applicable to non-CPM external pipeline leak detection systems discussed in Chapter 7, External and Intermittent Leak Detection System Types.

Because of the cost in terms of resources, time, and operational disruption, it is nearly impossible to use this approach to assist in tuning or to develop the kinds of performance maps discussed in Sections 9.1 and 9.2. Because of these issues, we tend to view commodity withdrawal testing as an essential method to perform limited confirmation of performance maps developed via other methods and to ensure that the fully implemented LDS is fully functional in an operational setting.

9.3.2 Field Point Edit-Based Testing of CPM Systems

A somewhat easier-to-implement alternative to commodity withdrawal testing utilizes field point modifications designed to represent the loss of commodity due to a leak. For a CPM system applied to a simple pipeline segment with inlet and outlet flow meters, pressure measurements, and temperature measurements, this would conceivably be achieved by applying a function in the SCADA system that would either artificially increase the upstream flow measurement or decrease the downstream measurement. With no other changes to the incoming field point data streams, this should cause an increase in the volume balance experienced by the CPM system. Assuming that the *VB* change is sufficiently large, this should trigger an alarm from the leak detection system.

As noted, this approach is generally easier to implement than a commodity withdrawal test. It avoids many of the complications of actually removing and handling commodity from the pipeline. However, the operator should note the following concerns with this methodology:

1. All of the concerns that apply to temporarily modifying field points in an actual withdrawal test obviously apply here as well. These include adjustment of custody transfer records as well as contamination of any long-term historical system data trends.
2. Modifying flow balance measurements alone may not be sufficient to test aspects of the LDS that go beyond simple leak detection. For example, in an RTTM capable of locating leaks, it may be necessary to alter upstream and downstream pressure measurements and to validate leak location capabilities. This is more likely to be required in pipeline systems with highly compressible media, such as natural gas, where pressure measurements may be key components of the leak detection methodology.
3. In larger pipeline systems with many internal flow, pressure, temperature, and other measurements, more sophisticated leak detection installations will utilize complex calculations ensuring internal consistency of a possible leak in one segment with the expected measurement changes in other pipeline segments before issuing an alarm. This means that the operator must modify even more field points in an effort to "fool" the LDS. If this is not done, then the LDS will (appropriately) tag the imbalance as a measurement problem.

4. This methodology is also much more difficult to apply to an operating rarefaction wave LDS, particularly if the rarefaction wave system uses dedicated components that are totally independent of the SCADA system.
5. As with withdrawal-based testing, this is a test on an operational system that limits the number of test points that can be collected.

Although generally easier to implement than commodity withdrawal testing, the limitations of dealing with an operating pipeline make this method complicated and limiting.

9.3.3 LDS Software-Based Testing

The previous two sections have illustrated two methods that are designed to perform LDS testing in the operating environment. In general, we observe that while commodity withdrawal and field point modification-based testing provide certain strengths in terms of providing confirmation that performance goals are met in a fully implemented operational setting, they are limited in terms of meeting other goals. This is primarily due to the potentially disruptive impact of performing tests or evaluations in an operating environment. This will tend to limit the ability of these methods to support development of detailed performance maps or as an assist in LDS tuning.

We now turn to a different approach, which is to utilize software-based offline testing. In the United States, API 1130 [1] permits CPM system testing through the use of editing CPM configuration parameters combined with accompanying software simulations designed to simulate commodity loss or a desired hydraulic condition. In this section, we discuss an offline software infrastructure that permits the calculation of performance metrics and maps as discussed in the previous sections of this chapter.

To perform our desired testing, we assume that we have available a copy of our LDS that can be run offline in an iterative fashion to evaluate a set of performance runs. For each performance run, we will configure the offline instance of the LDS through one or more sets of extended recorded or simulated (or both in combination) pipeline operational data. The cases we run will be divided into two categories:

1. Baseline case or cases designed to simulate normal operation of the LDS in the absence of a leak. The primary purpose of these cases is to determine the baseline false alarm rate under some specified tuning configuration.
2. Cases with LDS inputs consistent with imposed leaks but retaining the uncertainties, measurement errors, and modeling errors of the baseline cases. Imposed leak cases are designed to test leak detection sensitivity, leak location, instrument error analysis, and other features of the leak detection system, again under a desired leak detection configuration.

It is easiest to develop baseline performance by utilizing field data trends archived by the SCADA system to an appropriate historical database. It is essential for the scan frequency and detailed recorded data values to be retained by the database, and no form of lossy compression can be used to minimize storage. The archived data must also retain any and all data bits or information used to specify data quality if such information is used by the LDS.

A more refined methodology is to "strip" the error trends from the data by some means and then retain them in the form of field and model point error trends that do not incorporate the original ideal or "true" process values. This method provides the advantage of being able to utilize a new schedule of pipeline hydraulic operation that may be different from when the pipeline data trends were originally recorded. The noise is re-imposed when the leaks are simulated. In such a case, a hydraulic transient model would also simulate a new operational scenario, and the recorded errors would be added to the modeled scenario outputs. This frees the operator from being tied to the past operation. This approach may also be required in very compressible systems to ensure that physical transmission lags are properly addressed.

Extraction of errors may use very simple methods, such as simple subtraction of an idealized volume balance based on error-free inputs from the VB trend utilizing recorded inputs. Alternately, more sophisticated statistical methods based on state estimation approaches [2] can be used.

It is also potentially possible to evaluate performance using uncertainty-based "first principles," whereby the error behavior for all field instruments as well as model parameters is applied based on meter/instrument vendor information and proving records, and engineering judgment is used to drive the error stream. The error trends are then combined with the output of a pure pipeline hydraulic simulator, and the result is then fed to the offline LDS to develop baseline performance.

It is worth noting that the development of useful error probability distributions, random periodicity or autocorrelations, and other behaviors from first principles is nontrivial. This is because vendor information does not typically address long tails and outliers caused by data communications problems, stuck transmitters, random drift, and off-spec operation in complex devices like multipath ultrasonic flow meters. This can make it very difficult to develop reliable predictions of false alarms, as previously noted [3]. For this reason, the authors have tended to rely on judicious use and manipulation of in situ recorded signals that, if obtained over a long enough period, include sufficient error information to make reliable false alarm predictions.

Imposed leak cases utilize a schedule of imposed leaks as well as the underlying error information used in the baseline cases. In simple cases, the volume balance impact of the leaks can be combined with a recorded data or error stream without reference to any simulation process and then played directly to the offline LDS. In other cases (particularly those involving transient flow), the leaks are imposed as simulated offtakes in a pipeline transient simulator.

The simulator output is then perturbed with the error signals and the result is sent to the LDS. Each imposed error is run for a simulated period until it has effectively played through the evaluation period, t_{EVAL}, at which point the leak is labeled as being not detected, or until it is alarmed by the LDS and labeled as detected. Development of a detailed leak detection sensitivity map and other performance information is then performed by consolidating all of these cases.

Because of the need to impose a large number of leak cases, the operation of the performance analyzer places a significant premium on the ability of the LDS to run in the offline mode at a very high simulation rate. The ability to run cases either at a high simulated-to-real time ratio or on parallel processing architectures, or both, is a significant advantage. Fortunately, increases in the performance of both computer hardware and software are making this approach more and more feasible.

It should be clear that, even with current automation and hardware, the performance analysis can be complex and time-consuming. However, this type of analysis has proven to be feasible [4] and is slowly becoming part of the repertoire of commercially available pipeline leak detection systems.

9.4 LDS TUNING

In addition to performing testing, the software-based LDS performance analyzer described in the last section can be an invaluable for system tuning. LDS tuning refers to the process of developing an optimum set of leak detection parameters (such as a threshold or components used to calculate a threshold). Performing this process in an online system is often a very tedious and inefficient process, because: (1) a modern LDS may have a large number of adjustable parameters (such as the threshold value) that contribute to the performance; (2) the relationship of the parameters to the actual leak detection and location performance maps may be quite obscure; and (3) following a configuration adjustment, it will take time for changes in the baseline alarm rate to become apparent. In addition, online tuning involves trial-and-error tinkering with a complex operational support system, with inevitable impacts to the efficiency of the operation.

A software-based performance analyzer can improve this process considerably. The authors have found that as long as sufficient recorded data are available, the performance sensitivity curves obtained by this process are remarkably stable from year to year, as long as the pipeline installation/design and configuration parameters are not changed. Achieving this goal, however, assumes that the operational tuning data are recorded over a long enough period that they address a very large fraction of the important operational modes of the pipeline. For seasonably operated systems, this may require a year or more of data.

With respect to the baseline alarm rate, and as noted, the API 1149 Technical Report [3] indicates that prediction of false alarms is very difficult due to the importance of outliers, probability distribution fat tails, time-based data correlation, and the difficulty of obtaining this information. The authors have found that false alarm rates for an installed LDS are generally quite predictable if certain guidelines to prevent overfitting are followed, and as long as the error stream is well-grounded in data trends obtained during actual operation of the pipeline. If this kind of data is not available, then the user should approach the problem of tuning and predicting baseline alarm rates through the use of the performance analyzer with great caution.

As noted in the last paragraph, there is one issue that LDS users responsible for tuning should be very aware of, and that is the problem of model *overfitting* or *overtuning*. This phenomenon is common to complex classification systems with many tunable parameters, such as neural nets, but it is not limited to these systems alone. Overfitting is analogous to the problem of fitting a regression polynomial to a set of data points. A "better" fit to the data can always be achieved by applying more "tuning" constants or parameters as a result of increasing the order of the polynomial fit. However, such an improvement is likely to be achieved by effectively tuning to what is really random noise. Consequently, if the noise source is changed (ie, the tuned system is run against a new data set or period of pipeline operation), then the false alarms will increase and potentially skyrocket when compared to the number of alarms observed during the tuning process.

An example of how this works is shown in Fig. 9.6. Let's assume that we have two tuning parameters that are used to set or calculate the leak detection threshold. They could be measured volume balance and time, for example, but it could be any two tuning parameters that effectively partition the classification space for the leak and no leak categories. On the left, we see the result of using a high-order threshold scheme for our tuning set. This aggressive scheme correctly partitions the tuning space so that we capture every applied leak without experiencing any false alarms. Note that the partition boundary has a large number of twists and turns to accommodate irregularities observed in the tuning set. A less aggressive and lower-order scheme would have incorrectly missed four of the applied leaks and, in that sense, performed poorly in comparison to the higher-order threshold.

Now consider what happens if we test the high-order threshold on a new data set on the right side of Fig. 9.6. Because of new noise in the data, the pattern of perceived leaks has changed slightly. As a consequence, the zigs and zags of the high-order threshold are now in error. We now have three new false alarms and three new false negatives. Thus, application of the high-order threshold developed by the tuning set failed to predict the performance under the new data set because the high-order boundary was tuned to random variations in the data. This randomness changes in the test set. The new false negatives will barely be noticed in performance evaluations

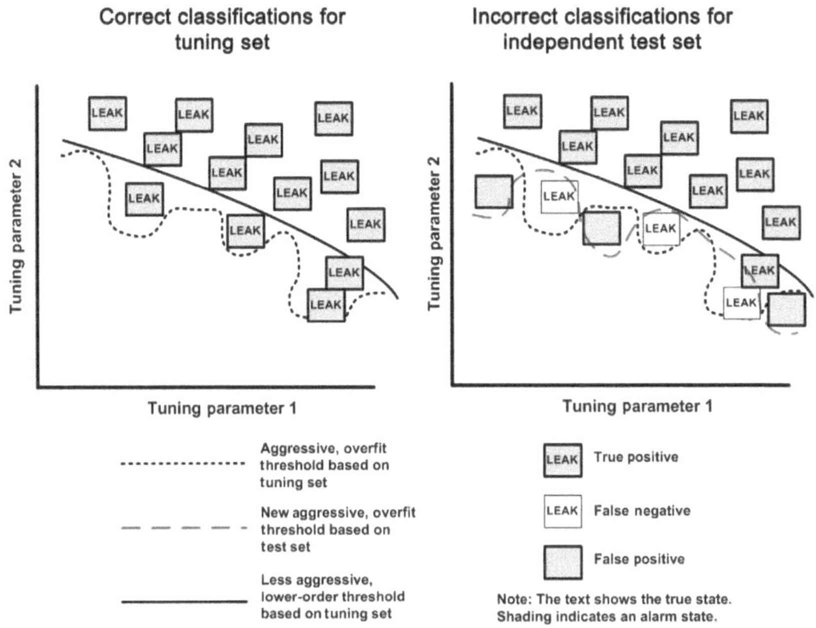

FIGURE 9.6 LDS threshold overfitting.

because, as we have seen previously, they will tend to shift the leak detection event confidence curves very slightly. The impact on false positives, however, may be noticed, especially if the number increases substantially. We could, of course, develop a new classifier boundary for the test data, but it will be subject to the same defects because it is set by new and unrepeatable randomness in the data.

Note that the performance under the less highly ordered threshold did not really budge: we still miss four real leaks and have no false positives. Thus, the less highly ordered threshold calculator developed by the tuning set has more *predictive* power than the highly ordered threshold.

What we need to do is *regularize* our decision boundary by penalizing unnecessary complexity. There are several methods discussed in the literature to achieve this, including ridge regression and Bayesian regularization, most of which are not supported by existing leak detection products [5,6]. However, one relatively easy method that can be used to avoid this unpleasant outcome is to use early stopping. We reserve a portion of our data as a test data set when using the LDS performance analyzer to tune the leak detection system. While tuning to the remainder of the data (the tuning set), we periodically check our performance against the test set. If the baseline alarm rate that occurs when the test data set is run through the performance analyzer starts to increase so that it is higher than the rate that was observed when tuning to the tuning data set, then the model has been overfit, and the tuning process should be stopped or repeated with a less aggressive or smaller set of tuning parameters.

As long as the cautions we have provided here are heeded, we feel that a well-designed LDS performance analyzer can be a great aid both in testing a pipeline leak detection system and in assisting in the LDS tuning process [7]. We note that performance analyzers have not been part of the tool sets provided by LDS providers in the past, but this situation is starting to change. We expect this trend to gain momentum in the future.

9.5 PERFORMANCE STANDARDS

Aside from a few relatively limited jurisdictions (see Chapter 12: Regulatory Requirements), there are virtually no accepted industry-wide standards or requirements for any of the LDS performance metrics listed in Table 9.1, or for the derived metrics discussed in the last section. However, it is worth noting the role of API Technical Report 1149 [3]. This guideline, which is primarily oriented around CPM systems, provides a mechanism for analyzing the performance of an internal LDS based on the expected instrument (and, if applicable, model) error and transient performance uncertainties. Much of API 1149 is dedicated to understanding and properly analyzing the uncertainties associated with various measurements, but the overall methodology recommended to utilize these errors is not inconsistent with the methodology described in Section 9.3.3. The principle differences are the emphasis we put on using in situ measurements of the baseline errors and the importance of the various performance metrics.

REFERENCES

[1] American Petroleum Institute Standard 1130. Computational pipeline monitoring for liquid pipelines; September 2007.

[2] Modisette JP. State estimation in online models. In: PSIG annual meeting, Galveston, Texas; 12 May–15 May 2009.

[3] American Petroleum Institute Technical Report 1149. Pipeline variable uncertainties and their effects on leak detectability; September 2015.

[4] Carpenter PS, Nicholas RE, Henrie ME. Accurately representing leak detection capability and determining risk. In: PSIG annual meeting, New Orleans, Louisiana, 26 October to 28 October 2005; 2003.

[5] Hagiwara K, Kuno K. Regularization learning and early stopping in linear networks. In: IEEE-INNS-ENNS International Joint Conference on Neural Networks (IJCNN'00), vol. 4, p. 4511; 2000.

[6] Dougherty ER. Probability and statistics for the engineering, computing and physical sciences. Prentice-Hall; 1990.

[7] Carpenter PS, Henrie ME, Nicholas RE, Liddell P. Automated validation and evaluation of pipeline leak detection system alarms. In: PSIG annual meeting, New Orleans, Prague, Czech Republic, 16 April–19 April 2013; 2013.

Chapter 10

Human Factor Considerations in Leak Detection

In this chapter we discuss the human factors involved in the leak detection system. For the purpose of this book, human factors include the design, implementation, and maintenance activities that are associated with the physical, mental, and work load aspects of how controllers interact with leak detection technologies within their working environment. In the first part of this chapter, we discuss the interaction between the technology side of leak detection and the people who must make decisions based on the outputs of that technology.

For us, human factors also encompasses the direct discovery or detection of pipeline commodity leaks and spills by people who are independent of the technology processes. Human leak detection systems comprise an important topic that is directly related to the holistic leak detection system approach that all pipelines should utilize. We discuss this in the second part of this chapter.

10.1 THE HUMAN–MACHINE SIGNAL DETECTION CONTROL LOOP

Let's begin by discussing the human factors element in the context of a complex hazard detecting system using signal detection theory and its interaction with industry-recommended control room management (CRM) practices in the United States of America (USA).

10.1.1 Diagnosing Alarms in the Face of Uncertainty

Leak detection systems are hazard detection systems. In line with our discussion in Chapter 5, Statistical Processing and Leak Detection, the leak detection system (LDS) obtains continuously updated information from the operating pipeline via one or more input channels that combine through various means to develop a real-valued hazard signal. The calculated hazard

Pipeline Leak Detection Handbook. DOI: http://dx.doi.org/10.1016/B978-0-12-802240-5.00010-8

signal is then compared to a threshold, at which time an alarm (or warning) is issued if the calculated result has met or exceeded the threshold [1].

However, this is not the end of the story. In the context of pipeline operations, leak detection systems include not only various technologies and software applications but the human element as well. The merging of technology and human entities forms a complex system. Detecting and responding to pipeline leaks therefore requires optimal integration of the human element within this system.

Key controller leak detection interaction requirements include monitoring and responding to changing conditions, events, and alarms as well as various leak detection information displays. We refer to this as the "Archimedean Point," or the position from which the human, in this case the controller, can take in the whole situation and determine the state of the system within the context of the alarm or warning state.

Leak detection systems are thus decision-making systems as the output of these systems results in further actions. The technology portion of the system calculates the probability that a leak may be present using the potentially limited input information available to it and then creates an alarm. This alarm contains elements of reliability, accuracy, and uncertainty. The human portion of the system must then make a diagnosis of the validity of the leak alarm using the information available to the LDS plus other operational information not available to the LDS, which in and of itself, may also not be fully reliable, accurate, and available. Another way of describing this process is that it is a signal detection system that can be modeled according to signal detection theory (SDT). SDT is based on the independent aspects of sensitivity and decision criterion performance [2].

To place SDT in the context of the human factors discussion, the National Transportation Safety Board (NTSB) studied 13 hazardous liquid pipeline accidents. Of these 13 incidents, the NTSB determined that 10 of these accidents involved a delay between recognizing that a leak had occurred and initiating efforts to reduce the effect of the leak [3]. The resulting delay was found to be a consequence of how the leak is graphically displayed, how alarms are managed, the level and adequacy of controller training and controller fatigue, as well as the leak detection system functions, features, and capabilities. Each of these identified elements are associated with the human factors portion of the LDS, SDT, and CRM systems.

To present an extended picture of the human factors portion of the leak detection system, this chapter provides a range of discussions on human factor–associated functions, features, and capabilities such as:

- the psychological, behavioral, and ergonomic aspects of the human factors portion of the system
- how leak alarms and statuses are presented to the controller
- how the human element component of the leak detection system interacts with the displayed information

- how these systems fit within regulatory CRM requirements
- how the pipeline owner/operator can merge or include leak alarms and status within the organization's control room determination and rationalization process
- how nonleak alarms impact the human element and factors that enable and hinder this portion of the leak detection system
- the level and quality of leak detection system diagnostics available to the controller

10.1.2 Human Factors in the Control Room

"Human factors," as used in this section, is a descriptive phrase addressing human and technology interaction. Let's consider the various aspects of this interchange, including psychological interactions, human behavior modes, and ergonomics.

Psychological aspects take into consideration the human experience associated with the interaction of these systems. Stated another way, they address how the control room controller interacts with their work environment. Some of the key elements of this environment include the leak detection system, supervisory control and data acquisition (SCADA) system, and other personnel who may be in the room, such as supervisors, other controllers, and so forth. The overall environment is highly dynamic and continually impacts and modifies the controller's work load.

Another factor that contributes to the environmental aspect of the system is the amount of information provided to the controller. SCADA systems, telecommunication systems, and leak detection systems can develop and transfer large quantities of data. As such, these systems can present more information than any individual can ever fully understand or possibly respond to. These systems have also been identified as hindering the ability of the controller to develop better comprehension of the fuller operational picture. The ability to fully understand what is occurring in one's environment is referred to as situational awareness. It has been shown that as the level of available information increases, there is a tipping point where the observer suffers information overload and can no longer develop a full situational awareness. When this occurs, more information increases the controller's psychological load and hinders rather than helps the decision-making process.

Leak detection systems are known to significantly contribute to the controller psychological load because they:

- Generate leak alarms that must be responded to in certain time frames
- Have an extensive range of different system status and alarm information that is available
- Require time-based, high-consequence responses based on uncertain knowledge

The leak detection system requires direct interaction from the controller regarding responding to leak alarms, system event notifications, completion of alarm attribution reports, potentially initiating spill response activities, and so forth. This interaction is referred to as the human-computer interaction process.

Human–computer interaction (HCI) research originated in the 1980s with a focus on understanding and improving computer system usability. One outcome of this research is identification that human-to-technology interaction is based on skill-based, rule-based, or knowledge-based behaviors. Of the three behavior types, human factors involved in leak detection deal predominately with rule-based behavior. That is, the controller applies rules learned from experience and training on how to respond to the leak detection system alarms, status, and event notifications.

Rule-based behavior is predominately grounded in the application of specific training and written procedures in response to a trigger event. In the case of the leak detection system, the controller would implement required procedures in response to a leak alarm trigger event. An example of this is shown in Fig. 10.1.

As outlined in Fig. 10.1, the sequence of events is serial in nature. Everything starts when the leak detection system determines that the rules to generate a leak alarm have been satisfied. Once the alarm occurs, the controller is to follow established procedures to make a determination of whether the leak alarm is valid or invalid, or if more information is required. Even in the best of times, the sequential and procedural-based response takes time.

This process is one of decision making based on uncertain information and can be modeled using SDT. SDT quantifies the human element of the leak detection system's ability to discern information-bearing patterns and to determine if the presented leak alarm is valid. Table 10.1 presents the four distinct decision states that are possible from the leak detection technology portion of the system.

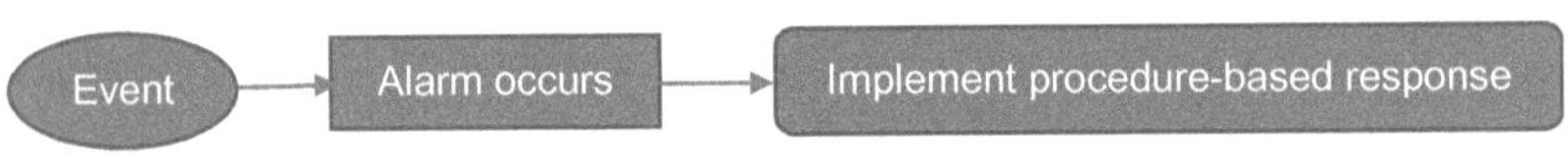

FIGURE 10.1 Rule-based process example.

TABLE 10.1 Leak Detection Technology Alarm Matrix

	Leak Present	No Leak Present
Leak Detection System Alarm	True alarm	False alarm
Leak Detection System No Alarm	Miss	True no alarm state

Based on the leak alarm state presented to the controller, the human element then must make a decision by selecting one of the states shown in Table 10.2. Note that the worst outcome occurs if the controller invalidates an alarm when there is an actual leak. In this situation, there is no remedial action (no pipeline shutdown, no field deployment, no field remediation, and a continued leak). We presume the leak will eventually be detected by other means but the resulting spill will be larger. We return to this issue later in this chapter.

As discussed previously in Chapter 5, Statistical Processing and Leak Detection, the goal of the combined LDS alarm/controller decision-making process is to determine if the leak is absent (hypotheses one (H1)) or present (hypotheses two (H2)). If we simplify the LDS operation by looking at it as a simple threshold, then the alarm process works as shown in the SDT model (Fig. 10.2).

The curve on the left is the outcome of the leak detection technology process assuming that no leak is present. This curve represents the normal "noise" associated with the overall process. The vertical line is the alarm threshold. The curve on the right incorporates all the noise associated with

TABLE 10.2 Leak Detection Human Element Alarm Matrix

	Controller Validates the Alarm	Controller Invalidates the Alarm
Leak Detection System True Alarm	True alarm	Miss/leak goes undetected
Leak Detection System False Alarm	False alarm/inappropriate responses	Correct rejection

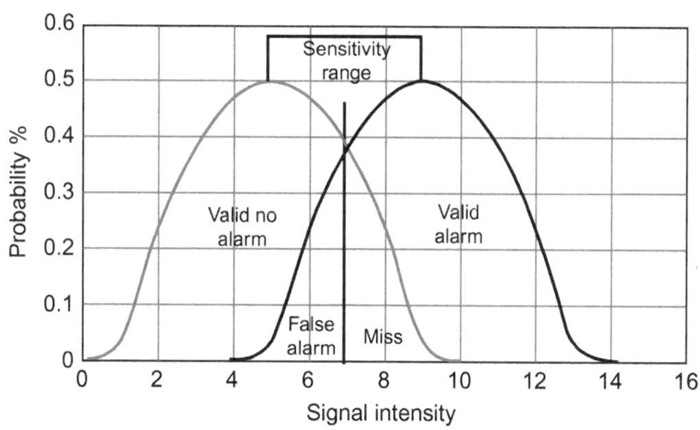

FIGURE 10.2 SDT decision curve.

the no-leak state plus the leak signature. The sensitivity of the system is the difference between the maximum points of these two curves.

The degree of overlap in this figure is an indication of the sensitivity of the systems to false positives and negatives. The relationship of the two curves is driven by field instrument quality, repeatability, reliability, and accuracy, as well as by leak detection algorithm system errors. It will be up to the controller to compensate for this uncertainty by bringing more information and reasoning regarding the problem of assessing the LDS alarm.

As noted previously, any response to leak detection alarms takes time. It also adds to the controller's workload and overall situational awareness requirements. Management of this workload requires consideration of the controller's ergonomic environment as well as the CRM regulatory requirements and industry-recognized best practices, as discussed later.

The ergonomic aspects of the human factor element involve how the person directly interacts with the system. In most operator control room environments, the first line of leak alarm determination typically resides with the controller. Occasionally, the first line of direct interaction is assigned to a leak detection specialist who monitors the leak detection system operation, performance, and operational events, rather than the operator.

The following section briefly discusses the ways these systems may be integrated into the control room environment. The variation of installations largely depends on the specific type of leak detection system that is used and the operator's philosophy and CRM program.

10.1.3 Data Display, Presentation, and Integration

Leak detection systems are generally vendor-supplied and come in many forms and types, with a range of different alarm and information display capabilities. Although some of the user interface design differences are driven by the vendor's unique desire for a specific look and feel, much of it is driven by the needs or desires of the pipeline operator.

Many interface design decisions are based on the fact that humans are visual entities. As such, we generally derive more information faster and at a greater degree of initial accuracy based on visually/graphically presented data and information rather than by reams of text-based tables. However, text is much easier to export to external applications for further analysis. Leak detection system developers and engineers have maximized flexibility by providing both visual and text-based alarm and system status information. That said, most systems provide initial alarm information in the form of visual displays.

Therefore, the context in which the system is operating is based on the fact that each leak detection vendor has a unique approach to presenting the system information to the end-user and on the need to reconcile that design with the end-user operating philosophy. The combination of the technology capabilities and operating philosophy contributes to or guides how these systems are implemented.

One display approach simply uses the vendor product as a standalone leak detection system. With this type of LDS configuration, acquisition of data, computation, and system status and alarm presentation are self-contained. As such, the interactions of the control room operator occur through the LDS human–machine interface (HMI). The HMI generally is self-contained and configured to show leak alarms and events as well as to provide a text-based alarm log. Depending on the type of leak detection system and vendor, the HMI may also provide other capabilities such a presentation of user-selected trends and profiles, field instrument status information, leak detection model status, and so forth. This system type requires the controller to monitor and control the pipeline and the separate leak detection system.

The standalone LDs requires the controller to monitor at least two HMIs. The challenges of locating these HMIs involve successful integration of separate and distinct ergonomic designs as well as operating philosophies. In general, borrowing from airplane instrument flying rules, the controller should scan these systems in a sequence that ensures they are always looking at the most meaningful information. But what is the most meaningful information? Meaningful information is derived from the specific and unique pipeline physical environment, control environment, and operating philosophy. Regardless of the environmental considerations, rule number one is to scan those instruments that maintain safe operation first and most frequently. The controller then extends the situational awareness scans to include the overall system operation, such as the leak detection system.

The advantage of this type of installation is that the leak detection system is not dependent on any other system. Consequently, a failure in one does not impact the other. The negative aspect is that it increases the controller's workload because it requires a completely different HMI to be taken into consideration.

Another type of leak detection system integration is to provide leak detection alarms and information as a SCADA contained component. A SCADA-based pipeline pressure and flow rate monitoring system, such as a deviation alarm system, is one such type. In this system design all leak alarms and leak detection system information are fully integrated within the operator's SCADA system alarm and status information, philosophy, and CRM plan.

An advantage of this approach is that the controller is interacting with a single HMI and alarm system. All information is available within a single context and ergonomic setting. The controller is not required to shift between a SCADA HMI and the leak detection system user interface for information gathering and analysis efforts. A negative aspect of this system, however, is that the leak detection system is fully dependent on the SCADA system. Any error or problem in the SCADA system can impact the leak detection system. This negative aspect is probably minimal because most SCADA systems are fully redundant and well-tested.

Another downside of the fully integrated system approach is that the range of leak detection analysis tools, such as detailed trends, profiles, and so forth, is usually not as extensive as that of dedicated leak detection system or hybrid leak detection system capabilities.

A third (and probably the most common) system installation type is the integrated hybrid system. This approach involves tightly coupling the leak detection system data gathering and derived information with the SCADA system. In this design, the leak detection system typically acquires field data from the SCADA system and provides any generated system status and alarms back to the SCADA system. From the point of view of the controller, system status and leak alarms will be displayed on both systems. However, in this configuration the two systems remain computationally independent.

Computationally independent leak detection systems require pipeline information inputs such as pressures, flow rates, temperatures, valve status, and so forth. To obtain the required inputs, the CPM system communicates with the SCADA system. As the SCADA system obtains the latest field information, it then transmits the required data to the CPM system. While the CPM system is receiving field information from SCADA, the leak detection system will send leak alarm and other system data to the SCADA system for display and annunciation.

Thus, these systems are tightly coupled in that information flows from SCADA to the CPM, and alarms and status information flow from the CPM to SCADA. Once the alarm and system status information has been transmitted to the SCADA system, this information is handled according to the pipeline operator's CRM plan.

Positive aspects of the tightly integrated system are that the controller receives all critical alarm information on the SCADA HMI. This supports an ergonomic design that does not require the controller to scan and be aware of different HMIs. This integration approach also capitalizes on the leak detection system's advanced analysis capabilities such as detailed information trends, profiles, and extensive modeling information.

The negative aspect of this approach is that the controller must interact with a second system for detailed leak alarm analysis. Although the leak alarm is annunciated via the SCADA system, if a detailed analysis is required, then this occurs on the leak detection HMI. While engaged in this analysis, the controller could lose overall situational awareness of pipeline operations by being engrossed in analyzing the leak alarm supporting information. Examples of this type of situational awareness loss can be found in airline and pipeline incident investigations. Any time when the controller is not focused on the SCADA HMI, there is an opportunity to lose full pipeline operational situational awareness.

Integration of pipeline leak detection systems into the SCADA control room environment is also complicated by the number of leak detection systems that are deployed. Often, operators utilize two or more different leak detection technologies on a single pipeline system. As an example, the pipeline may have a CPM system that is tightly coupled with the SCADA system. At the same time, they may have a pressure and flow deviation leak detection application monitoring for major pipeline failures as part of the SCADA system alarm and monitoring function. Alternately, they may have a rarefaction wave system installed across a short section of high-consequence area (HCA) pipeline. In this situation, the control room environment has multiple leak detection systems that are integrated into the environment in different ways. The CPM would be tightly integrated with the SCADA system and the pressure/flow deviation alarm system would be part of the SCADA system, but the rarefaction wave system could be a separate system with its own HMI.

The decision on how to integrate the selected leak detection system or systems into the control room environment should be based on:

- The pipeline-specific CRM plan, as discussed in the next section
- The type of leak detection system or systems under consideration
- The functionality available within the leak detection system

The following section discusses how human factors involved in leak detection are a function of CRM, regulatory requirements, industry standards, and recommended practices.

10.1.4 CRM Regulatory Requirements, Industry Standards, and Recommended Practices

This section presents a discussion of human factors involved in leak detection in the context of CRM regulatory requirements, industry standards, and industry best practices. In the United States, federal and state regulatory agencies have a long history of establishing guidelines and requirements for safe operation of hazardous liquid and natural gas pipelines. As an example, United States Public Law 109-468-Dec. 29, 2006, Section 60137, requires the following: "... each operator of a gas or hazardous liquid pipeline develop, implement, and submit ... a human factors management plan designed to reduce risks associated with human factors ..." [4]. Other countries, such as Germany and the United Kingdom, have taken major steps in regulating hazardous material pipelines to increase the safe operation of these symptoms as well.

USA Federal regulatory human factors requirements are found in United States 49 CFR (Code of Federal Regulations) Part 192 and US 49 CFR Part 195.

Portions of these CFRs require the pipeline operator of natural gas, other gas, and hazardous liquid pipelines to have specific requirements for safety-related alarms, such as LDS leak alarms. Such requirements include the need to ensure that the alarms are accurate, are initially and periodically reviewed; they must also support safe pipeline operations. Other regulatory requirements define how the alarms are monitored and how the owner must ensure that a management-of-change process is in place and followed. As noted previously, leak alarms are safety-related alarms and are included within the scope of these regulations.

Regulatory requirements also support and are part of the control room operator rule-based behavior. That is, regulations require a written alarm management plan that includes and provides for effective operator response to these leak alarms. The alarm management plan is inclusive of the corporate policy and procedures, which constitute the leak detection system's human factors rule-based behavior.

PHMSA has also issued the Pipeline Safety: Control Room Management/ Human Factors Rule [5]. This rule requires several distinct actions, such as:

- Have and follow written CRM procedures
- SCADA displays must meet API RP 1165
- Implement measures to prevent fatigue
- Develop, implement, and maintain a SCADA alarm management plan

Other nations have leak detection regulatory requirements that are applicable within their regions, such as:

- Germany: Technische Regel fur Rohrfernleitungen (TRFL)
- Brazil: ANP's Technical Regulation of Pipelines for the Transport of Petroleum, its by-products, and Natural Gas (RTDT)
- Great Britain: Pipelines Safety Regulations

These regulations are not as explicit about requirements for human factors. Reference Chapter 12, Regulatory Requirements for a broader discussion of various international leak detection regulations.

The oil and gas industries have also developed and institutionalized the following standards and recommended best practices:

- ANSI (American National Standards Institute)/ISA (The Instrument, Systems, and Automations Society) 18.2-2009
- API (American Petroleum Institute) Recommended Practice 1165
- API Recommended Practice 1167
- API Recommended Practice 1175

The referenced standards and best practices are designed and structured to provide guidance regarding the development, design, installation, and management of control system displays, alarm systems, and a leak detection program. The focus is to structure the identification, selection, display, response, and maintenance of safety-related leak alarms. Thus, those who are responsible for the monitoring of safety system alarms can do so in an efficient and effective manner.

As noted previously, issues regarding human factors in the control room, such as creating delays to assist in determining when a leak has occurred, have been found to contribute to and exacerbate the effects associated with leaks and resulting spills. The identified regulations, standards, and recommended practices are intended to provide a consistent structure for how owners/operators monitor, manage, and respond to safety system alarms, including leak detection systems leak alarms.

The following section expands on the control room alarm management processes associated with the safety-related leak alarms.

10.1.5 Alarm Management Overview

As presented in API 1167:

> *. . .operators utilize "leak detection systems" systems to detect and alarm possible leaks. Such alarms have high importance. Depending on the sophistication of the leak detection method employed, "false alarms" may be generated. These are often due to combinations of communication problems, sensor calibration metering and telemetry uncertainty, and pipeline transients. [6]*

Modern technology and telecommunication systems have advanced to a point where it is all too easy to send an extensive number of alarms and statuses to the controller. A fundamental issue with the controller is flooded with alarms and status information, thus contributing to a loss of situational awareness. In short, displaying and/or annunciating too many alarms and statuses can potentially cause controllers to experience information overload, which significantly detracts from their ability to effectively and efficiently monitor and operate the pipeline. A worst-case outcome of this is that safety system critical alarms are not responded to in a timely manner. It is imperative for annunciated indicators to be presented such that they can be quickly and easily identified.

It is an industry best practice and, within the United States, a regulatory requirement (US Code of Federal Regulations 195.446), that operators must have a CRM plan that includes an alarm management plan. Under this regulation, the operator must have and follow a set of CRM procedures. In conjunction with the regulatory environment, the American Petroleum Institute developed an alarm management recommended practice, API 1167.

As outlined in API 1167, an alarm management plan is inclusive of a well thought-out and developed alarm philosophy. The operator alarm philosophy becomes the guiding force behind the control room alarm management and a significant contribution to the overall control room workload management requirements. It also defines what constitutes alarms versus other data, such as system status or system information, as well as how to define, design, implement, maintain, monitor, and test an alarm system.

The alarm management philosophy becomes a structural component of the overall CRM program. This philosophy provides the framework that supports required tasks such as alarm documentation and rationalization. The alarm documentation and rationalization (D&R) is a structured process approach by which alarms are determined, prioritized, and documented. The D&R process also defines the required alarm response times.

The overarching objective of these regulations, standards, and best practices is to ensure that the controller receives the essential information to maintain a clear situational awareness of pipeline operation. Information that is not required to support this need is placed at a much lower level or suppressed such that it does not create information overload.

10.1.6 Balancing Sensitivity and False Alarms

Leak detection systems derive commodity imbalances and present leak alarms to the operator. The largest percentage of these alarms is not associated with a valid commodity release. These are referred to in the pipeline industry as "false alarms," "false positives," or "nonleak" alarms. Another way to look at false alarms is that these are alarms that the controller receives that do not correctly and accurately reflect the pipeline's actual state, operating conditions, or system status.

The generation of leak detection system false alarms are a result of many things, such as the type of leak detection system in use, system tuning, hydraulic transients caused by pump startups and shutdowns, valve position changes, relief actions and other normal operating changes, instrument errors, telecommunication uncertainty, modeling errors, and so forth. As an example of how the type of leak detection system selection impacts the level of false alarms, experience demonstrates that properly tuned rarefaction wave systems are less prone to false alarms than other internal leak detection systems. This is associated with uncertainties regarding inputs to the system and system timing aspects. As discussed in Chapter 6, Rarefaction Wave and Deviation Alarm Systems, these systems minimize the likelihood of false alarms because they will only trigger if two specifically characterized pressure disturbances occur within a small window of time.

As compared to negative pressure wave systems, mass balance systems tend to generate more false alarms. This occurs as these systems generally require a significantly increased number of inputs, many more pressure sensors, flow meters, temperature sensors, and so forth. Each of these field inputs have a level of uncertainty and error that may not be fully independent. As such, the level of uncertainty and error that the rarefaction wave system must deal with is significantly less than the that of the mass balance systems, which directly impacts the number of false alarms which may be generated.

As noted, false alarms are also a function of pipeline transient conditions. As pipeline operation transitions from one state, such as a steady state, to another state, such as a shutdown state, it is in hydraulic transition. Operational changes, such as starting and stopping pumps or compressors, are other sources of transient hydraulics. When this occurs, the pipeline starts to pack or unpack due to the dynamics of the changing pipeline state. To some leak detection systems, a line that is unpacking can appear to have a leak. This would result in the generation of one or more false alarms. Mass balance systems that use real-time transient models are designed to cancel out these effects. However, certain hydraulic states, such as slack line or other multiphase conditions, can be difficult to model accurately.

Consequently, even leak detection systems that use RTTMs may exhibit higher rates of false alarms when these conditions are present.

Another source of false alarms is bad or noisy field instrumentation. Because all internal-based leak detection systems utilize field data, such as flow meters, pressure instruments, and temperature instruments, the ability to correctly detect a leak is grounded in the reliability, repeatability, and availability of these data. If the field instrument information changes rapidly or drifts over time, then the internal leak detection system may conclude that a leak is present rather than that the system is dealing with bad information.

The point is that there are many reasons why a leak detection system may generate a false alarm. The occurrence of false alarms is usually viewed in a negative light, but they may also be viewed positively under some circumstances.

One reason why false leak alarms are viewed negatively is that they require controller time to respond and analyze the leak indication. This adds to the control room work load that could have been applied to other efforts when the leak alarm is false.

Another reason why false alarms are viewed in a negative light is that they erode controller confidence in the leak detection system. If the leak detection system generates many false alarms, then it can be viewed as an instance of Aesop's fable *The Boy Who Cried Wolf*. In this fable, the shepherd boy repeatedly cries out that the wolf is attacking when no wolf is present. The villagers stop believing the shepherd boy due to his large number of false alarms. Consequently, when a wolf does attack, the villagers do not respond and the sheep are lost. The same can and does occur in control room situations when an alarm repeatedly goes off but provides a false indication. After some time, the controller can be desensitized to the alarm and fail to take required action when the "valid" alarm occurs.

There is also the counter view that having a few false alarms can be regarded as beneficial. The primary driver of this view involves maintaining the controllers skill sets. Another perceived benefit is that although an aggressively tuned leak detection system generates more false alarms, there is no doubt that it will catch smaller leaks. Note that this places a burden on the controller to distinguish between true and false alarms.

Minimizing the potential negative impacts associated with false alarms involves finding a balance. There is the distraction that occurs when there are too many. However, there are the benefits of increased leak detection sensitivity resulting from aggressive tuning, and the refreshment of controller skills that occurs through occasional analysis of false positives. Achieving this balance requires a complete leak detection program. Looking at part of the system while ignoring the rest tends to develop single-factor optimization with an overall degradation in system performance.

This is a classic constrained optimization problem. That is, various technology, physical, telecommunication, and modeling constraints exist that prevent one from achieving a "perfect" state of detecting all leaks with no false alarms. Rather, one must fully understand the restrictions and constraints that exist within the parts of the system so that an optimized LDS can be designed. This means that the LDS must be designed to balance the various constraints to meet regulatory requirements as well as the organization's leak detection policy and LDS technology limitations.

As part of the balancing process, the system engineer should take into consideration key elements such as understanding and documenting all regulatory and industry standard requirements. This is achieved by surveying all applicable regulatory, governmental, and industry standards and company internal requirements, as well as any additional operational, performance, and other engineering needs for the system. This activity develops and documents the regulatory industry standards foundation that supports all other processes, such as tuning and human factors requirements. The LDS engineer will use the regulatory industry standards and other formal requirements, in conjunction with operational, performance, and other system needs, to develop a written leak detection system functional requirements document. Included in this document would be items such as:

- Detectable leak size policy. Although the objective is to detect smaller leaks with few false alarms, the owner/operator needs to develop a realistic detectable leak size target. This target may be set by regulations as well.
- Identification of the type and number of leak detection systems to be deployed. Each leak detection system has strengths and weaknesses. By deploying complementary leak detection systems, a higher probability of detecting the targeted leak size is possible.
- Identification of human factors requirements for the control room to support the leak detection system.
- Specification of the alarm management requirements and classifications.
- Identification of roles and responsibilities with respect to management of leak alarms.
- Survey the pipeline design and operation and determine if the selected leak detection systems can support the organization leak detection policy and functional system needs.
- Development of a leak detection system instrument maintenance plan. The reliability, repeatability, and capability of the LDS are based on quality of the field instrument data. Improving the maintenance of these field devices will enhance the performance capabilities of the leak detection system.

Given the fact that false alarms will occur, it is necessary that clear unequivocal guidelines should be provided to the pipeline controller on the actions to be taken in response to a leak alarm. These guidelines should be such that the pipeline controller can take the appropriate action quickly. Since an alarm may occur during complex pipeline operations requiring the controller's focus, that individual should not have to shift focus to evaluating the leak alarm. An example of a clear guideline is: "If the leak alarm cannot be clearing determined not to be a leak within 10 minutes, the pipeline should be shut down." An alternative guideline might be: "Upon the receipt of a leak alarm, the on-call leak detection engineer should be immediately notified and that person will take responsibility for responding to the leak alarm."

10.1.7 Training

Training is a key human factor for operators/controllers and LDS support personnel. Within the United States, leak detection system training is also a regulatory requirement, as outlined in 49 CFR 195.444. When considering the range of personnel who work with the leak detection system, such as controllers, engineers, analysts, technicians, and so forth, there is a corresponding wide range of required training. We can look at training from a content requirement perspective as well as a frequency perspective.

From a content perspective, the operator should craft a training program that is specific to each job classification and its linkage to the leak detection system. The objective of each job classification training program is to transfer the knowledge and skills that allow each job classification to directly interact with the system, understand all required technical aspects of the system applicable to the job, and understand how their positions fit within the overall leak detection system team.

Although the scope of training is technology-specific, some general guidelines apply to all. First, training should be on the system that has been or will be deployed within the operating environment. Generic training is usually not adequate to meet the needs.

Second, training should include theory as well as hands-on training. As an example, controllers should receive training on the pipeline hydraulic theory that supports leak alarm analysis. They should also receive hands-on training whereby they apply the theory to real recorded leak events or simulated leak events. This allows them to apply the theory to practice.

Third, training should always include all operator policies and procedures. All personnel must understand what policies apply and how to use existing procedures.

The training effectiveness should also be measured. You cannot improve something if you cannot measure it. One should always strive to improve the training program over time. To achieve this, you need to measure the training results. Some key training metrics that can be developed specific to the operating environment include: (1) percentage of correct nonleak alarm analysis; (2) percentage of correct valid leak alarm analysis; (3) length of time to develop a leak alarm attribution for the same alarm; and (4) student evaluations of the training class. Depending on the leak detection system and operational environment, other specific metrics may be available.

Leak detection system training is not a one-time event. The training program should include original or initial training, refresher training, and change management training. Whenever a new leak detection system is deployed or when a new person transitions into a role directly associated with the leak detection system, initial training is required. Following that, and usually on an annual basis, all roles should receive refresher training. Refresher training strengthens the individual knowledge base and skill sets. Finally, as part of any change management program, if major leak detection system changes have been implemented, then the affected personnel should receive corresponding training.

An effective and continuously improving leak detection training program is a primary contributor to risk mitigation. Highly trained and skilled personnel are more likely to quickly and accurately diagnose the leak detection system alarm and take appropriate actions. This capability reduces the potential negative consequences of a commodity release.

10.1.8 Human Factors Summary

Leak detection is a complex process that merges technology and human processing elements. The human factors portion of this system is a critical component that must be optimized to avoid contributing to slower responses to leaks alarms and subsequent spill damage.

Human factors encompass the design, implementation, training, and recurring maintenance associated with the physical, mental, and work-load aspects of how the control room operator interacts with the technological leak detection–based systems within their working environment. These interactions include psychological, environmental, and ergonomic issues. There are also regulatory requirements that must be addressed, as well as industry standards and best practices that should be considered.

Designing, implementing, and maintaining the human factors portion of the leak detection system is an "evergreen" process. Foundationally, the system design, operation, and maintenance are driven by the operator's leak detection policy and supporting procedures. The policy should provide clear direction on how the system integrates humans and technology into a cohesive whole and guides performance goals, such as specifying the smallest detectable leak size, controller leak alarm acknowledgment, analysis and response times, as well as the maximum number of false alarms. Each of these guiding objectives is a key system design and maintenance input associated with human factors management. The system objectives help to determine how leak alarms are displayed, as well as the processes and procedures that are required to respond to them.

10.2 DIRECT OBSERVATION LEAK DETECTION

In Chapter 13, Leak Detection and Risk-Based Integrity Management, we discuss how most leak, spill, and rupture events in the United States (and probably in most of the developed world) are detected by people, with little to no leverage provided by technology. Further, as noted previously in Chapter 7, External and Intermittent Leak Detection System Types, people are efficient at ultimately detecting smaller leaks that are often missed by technology. On the other hand, Chapter 13 also demonstrates that technology-based leak detection systems often excel for high-volume liquid commodity spills, and can efficiently and quickly detect releases that occur at high flow.

It is clear that the advantages and disadvantages of leak detection by humans, or *direct observation*, do not necessarily cover the same ground when compared to the pros and cons provided by leak detection technology solutions. In this section, we use some simple models to analyze the reasons why. We begin by discussing commodity physical release models.

10.2.1 Physical Release Models

There are only two ways in which a person is going to detect a leak without the aid of leak detection technology. One way is for the controller to independently notice (via the SCADA system inputs) that there is an imbalance in flows or some unusual transient, possibly some set of device alarms, that is possibly indicative of a leak. We do not discuss this possibility further here, not because it is not important. Rather, this is because the thought processes used by the controller will duplicate to a great extent the means by which an internal LDS would detect a leak.

The other way is for the commodity itself to be observed once it has escaped the pipe. This is direct observation. Such detections automatically place people in special cases of external leak detection systems (see Chapter 7: External and Intermittent Leak Detection System Types). To get a handle on how efficiently this can work, we need to develop simple illustrative or explanatory physical models that broadly describe how a commodity behaves from the start of the leak until the point when it is observed. Chapter 7 noted that the migration of spilled liquid commodity is a complex issue; therefore, to get a handle on issues that influence direct observation, we only analyze two types of simplified releases in this section. The first type is a low-flow liquid commodity with low vapor pressure spilled onto the ground from an aboveground pipe or other components. The second is a low-flow LVP commodity released from a high-pressure source into homogeneous soil from a buried pipe. These low-flow cases are particularly challenging for internal leak detection technology approaches, such as mass balance, RTTM, or rarefaction wave approaches (or, for that matter, a pipeline controller limited to data viewable only via the SCADA system).

We do not look at more complicated releases involving complex underground soil geometries, spills involving HVP liquids, or discharges of gas-phase commodities. This is partially due to the fact that it is not feasible in the limited space available for this book to address all of the possible loss cases that can arise. We also do not consider the spread of oil slicks on water (again, because this is a fairly large topic) except to note that water spills tend to spread and diffuse rapidly, and thus they tend to have a greater scope of contamination. They also tend to impact many separate parties at the same time. Detection of HVP and gas commodities is often expedited by the fact that these fluids are often flammable (bright!) or explosive (loud!), often resulting in rapid detection over significant distances by...people.

Before we start, let's keep in mind that what we are looking for are physical descriptions that are conservative with respect to discovery. That is, we want models that will tend to minimize the probability or rate of detection on the part of site observers while still remaining physically reasonable.

Let's start by considering a liquid hydrocarbon spill directly onto solid ground. The ground may be either impermeable (such as a concrete pad) or permeable (soil, sand, or gravel). A solution to the equations that describe viscous gravity currents associated with pool spreading due to a continuing leak of flow, q_{Leak}, on a horizontal impermeable surface is provided in [7]:

$$R_{\text{Spill}}(t, q_{\text{Leak}}) = 0.715\left(\frac{g}{3v}q_{\text{Leak}}^{3}\right)^{1/8} t^{1/2}$$

EQUATION 10.1 Radius of Liquid Spill to Impermeable Surface

where R_{Spill} is the spread radius of the spilled commodity on the ground, t is the time after leak onset, g is the acceleration of gravity, and v is the kinematic viscosity of the commodity. Because we presume the leak is from an above-ground pipe or component, we can assume that the rate could potentially be calculated based on the outlet diameter and an orifice coefficient if the pipe internal pressure is known. Note that Eq. (10.1) does not include the effects of evaporation, which would presumably act to limit the size of the spill and constrain its usefulness for HVP commodities. If the spill is on a permeable surface with saturated permeability constant k_{Soil} (units of distance squared) for the porous soil, then the spilled commodity pool will transition to draining behavior and stabilize for times $t > t_{\text{T}}$ [8], where:

$$t_{\text{T}} = 0.697\left[\frac{6qv^5}{\pi g^5 k_{\text{Soil}}^4}\right]^{1/4}$$

EQUATION 10.2 Spill Draining Behavior Transition Time

If $t > t_{\text{T}}$, then the spill radius is constant. Note that k_{Soil} is a strong function of the soil type. The constant maximum value of the radius is given by Eq. (10.1), with t set equal to t_{T}. Note that once the spill radius stabilizes in the face of a continuing leak, this implies that the entire flow of the leaking commodity now drains to a continuously expanding bulb of soil beneath the spill source, where the extent of the contaminated soil bulb is a function of the soil porosity fraction $\varepsilon_{\text{Soil}}$. This implies that a considerable amount of soil could require remediation under these circumstances.

Now, let's consider a leak from a pipe buried in dry soil. Such a leak could arise from a pipe rupture or a failed corrosion defect. As discussed in Chapter 7, External and Intermittent Leak Detection System Types, flow of a commodity into the soil will be influenced by the pressure of the source and the orifice size of the hole in the pipe. Small orifices tend to be associated with downward gravitational flows, whereas larger leaks into high-permeability soil tend to have more spherical intrusion fronts and minimum gravitational distortion. Let's consider a model for the latter. If the soil is well packed around the pipe and homogenous in properties and extent, then we have a Darcy flow condition, where the local component of the subsurface commodity flow velocity u_{DF} in coordinate direction i beneath the surface of the soil is given by:

$$u_{\mathrm{DF},i} = \frac{k_{\mathrm{Soil}}}{\mu}\frac{\partial p}{\partial x_i}$$

EQUATION 10.3 Liquid Commodity Darcy Flow Velocity from Buried Leak Source

In this equation, μ is the commodity viscosity and p is the local commodity pressure. If we consider a transient situation, then we can use the unsteady Richards equation [9]. This equation is expressed in several forms, one of which is the matric head form:

$$C_{\mathrm{S}}(\psi_{\mathrm{C}})\frac{\partial \psi_{\mathrm{C}}}{\partial t} = \nabla\cdot\left[K_{\mathrm{S}}(\psi_{\mathrm{C}})\nabla\psi_{\mathrm{C}}\right] + \frac{\partial K_{\mathrm{S}}(\psi_{\mathrm{C}})}{\partial z}$$

EQUATION 10.4 Richards Equation for Diffusive Darcy Flow

where ψ_{C} is the liquid commodity matric or tension head (units of length), z is the vertical distance coordinate, and K_{S} is the hydraulic conductivity (length/time):

$$K_{\mathrm{S}}(\psi_{\mathrm{C}}) = \begin{cases} \rho_{\mathrm{C}} g \dfrac{k_{\mathrm{Soil}}}{\mu_{\mathrm{C}}} & \psi_{\mathrm{C}} \geq 0 \\ \ll \rho_{\mathrm{C}} g \dfrac{k_{\mathrm{Soil}}}{\mu_{\mathrm{C}}} & \psi_{\mathrm{C}} < 0 \end{cases}$$

EQUATION 10.5 Richards Equation Hydraulic Conductivity

Parameter μ_{C} is the commodity viscosity and C_{S} is the rate of change of saturation with respect the hydraulic head function (1/length):

$$C_{\mathrm{S}}(\psi_{\mathrm{C}}) = \frac{d\theta_{\mathrm{C}}}{d\psi_{\mathrm{C}}}(\psi_{\mathrm{C}}) = \begin{cases} \rho_{\mathrm{C}} g \dfrac{\varepsilon_{\mathrm{Soil}}}{B_{\mathrm{C}}} & \psi_{\mathrm{C}} \geq 0 \\ \gg \rho_{\mathrm{C}} g \dfrac{\varepsilon_{\mathrm{Soil}}}{B_{\mathrm{C}}} & \psi_{\mathrm{C}} < 0 \end{cases}$$

EQUATION 10.6 Richards Equation Saturation Function

Here, θ_C is the dimensionless commodity fraction, B_C is the commodity bulk modulus, and ε_{Soil} is the soil porosity. If the soil is fully saturated with commodity, then the matric head is positive and equivalent to the pressure head. However, if the soil is not saturated (ie, has some remaining pore space still containing air), then the tension head can be negative, primarily due to capillary suction resulting from attraction between the commodity and the soil. This suction increases (ie, the tension head becomes more negative) as the soil becomes less saturated with commodity.

This equation also assumes that the soil is dry; it neglects the impact of any water in the medium surrounding the pipe. It is thus inappropriate for any analysis of commodity leakage into water-saturated soil, as is the case if the pipe is below the water table. Under these circumstances, more complex multi-component equations would be required, and the form of the equations would be influenced by whether or not the water and commodity were miscible.

If the matric head is very high, as would be the case for a leak from a buried pipe operating at pressure, then we can neglect the vertical derivative on the soil conductivity, or:

$$\frac{\partial \psi_C}{\partial t} = \frac{1}{C_S(\psi_C)} \nabla \cdot \left[K_S(\psi_C) \nabla \psi_C \right]$$

EQUATION 10.7 Richards Equation for Diffusive Darcy Flow (Gravity Neglected)

This equation is highly nonlinear and typically requires numerical solution. It also has a split personality. For a given soil and commodity combination, the relative commodity capacity function and soil conductivities are both presumed to be functions of the tension head and, as implied by these relations, these parameters change dramatically when the oil transitions between unsaturated and saturated states. We previously noted that when the matric head is low and the soil is not saturated, soil pore spaces are not filled and the fluid travels with high resistance because it will move in the form of very thin films on the soil particles. This means that an unsaturated fluid will have high available capacity and very low permeability relative to a situation when the soil is saturated. Consequently, it is difficult for the fluid to move forward into dry soil from a high-pressure source, such as a leaking pipe. When it does so, the pore spaces must completely fill to occupy more volume, and it will move slowly into the soil with a very distinct saturated fluid front.

For a point source in an infinite medium, an analysis of an expending semi-hemispherical saturated infiltration front in a porous medium has been presented [10], which we adapt here to address a fully spherical infiltration front. We assume the following: (1) the liquid leak orifice is small enough to be considered as a point source, and the pipe geometry does not impact the solution; (2) the shape of the advancing liquid front is therefore an

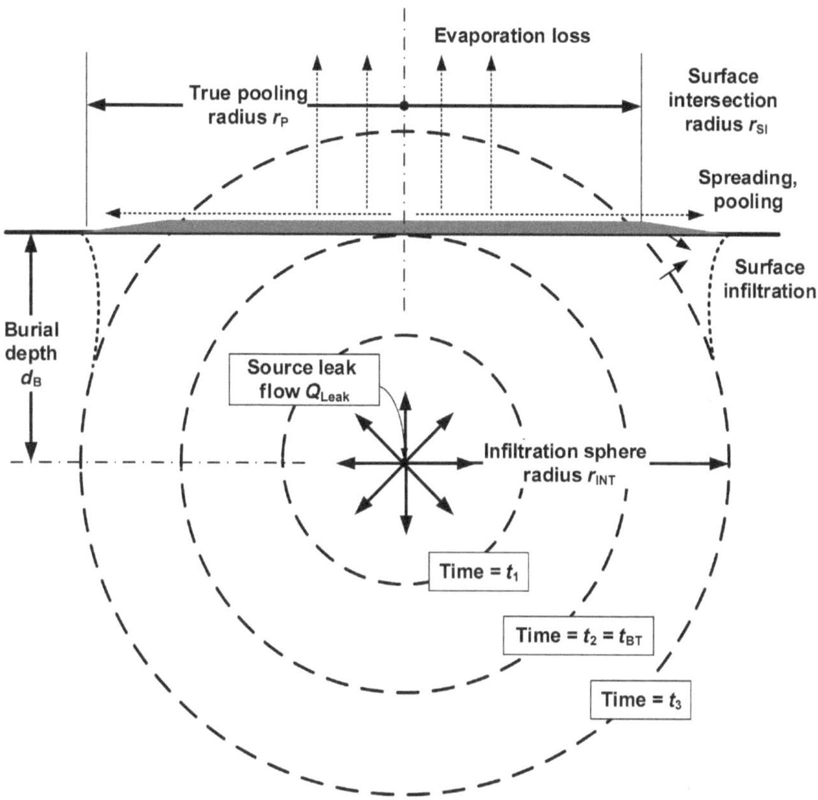

FIGURE 10.3 Spherical infiltration front from buried pipeline leak.

expanding sphere with distinct concentration discontinuity across the front; (3) liquid flows radially from the source to the advancing front and the radial velocity is uniform along the advancing front; (4) the pressure gradient is radial; (5) gravitational effects can be neglected; (6) the soil resistance is high and the leak is relatively small, so we can neglect orifice effects across the hole; and (7) we neglect surface effects of evaporation and spreading once the front breaks the surface (more about this later).

Refer to Fig. 10.3. We see a leak source initiated at time t_0 and buried at depth d_B, with a series of successive intrusion spheres and progressive times t_1, t_2, and t_3. Let's first consider the time before the intrusion sphere reaches the surface. Because the volume inside any particular intrusion sphere of radius r_{INT} is saturated, we can define the leak flow rate as:

$$q_{Leak}(t) = 4\pi\varepsilon_{Soil} r_{INT}^2 u_{INT} = 4\pi\varepsilon_{Soil} r_{INT}^2 \frac{dr_{INT}}{dt}$$

EQUATION 10.8 Leakage Flow Rate from Pressurized Buried Source where u_{INT} is the radial velocity, which, from Darcy's law, allows us to express the pressure gradient as:

$$\frac{dp}{dr} = \frac{\mu q_{\mathrm{Leak}}(t)}{k_{\mathrm{Soil}} 4\pi r_{\mathrm{INT}}^2}$$

EQUATION 10.9 Pressure Gradient for Spherically Expanding Buried Commodity Source

The parameter μ is again the commodity viscosity. Neglecting the pressure drop across the leak orifice, we assume that the pressure at the buried source ($r_{\mathrm{INT}} = r_{\mathrm{S}}$, the radius of the leak orifice) is the pipeline operating pressure p_{PL} and that the pressure at r_{NT} is atmospheric pressure p_{ATM}. Integrating both sides of this equation and assuming that $q_{\mathrm{LEAK}}(t)$ is constant as a function of radius yields:

$$p_{\mathrm{PL}} - p_{\mathrm{ATM}} = \frac{\mu q_{\mathrm{Leak}}(t)}{k_{\mathrm{Soil}} 4\pi} \left(\frac{1}{r_{\mathrm{S}}} - \frac{1}{r_{\mathrm{INT}}} \right)$$

EQUATION 10.10 Integrated Pressure Gradient Equation (Darcy Infiltration Sphere)

We can combine Eq. (10.8) with Eq. (10.10) to obtain:

$$r_{\mathrm{INT}}^2 \frac{dr_{\mathrm{INT}}}{dt} \left(\frac{1}{r_{\mathrm{S}}} - \frac{1}{r_{\mathrm{INT}}} \right) = \frac{k_{\mathrm{Soil}}}{\mu \varepsilon_{\mathrm{Soil}}} (p_{\mathrm{PL}} - p_{\mathrm{ATM}})$$

EQUATION 10.11 Infiltration Sphere Pressure Gradient Differential Equation

Integrating this gives us:

$$\frac{1}{3r_{\mathrm{S}}} (r_{\mathrm{INT}}^3 - r_{\mathrm{S}}^3) - \frac{1}{2} (r_{\mathrm{INT}}^2 - r_{\mathrm{S}}^2) = \frac{k_{\mathrm{Soil}}}{\mu \varepsilon_{\mathrm{Soil}}} (p_{\mathrm{PL}} - p_{\mathrm{ATM}}) t$$

EQUATION 10.12 Infiltration Sphere Radius vs. Time

Over long periods of time, this simplifies to:

$$r_{\mathrm{INT}} \approx \left[3r_{\mathrm{S}} \frac{k_{\mathrm{Soil}}}{\mu \varepsilon_{\mathrm{Soil}}} (p_{\mathrm{PL}} - p_{\mathrm{ATM}}) \right]^{1/3} t^{1/3}$$

EQUATION 10.13 Simplified Infiltration Sphere Radius vs. Time Solution

If we now substitute this back into Eq. (10.8), then we find that the time components of the radial and velocity terms cancel. Therefore, over long periods the flow must be relatively constant or:

$$q_{\mathrm{Leak}} = 4\pi \varepsilon_{\mathrm{Soil}} r_{\mathrm{INT}}^2 \frac{dr_{\mathrm{INT}}}{dt} \approx 4\pi \frac{k_{\mathrm{Soil}}}{\mu} (p_{\mathrm{PL}} - p_{\mathrm{ATM}}) r_{\mathrm{S}}$$

EQUATION 10.14 Infiltration Sphere Flow Rate

We can now combine the last two equations to determine the breakthrough time, t_{BT}, that it takes the intrusion sphere to reach the ground surface:

$$t_{\mathrm{BT}} \approx \frac{4}{3}\pi\varepsilon_{\mathrm{Soil}}\frac{d_{\mathrm{B}}^{3}}{q_{\mathrm{Leak}}}$$

EQUATION 10.15 Infiltration Sphere Breakthrough Time

Our exercise has provided a model that is consistent with theories of saturated and unsaturated liquid commodity flow in permeable media while providing a reasonable estimate of the time it will take for a small underground leak to reach the surface. Now, let's refer back to Fig. 10.3 and consider what will happen after the intrusion sphere reaches the surface. Let's pretend that the intrusion sphere simply ignores the ground surface discontinuity and moves through it. In that case, we can use geometry and Eqs. (10.13 and 10.14) to describe an intersecting circle on the ground defining some observable wetted pool of commodity with radius r_{SI} defined by:

$$r_{\mathrm{SI}} = \left(r_{\mathrm{INT}}^{2} - d_{\mathrm{B}}^{2}\right)^{1/2} = \begin{cases} 0 & t \le t_{\mathrm{BT}} \\ \left[\left(\dfrac{3q_{\mathrm{Leak}}}{4\pi\varepsilon_{\mathrm{Soil}}}\right)^{2/3} t^{2/3} - d_{\mathrm{B}}^{2}\right]^{1/2} & t > t_{\mathrm{BT}} \end{cases}$$

EQUATION 10.16 Above Ground Wetted Pool Radius from Small Subsurface Leak

Experimental verification of the infiltration front model are provided in [10], and, for analogous two-dimensional cylindrical infiltration line source, in [11].

In reality, the intrusion region in the soil will, of course, cease to be spherical once the intrusion sphere radius intersects the ground profile at $t = t_{\mathrm{BT}}$. One reason for this is that the effective pressure gradient will become limited by the distance between the buried leak source and the ground, causing the radial streamlines to bend upward, thus increasing the actual wetted pool radius r_{P}. Another reason is that if there is only limited evaporation, then the leaked commodity on the ground will accumulate and flow radially away from the center of the pool. Although the spreading commodity will tend to re-infiltrate the soil due to gravity current effects, the net effect will be to increase the size of the pool. In fact, virtually all real-world effects will tend to increase the pool size so that $r_{\mathrm{SI}} < r_{\mathrm{P}}$. This means that from a human-only mediated leak detection point of view, r_{SI} should suffice as a conservatively small estimator of the observable size of the spill pool.

Fig. 10.4 shows predicted spill diameters using our very simple models for a relatively small leak at 0.5% of 15,000 BPD nominal flow based on above-ground and buried pipe leaks. We see that for above-ground leaks, spills on an impermeable surface such as a concrete pad produce a spill with a sizable diameter, which would be expected to be quickly identified assuming there is an observer around to detect it. However, if the spill is on a

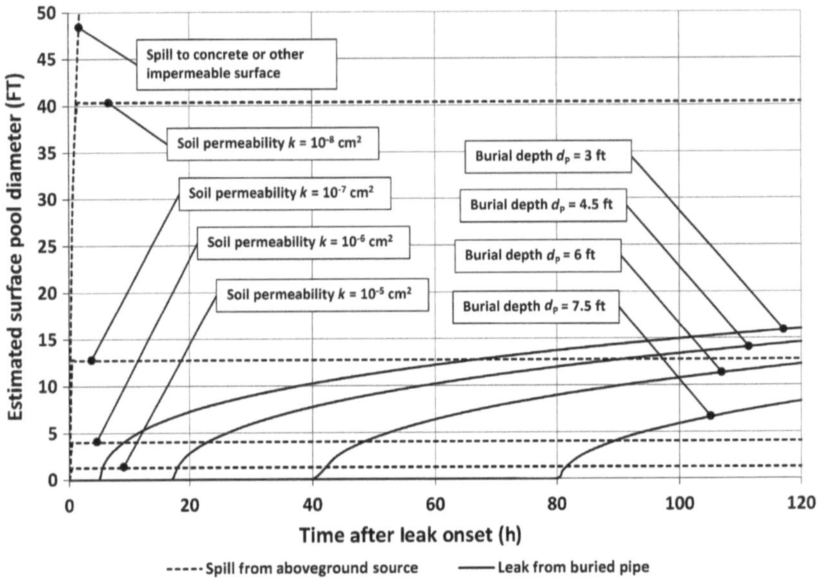

FIGURE 10.4 Estimated pool diameter versus time for a 0.5% nominal flow leak from a 15,000 BPD pipeline.

permeable medium, such as well-sorted sand or gravel with low permeability in the range of 10^{-5} to 10^{-6}, then we can see that the spill will be limited to very small diameters because it will tend to soak into the soil. Note that the small diameter may not be at all indicative of the very large volume of contaminated soil beneath the observed spill pool.

A leak from a buried pipe has a somewhat different detection problem in that, in the absence of discovery by some other mechanism (such as leak detection technology), the leak is occulted or concealed until the infiltration sphere reaches the surface or it causes a significant change in vegetation above the spill. Even then, our conservative estimate of the pool diameter grows slowly and, again, may not be indicative of the potentially large volume of contaminated soil inside the occultation sphere.

Darcy diffusion in the soil is but one means by which oil occultation can occur. Other possibilities include spills that route to drainage ditches, ravines, and storm drains, capping of the oil due to an impenetrable barrier (such as a parking lot) atop the pipeline, preferential flow through low-permeability soils below the surface, preferential migration of the oil along the pipe, and (of course) flow of the commodity to water. Also note that if the leak is beneath or just above the water table, then the flow of commodity will be either preferentially within the water (if the commodity is polar), or above/below it (if nonpolar). Finally, a high-pressure leak of this size from shallow burial at the top of a pipe has a distinct likelihood of violating the integrity of the soil over the pipe, with a resulting cracking or rupturing of the soil and consequent short-circuit of the flow to the surface. In this case, the leak will behave more like a surface spill. The important point to remember is that all of these cases ultimately and qualitatively behave like our simplified models: they eventually become detectable by observers, with some lag.

10.2.2 Detection of Leaks by the Public

Direct detection of spills by people can be accidental (members of the public, roving operating company staff doing their jobs), on purpose (ground patrols, air patrols), or mixed (emergency responders). Let's use the physical release models described here to model direct detection of the leaks by members of the public in the vicinity of the spill. Fig. 10.5 shows a view of the visible spill pool from above. In the vicinity of the spill is a population of human observers going about their business. Many of the observers are stationary; they are in their homes or places of business, and they are not moving. Others are mobile. Although the figure shows them as pedestrians, they could be moving about in vehicles of some kind.

Let's assume for illustrative purposes that the population density, ρ_{HO}, of all observers is known, and that we can separate them into two categories using parameter f_M, representing a number between zero and one, which we

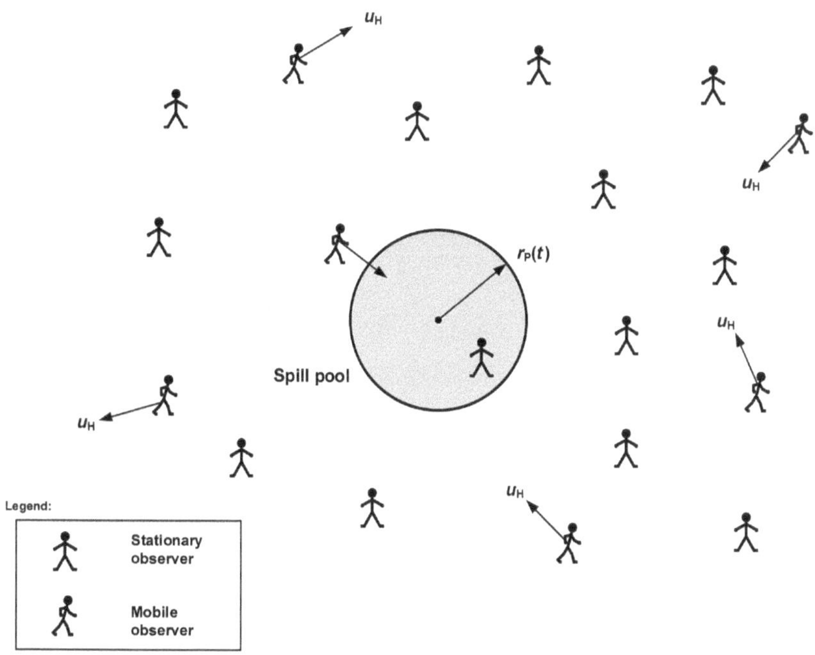

FIGURE 10.5 Human observers in the vicinity of a pipeline spill.

will refer to as the mobile fraction. The velocity of stationary observers is, of course, zero, and the average velocity of mobile observers is u_H. We further assume that the velocity vector for the mobile observers is randomly oriented. From the last section, we already know that the radius of the spill pool grows over time. It is reasonable to assume that the spill is potentially detected every time a new observer becomes aware of it. Here, we conservatively assume that an observer detects the spill by seeing the commodity, smelling the commodity, or feeling the change in traction as a result of stepping into it. In reality, the observable radius is probably somewhat larger than the pool radius, but we conservatively assume that the observer has to actually be inside the surface pool radius to detect the spill.

The rate R_{sc}, at which new observer-based channels n_{OBS} are opened, is equal to the rate $\dot{n}_{SO}$, at which the expanding radius of the spill pool encompasses members of the stationary population plus the rate $\dot{n}_{MO,IN}$, at which members of the mobile population enter the pool (people continuing to stay inside or exiting the pool are conservatively assumed to have had their chance and will have no further opportunity to become aware of the spill):

$$R_{SC} = \frac{dn_{OBS}}{dt} = \dot{n}_{SO} + \dot{n}_{MO,IN}$$

EQUATION 10.17 Spill Pool Detection Definition

The rate at which stationary observers are overtaken by the expansion of the pool is:

$$\dot{n}_{SO} = (1 - f_M)\rho_{HO}\frac{dA_P}{dt} = 2\pi(1 - f_M)\rho_{HO} r_P \frac{dr_P}{dt}$$

EQUATION 10.18 Stationary Observer Spill Pool Detections

where A_P is the spill pool area and r_P is the pool radius, estimated by r_{SI}, as discussed in the last section. The rate at which members of the mobile population enter the pool is likewise given by:

$$\dot{n}_{MO,IN} = \frac{1}{2} f_M \rho_{HO}(2\pi r_P) u_{IN} = f_M \rho_{HO} \pi r_P u_{IN}$$

EQUATION 10.19 Mobile Observer Spill Pool Detections

The factor of ½ is because only half of the population on the spill boundary is entering the spill at any time (the rest are exiting the spill), and u_{IN} is the entrance velocity of the observers normal to the spill boundary. This velocity is:

$$u_{IN} = \frac{u_H}{\pi}\int_{-\pi/2}^{\pi/2} \cos\theta \cdot d\theta = 2\frac{u_H}{\pi}$$

EQUATION 10.20

The mobile observer detection rate therefore simplifies to:

$$\dot{n}_{\mathrm{MO,IN}} = f_{\mathrm{M}} \rho_{\mathrm{HO}} 2 r_{\mathrm{P}} u_{\mathrm{H}}$$

EQUATION 10.21 Mobile Observer Spill Pool Detections Based on Observer Velocity

And the spill pool detection rate becomes:

$$R_{\mathrm{SC}} = 2 r_{\mathrm{P}} \rho_{\mathrm{HO}} \left(\pi (1 - f_{\mathrm{M}}) \frac{dr_{\mathrm{P}}}{dt} + f_{\mathrm{M}} u_{\mathrm{H}} \right)$$

EQUATION 10.22 Spill Pool Detection Rate Equation

Let's consider the terms of this equation. At any time, the mobile fraction of the population, f_{M}, in the vicinity of the pipeline right-of-way is likely to be low, probably approximately a few percent. Most of the population will be in their homes or at work or, for a good fraction of the day, asleep. Leaks are most likely to be detected by observers moving at relatively low velocity who are not occupied by driving activities; therefore, for u_{H}, we are talking about pedestrian velocities of approximately 2 or 3 miles/h (approximately 3½ feet/s). However, if we refer back to Fig. 10.4, the rate at which a relatively small spill expands on a permeable soil substrate is much lower, approximately 10^{-5} feet/s. If we plug these numbers back into Eq. (10.22), then it rapidly becomes clear that for very small leaks onto permeable soil, the first term is insignificant. Therefore, the observation opportunity rate is set by the number of mobile observers or:

$$R_{\mathrm{SC}} \approx 2 r_{\mathrm{P}} f_{\mathrm{M}} \rho_{\mathrm{HO}} u_{\mathrm{H}}$$

EQUATION 10.23 Simplified Spill Pool Detection Rate

In effect, we have recapitulated our discussion of Chapter 2, Pipeline Leak Detection Basics, where we pointed out that mobile external observers were more efficient at detecting free riders exiting the highway system than stationary observers could ever be. Over a period time, t, the number of potential reporting opportunities, N_{SC}, is then:

$$N_{\mathrm{SC}} \approx 2 f_{\mathrm{M}} \rho_{\mathrm{HO}} u_{\mathrm{H}} \int_0^t r_{\mathrm{P}}(t) dt$$

EQUATION 10.24 Cumulative Spill Pool Reporting Opportunities Definition

Just because an observer has the opportunity to observe a spill does not mean that the observer will notice it or do anything about it. Consequently, we now define two more terms: the spill surprise parameter (p_{SS}) and the vigilance (p_{VIG}). The spill surprise is a probability that defines whether or not the spill will actually be noticed. If you step in a 6-inch-diameter oil pool, for example, you might not think much of it; however, a 10-foot pool is more likely to get your attention. Thus, we can assume that this parameter is a cumulative probability distribution that is a function of the diameter of the spill pool. We really do not have a lot of data to back up this parameter (the distribution could be normal or log-normal, or something else, for example). Here, we will for the sake of argument assume a simple exponential distribution:

$$p_{SS} = 1 - e^{2r_P/D_{ATT}}$$

EQUATION 10.25 Observer Spill Pool Surprise Parameter

where D_{ATT} is the spill pool surprise or diameter (the average pool diameter that will actually be noticed as unusual by members of the public).

The vigilance describes the willingness of people to report the spill to the authorities. In developed nations, we would expect vigilance to be well above 50%, but probably not 100%. In other places without well-developed institutions and social norms, this number may be considerably lower. In other words, there will always be a residual core of the population that will not report anything to anybody. Note that we are ignoring reporting lag in this equation. For developed nations in the cell phone era, this is probably not unreasonable; however, this assumption may not apply everywhere.

For each spill observation opportunity defined in Eq. (10.24), the probability (p_{RPT}) that the opportunity will be reported as a spill is then:

$$p_{RPT} = p_{VIG}(1 - e^{2r_P(t)/D_{ATT}})$$

EQUATION 10.26 Observer Spill Pool Vigilance Parameter

We now have an equation that defines the total number of reporting opportunities, N_{SC} (Eq. 10.24), and an equation defining the probability that each opportunity will actually be reported (Eq. 10.26). This gives us a succession of Bernoulli trials, and we can thus calculate the cumulative probability p_{PUB} that the spill will be reported as a result of a succession of observations, each of which occurs with probability p_{RPT}:

$$p_{PUB} = 1 - (1 - p_{RPT})^{\mathrm{Int}(N_{SC})}(1 - \mathrm{Mod}(N_{SC})p_{RPT})$$

EQUATION 10.27 Cumulative Probability of Spill Detection by Direct Observation

where the functions Int and Mod return the integer and fraction portions of a number, respectively.

Fig. 10.6 shows an illustrative example of the estimated cumulative probability of detection by the public for a very small leak of 0.1% of nominal flow from a buried pipe. The nominal flow in this hypothetical pipeline is 20,000 BPD and the soil porosity is 40%. Vigilance is set to 75%, the surprise diameter is 5 feet, and the mobile fraction is 5%. The leak originates 3.4 feet beneath the surface. We see that the spill is undetectable by site

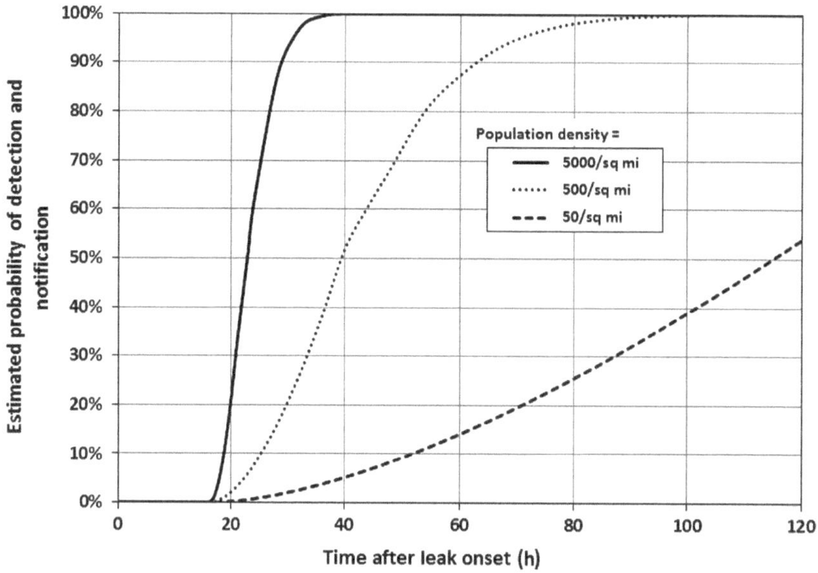

FIGURE 10.6 Estimated spill detections by the public from a buried pipe leak.

observers for a period of 17 h as the commodity works its way to the surface. Once it does, however, the detection probability is strongly dependent on the population density; the probability increases fairly quickly for a high population density, as might apply in an urban area.

In a low-density rural area, however, the spill detection probability curve increases much more gradually. Note that this means detection is obviously slow when compared to other methods due to the occultation period, and it can take considerable time when the population density is low. However, when the spill becomes detectable, detections by human observers ultimately approach 100% in efficiency. This is as compared to CPM systems, which typically are constrained at the low end by statistical limitations inherent in the data analysis process, as is discussed in Chapter 5, Statistical Processing and Leak Detection and Chapter 9, Leak Detection Performance, Testing and Tuning (see Figures 9.1 (a) and (b)). This is an important difference between these approaches, and we return to this topic in Chapter 13, Leak Detection and Risk-Based Integrity Management.

Other models involving detections of leaks by humans at the leak site, such as might occur through station or site pipeline personnel, ground patrols, air patrols, and other mechanisms, or for more complex escaped commodity migration models, can be developed using approaches similar to the one used here for detections by the public.

REFERENCES

[1] Balaud M. Likelihood alarm systems: the impact of the base rater of critical events, the cost of alarm validity information, and the number of stages on operator's performance. University of Berlin, Nov. 11, 2015.

[2] Green DM, Swets JA. Signal detection theory and psychophysics. New York: Wiley; 1964.

[3] National Transportation Safety Board. Supervisory control and data acquisition (SCADA) in liquid pipelines. Safety Study NTSB/SS-05-02, Nov. 29, 2005.

[4] Public Law 109-468-Dec. 29, 2006, Pipeline Inspection, Protection, Enforcement, and Safety Act of 2006.

[5] Pipeline and Hazardous Materials Safety Administration. Pipeline safety: control room management/human factors. Final Rule, Jan. 26, 2010. https://federalregister.gov/a/E9-28469.

[6] American Petroleum Institute. Pipeline SCADA alarm management. API Recommended Practice 1167, Dec 01, 2010.

[7] Huppert HE. Gravity currents: a personal perspective. J Fluid Mech 2006;554:299–322.

[8] Grimaz S, Allen S, Stewart J, Dolcetti G. Predictive evaluation of surface spreading extent for the case of accidental spillage of oil on the ground. AIDIC Chem. Eng. Trans. 11, June 2007, p. 389–394.

[9] Celia MA, Bouloutas ET, Zarba RL. A general mass-conservative numerical solution for the unsaturated flow equation. Water Resour Res 1990;26(7):1483–96.

[10] Xiao J, Stone HA, Attinger D. Source-like solution for radial imbibition into a homogeneous semi-infinite porous medium. Langmuir 2012;28(9):4208–12.

[11] Hosseinalipour SM, Aghakhan H. Numerical & experimental study of flow from a leaking buried pipe in an unsaturated porous media. Int J Math Comput Phys Electr Comput Eng 2011;5(6).

Chapter 11

Implementation and Installation of Pipeline Leak Detection Systems

"Pipeline operators function today in an environment where leak detection systems are seen as a condition of operation mandate" [1]. However, there are many different types of leak detection systems with corresponding strengths and weaknesses. Identifying which leak detection system is best for a specific pipeline environment is not easy because no two pipeline environments are the same.

When considering the variations in pipeline environments, the one constant is that they are all unique. They vary in their physical characteristics, such as length, pipe diameter, pipe wall thickness, type of pipe material used, internal roughness coefficients, location of pump or compressor stations, and others. Furthermore, each pipeline has specific operating conditions such as batched, intermittent, continuous flow, the presence of slack or not, blending, product fluid characteristics, and regulatory environment, to name a few examples. The distinctiveness of each pipeline as well as the operator's policies and procedures guide the final selection of the most appropriate leak detection system or systems.

In this chapter, we discuss the overall process of installing a new leak detection system, or replacing an existing system, and even augmenting an existing system. As indicated in Fig. 11.1, the implementation process starts with defining the system's functional and performance requirements and ends with final testing and system verification, also known as the commissioning process.

As discussed in the next sections, the implementation process is interactive and iterative. As the team responsible for implementing the project proceeds, they may have to return to earlier steps to adjust or modify requirements and specifications established early in the process. Those involved in the effort should be prepared for this and include it as a contingency in their overall plan.

Pipeline Leak Detection Handbook. DOI: http://dx.doi.org/10.1016/B978-0-12-802240-5.00011-X

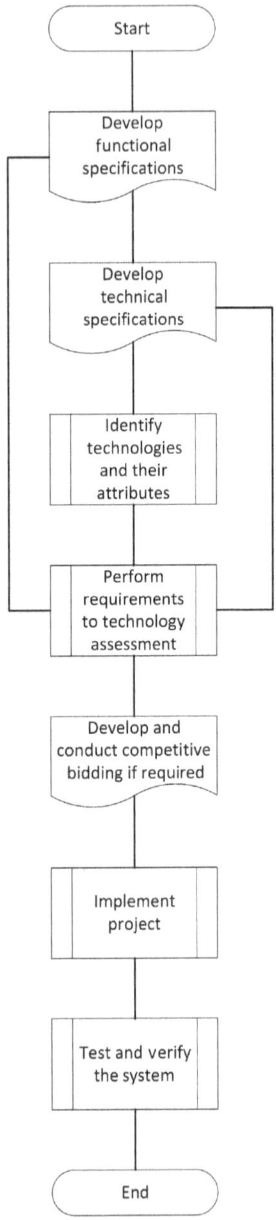

FIGURE 11.1 Overall implementation flow chart.

11.1 PERFORMANCE REQUIREMENT SPECIFICATION

As you embark on a leak detection technology project, a set of specifications or metrics at the outset is necessary. Developing the full range of functional and technical performance requirements involves data gathering and specification development. Fig. 11.2 provides a flow chart of this process.

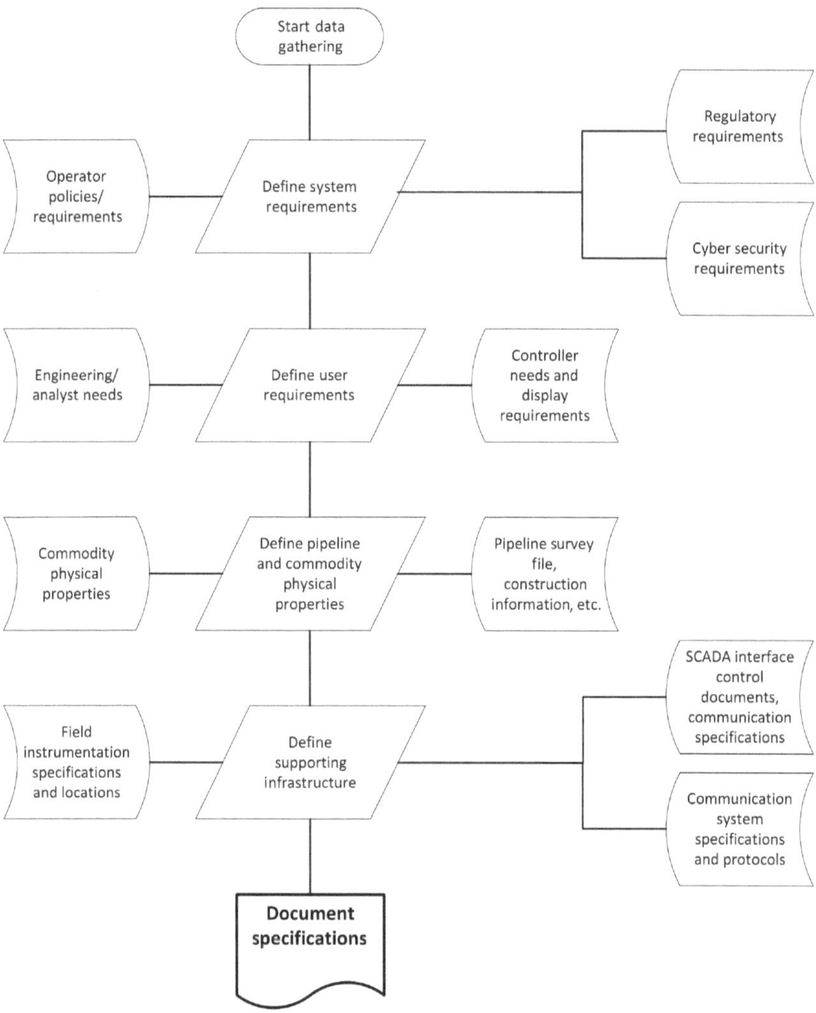

FIGURE 11.2 Specification development flow chart.

The development of the functional and technical specifications involves gathering data from several sources. These data becomes the substance and foundation for the rest of the project. But to start and to ensure a common understanding we start with answering the following three questions. (1) What is the difference between functional and technical specifications? (2) Are distinct documents required for each of these items? (3) Moreover, why is it important to write it all down?

So, what is the difference between functional and technical specifications? Depending on your research, the answer you find may be "really nothing" or "major difference." From our view we consider the documents to be individually distinct and mutually supportive. Each serves a required function and both are essential in defining the system.

The functional design or specification is the information and data that define the system behavior. Sometimes you will see this document described as a system requirement document. The document clearly lays out how the user will interact with the system as well as the system inputs and outputs. The documentation should identify and define all system level functions. These detailed and defined functions, once met, will fulfill the stakeholder's needs and requirements. This documents the "what" of the system.

The technical specifications document details the required performance of the system as well as all human machine interface (HMI) requirements. It also provides details on how the system will operate "under the hood," such as specific details on how the vendor system will interface with the supervisory control and data acquisition (SCADA) system or other data acquisition systems, data storage requirements, bandwidth allocations, and so forth. The operative word is "detailed," indicating in-depth information on how the leak detection system (LDS) will work.

As an example, functional specifications may state that the LDS must estimate leak location. This is a functional requirement, but it lacks explicit details regarding how it occurs. The technical specification document details how this functional requirement is defined. It could include requirements, such as the estimated leak location should be within ± 5 miles of the actual leak location with a 95% confidence for all leaks larger than 250 BPD. The technical specifications document is also the link between system functional requirements and acceptance criteria.

Are both of these documents specifically required? The answer is yes, the information in both of these documents is required. However, should they be separate documents? From a purest perspective, the answer is yes. Yet, successful project implementation occurs with one document that includes both sets of information. The decision of a single or two documents is specific to each system operator's policies, procedures, and project team decisions. Often, the functional specification forms the basis for the project justification and a mandate for the project team. The technical specification may be developed by the project team to define how the system will be implemented and to establish technical success criteria.

Why take on the effort of developing these documents? Because the resulting documentation defines all functional requirements and technical specifications. This helps ensure that the implementation efforts and requirements are:

- Traceable—each system function and capability can be traced to a requirement
- Unambiguous—all participants in the process understand what is to be delivered
- Measurable—the system performance has specific metrics that can be and will be measured
- Testable—all testable items are linked to measurable system requirements
- Feasible—this ensures that the desired functions or requirements can be achieved within the specific pipeline environment

With all of this, where to start? As shown in Fig. 11.2, we start by identifying the high-level requirements that may exist within the regulatory framework and the owner's policies and internal requirements.

Once identification of regulatory and owner requirements is complete, one needs to gather and document user requirements, including those associated with the controllers, leak detection engineers, and system analysts. This data-gathering step clearly defines how users will directly interact with the leak detection system as well as defines all specific controller leak alarm and leak system monitoring requirements and procedures.

Similarly, the needs and requirements of other leak detection system users, such as the system analysts and leak detection engineers, should be specified. Items to consider are specific requirements associated with working with the system on a daily basis, when a leak alarm or abnormal system alerts occurs, and any daily, weekly, monthly, or annual system maintenance requirements.

Once we understand and have documented the underlying regulatory, owner, and user requirements, we move into obtaining and documenting the unique pipeline physical and commodity requirements. Because not all leak detection technologies are applicable to all physical environments, it is essential to define the pipeline and commodity physical details and operating states. As an example, if the pipeline always operates in the slack or multiphase region, then this would eliminate rarefaction wave leak detection systems. Thus, it is essential to detail the pipeline and commodity specifications as part of this process.

It is also essential to define the technical specification of any field instruments, telecommunication systems, SCADA systems, or data historian systems that will interact with the leak detection system. Leak detection system outputs are only as good as their inputs. As such, it is imperative for the operator to understand any limitations of the existing infrastructure before the selection of an appropriate leak detection technology occurs. This data-gathering and analysis process may identify that the current infrastructure does not support the overall system functional requirements and technical specifications. If the infrastructure will not support the identified system, the owner must either upgrade the infrastructure or modify the requirements and specifications.

Once this information is gathered, the operator's team can finalize the requirements and specification development efforts. The resulting documentation is an essential element to determining the type, but not vendor, of LDS (or LDSs) that will meet these requirements. Section 11.2 discusses the process of selecting an appropriate leak detection technology or technologies.

11.2 LEAK DETECTION TECHNOLOGY/METHODOLOGY DECISION

Once the functional and technical specifications are completed, then there is the question regarding what types of leak detection systems exist that will meet these needs. This is not the same as the question regarding what leak detection vendor can provide a system that meets our needs. One must start by identifying the type of leak detection system or systems that will meet the needs and then proceed to vendor selection, not the other way around.

There are different approaches applicable to identifying the most appropriate leak detection technology for a unique pipeline environment. One method is the group meeting process. In this approach, a group of knowledgeable individuals and stakeholders meet to discuss the system needs and to identify the leak detection technologies that would be applicable. This approach relies on the participants' knowledge of the range of potential technologies, including the positive and negative benefits of each. On the positive side, with a set of very knowledgeable participants, this approach can identify a system or system type quickly. On the negative side, this approach may experience problems because of the following:

- The participants may not have an extensive knowledge of all available leak detection technologies.
- The participants do not have a full understanding of each technology's features, functions, and capabilities.
- One or more of the participants may have a preferred vendor or technological solution that they want to see implemented, also known as the "sacred cow" solution. Rather than selecting the best technological solution, the team ends up agreeing to the sacred cow.
- There is a lack of detailed understanding regarding how the pipeline's unique features may influence the functionality of each leak detection.
- Groupthink may occur. Groupthink is a pattern of thought characterized by self-deception and forced manufacture of consent. With groupthink,

everyone agrees to a selection because it appears that everyone else knows more and everyone else wants the identified selection. It often turns out most participants were assuming the same points and went along with the group instead of voicing their true desires.

Another approach is reliance on the in-house technical expert to make a recommendation. This method assigns the responsibility to a single person. While it may appear that this is a simpler approach it has all the potential positive and negative aspects of the group selection method except for the groupthink issue.

A better approach is a structured method that takes into consideration not only the functional and technical requirements but also items such as the following:

- What technologies are available?
- Is the technology available for and used in similar pipeline situations?
- Is the technology applicable to the operator's specific pipeline operations?
- What is the expectation that the potential technology will provide improved leak detection?
- What is the predicted life cycle cost? Is it justifiable in relationship to the remaining years of service of the technology?
- What is the age and condition of the potential technology?
- Is the technology compatible with the operator's existing installed systems?
- Is the system practical and feasible in terms of engineering and other operational aspects?
- Will there be environmental impacts of the selected technology and, if so, will these offset any anticipated environmental benefits?

One way to look at this stage of the analysis is like a funnel. The requirements and specifications form the boundary areas of the funnel. In the beginning, the funnel is wide open regarding the potential technologies, application methods, and approaches. At each stage of analysis, the funnel becomes smaller until selection of the final technology or technologies occurs.

Let us go back to the top of the funnel. At this stage, we have defined all the system requirements and specifications. This provides a boundary to work with. Now, we need to identify the technologies that may meet the identified needs.

The system requirements drive selection of the potential technologies. As an example, if the requirement were to provide a rapid response leak detection system over a very small area, then available technologies would be different than if the requirement were a leak detection system that encompassed the full pipeline and looked for very small leaks.

An approach to the technology assessment is to review various technology assessments published in academic and industry works, communication with industry peers, and/or the use of consultants who are experts in this field. At this stage, the analysis is not considering specific vendors. Rather, the focus is on the types of technology that may meet the defined requirements and specifications.

Once you have selected a technology or sets of technologies, two key questions are: (1) is the technology used in similar situations? and (2) is the technology available for use in your operations?

The objective of answering the first question is to ensure that successful deployment of the identified technology has occurred in a similar situation. There are situations in which the leak detection system's physics or approach appears to be fully compatible with the requirements. Yet, the theory underlying that system (or other issues) has limited the number of implementations. You would like to avoid spending a lot of time and energy supporting research and development when a commercially available system would meet all the requirements. This first question is intended to provide that focus.

The intent of answering the second question is to ensure that the technology is actually commercially available. Engineers and researchers continue to identify and publish papers on new approaches or methods. Although the research indicates that the concepts could add value, there is no commercially available product. We have identified a similar situation in the discussion on rarefaction waves. In Chapter 6, Rarefaction Wave and Deviation Alarm Systems, we discuss how the merging of the flow and pressure rarefaction waves could enhance the system capability; however, at the time of writing, there were no commercial systems that support this method. As you are evaluating technologies, you need to ensure that the technologies actually exist and are commercially available to you.

Now that you have identified a technology or sets of technologies, you need to determine if the technology is transferable to your specific pipeline and operations. If you are a small pipeline operator and one of the identified technologies requires dedicated analysts and engineering support 24 h/day, 7 days/week, 365 days/year, then you may determine that this technology is not realistically transferable to your operations. Other limiting issues could be that the identified technology requires field data at a rate that the current system will never support. Or an issue may be that the technology requires equipment installations where there is no existing supporting infrastructure. It really does not matter how great the technology is if it will not merge into the existing infrastructure; it will not be functional.

At this point, you should have a good idea of the technology or set of technologies that appear to meet your requirements. If so, then you are well on your way to making a preferred technology selection. Alternatively, you may have determined that no commercially available system can meet the defined requirements and specifications. If you have concluded that nothing appears to be available, then you must circle back to the requirements and specifications and see what can be changed or modified. From beginning to the end, this is an iterative process; as you gather information, you may need to revisit and perhaps change earlier decisions.

It is now time to ask the following question: is there a reasonable expectation that the selected technology will meet your defined leak detection requirements? To answer this question, you must analyze the selected technology (or technologies) in light of key issues such as available pipeline infrastructure (see Chapter 8: Leak Detection System Infrastructure) and the required LDS reliability, sensitivity, accuracy, and robustness.

We borrow from API 1130 [2] to define reliability, sensitivity, accuracy, and robustness:

- Reliability is defined as a measure of the system ability to render accurate decisions about the possible existence of a leak while operating within the pipeline and operational envelope. This is viewed as the ratio or frequency of false alarms to valid alarms under all defined operational states. If the system generates nearly continuous false alarms, then its reliability may not meet the intended system installation requirements. Refer to Chapter 5, Statistical Processing and Leak Detection and Chapter 9, Leak Detection Performance, Testing, and Tuning for more discussions about reliability and sensitivity.
- Sensitivity is defined as a composite measure of the size of a leak that a system is capable of detecting and the time required to issue an alarm in the event that a leak of that size occurs. Although sensitivity metric testing approaches are available and used on internal leak detection systems, no corresponding and universally accepted testing method or set of methods exists for external leak detection systems.
- Accuracy applies to the validity of leak parameters estimates, if provided, such as leak flow rate, total volume lost, type of fluid lost, and leak location. When comparing the leak detection system outputs to actual or simulated leaks, accurate system outputs will closely or exactly match the actual leak parameters.
- Robustness is defined as a measure of the system's ability to continue to function and provide useful information even under changing pipeline conditions (ie, transients) or in conditions in which data are lost or suspect. A robust system will continue to function under less than ideal conditions.

Note the distinction between reliability and robustness: reliability is a measure of performance within a specified operations envelope and robustness is a measure of the effective size of the operational envelope. For example, regarding the previous selection criteria, the feature "Be minimally impacted by communication outages or by data failures but provide alarms based on a degraded mode of operation" would be a robustness consideration [2].

We can now start to focus on technology selection considerations such as the maturity and condition of the apparent preferred technology. Is this a brand new technology that is closer to research and development or a mature technology with a broad installed base? If the technology is relatively new and untested, then the risk level (both initially and possibly throughout the life of the system) is much higher than that for a mature LDS. Conversely, is the technology at the stage where it may become obsolete, so that you might end up with an orphaned application? Ideally, the technology will be both mature and well supported and designed so that it may incorporate technical advances as they are carefully integrated into the LDS for many years to come.

A critical consideration is the ability of the preferred technology to integrate with the operator's existing infrastructure. Will you have to develop new interfaces, or will the current SCADA system, telecommunication system, and field devices provide the capabilities to link to the new technology? What about the new technology operating system? Is it a match for the rest of the current infrastructure systems? Can the field instrumentation data update rates support the new technology requirements? Evaluation of the whole system, as a system, must occur. Section 11.3 provides further discussion.

While evaluating how well the new technology will interact with the current infrastructure system, you must also take into consideration all operational impacts. These could include the ergonomics of the system as well as the calibration, tuning, and testing of all field devices that are directly connected to the technology and the leak detection application or technology.

Unique and major evaluation considerations, when retrofitting an existing pipeline with an external LDS technology, are concerned with what potential environmental impacts and risks the existing infrastructure will be subjected to. If you need to bury cable or a series of sensors along the pipeline, then do existing right-of-way agreements allow this? Will there be a negative environmental impact associated with the construct work, such as trenching? If so, then will the new leak detection technology benefits outweigh the potential negative environmental impacts?

Once you've answered the previous questions you should have a good idea of what leak detection technology or technologies are the most appropriate for the defined requirements. However, thus far, we have not taken costs into consideration. Leak detection systems have extended life spans, typically 10 years or longer. During this period, they all require "care and feeding." There will be daily support costs, upgrade costs, maintenance costs, and so forth. There is no such thing as an "install and forget" LDS that will continue to meet all requirements.

Therefore, a life cycle cost estimate should be prepared for the range of potential systems. The life cycle cost estimate will provide a common financial comparison. Deriving life cycle costs for competing technologies can help direct the final selection toward the most economical system capable of meeting the system requirements.

In summary, it is best to focus on determining the technology that is most appropriate for your requirements and marries well with your pipeline infrastructure and operating conditions. We recommend a structured approach to answering this question. Using a structured method provides traceability to the process as well as clear justification for the selection.

11.3 LDS SYSTEM INTEGRATION REQUIREMENTS

In this section, we provide insights and details regarding how the LDS exists within the overall system and supports the broader organizational requirements. LDSs, regardless of whether they are external or internal systems, are not technology islands. Instead, they must integrate with the existing pipeline infrastructure at some level.

From a minimalistic integration view, the leak detection systems must provide a leak alarm output to the controller. While providing just a leak alarm is possible, in reality, the interaction between the LDS and the operator infrastructure tends to be much broader and involves more data than a single status point.

The details of how the leak detection system must interface with the operator's technology and operational systems are pipeline- and operational-specific. Yet, some common aspects of this integration are technology independent. The following sections discuss some of these characteristics.

11.3.1 External Leak Detection Integration Requirements

External leak detection systems, as discussed in Chapter 7, External and Intermittent Leak Detection System Types, are technologies that typically detect the commodity after it has left the pipeline pressure boundaries. These can be point-specific or pipeline-wide detection methods. Regardless of the technology selected, the engineer must take into consideration that each field site will:

- require a source of electrical power
- need a shelter (even if it is just a weather-proof box on a pole) to protect the electronic equipment
- have a data transfer link between the external LDS and the SCADA or controller HMI system

These are minimal requirements for any external installed leak detection system. Other considerations are technology-dependent. As an example, if the technology selection involves sensing cables over a long distance, then multiple field sites and supporting infrastructure will be required. Implementation of these external leak detection locations may even require development of "evergreen" sites. Evergreen sites are those that lack supporting infrastructures such as electrical power, equipment shelter, and telecommunication systems.

External leak detection systems also generally transfer leak alarm and system status information to the controller. A common approach is to link the leak alarm status bit to a local programmable logic controller (PLC) or other type of data concentrator. These field devices are usually part of the SCADA system, which ultimately displays the alarm to the controller. Transferring additional system status information, such as running or system fault, may occur in this manner as well. The data transfer from the external leak detection system to the PLC, into the SCADA system, and display on the controller's HMI.

Alternatively, the transferring external LDS may transmit alarm and status information directly to the SCADA system over the telecommunications network. In this case, an interface control document (ICD) is required to define the message structure, data transfer rates and methods, and so forth.

In summary, the data presented to the controller and the type of external leak detection technology will drive the interface details.

11.3.2 Internal Leak Detection Integration Requirements

Integrating an internal LDS into the operator's existing infrastructure is usually more complex than for an external LDS. This complexity is a direct result of the fact that all internal LDSs are dependent on frequently sampled real-time field data. Chapter 8, Leak Detection System Infrastructure discusses the infrastructure requirements.

A detailed description of how to interface field data to the LDS is an essential part of implementing an internal leak detection system. This requires development of an ICD which provides specific details such as:

- Exact message format structure
- Exact details on each bit within the message structure
- What is the underlying communication structure? As an example, the communication from the SCADA to the leak detection system may be based on an existing standard such as Modbus, TCP/IP, OLE for Process Control (OPC), and so forth.
- Specific details on data quality bit definitions
- Analog ranges on a per-device level or at least on a common device type level
- Alarm limit details for each device with low, low–low, high, and high–high alarm limits
- Flow meter flow rate details such as gallons per minute, over range limits, and others
- Flow meter accumulator rollover values

- Analog instrument dead bands
- Field device update periodicity rates

Interface redundancy requirements are another key ICD element. Defining the redundancy details depends on the previously defined system requirements, physical environment, telecommunication infrastructure, and local area network details. This portion of the ICD will provide information such as:

- where the redundant systems will be located
- the telecommunication infrastructure and its redundancy level
- the local area network infrastructure and its redundancy level
- what the system component monitors and controls in the selection of the prime and backup leak detection application
- how the system fails from prime to backup in all failure scenarios
- the time limit between the prime system failure and backup becoming primary
- the capability that is provided to force the prime to backup (and vice versa)
- how the system returns to the "normal" prime system after a failure

These lists are not all-inclusive, but they do illustrate the level of detailed information to be included in the ICD. It is almost impossible to provide too much detail in this document.

There are also system integration requirements defining how and what data must be shared between the LDS and other operator systems, such as a data historian, a graphical information system, maintenance system, and so forth. For each point when the leak detection system will share data with a different system, a specific and unique ICD is required. The specific ICD information is unique to the actual systems involved but, generally, the details in the previous lists are illustrative.

Another very useful and necessary document is the controller interface document. This document details how the controller and/or leak detection analysts will directly interface with the system. This document must take into account the operator's control room standards, procedures, and training requirements. It is an effective approach to documenting the controller interface to develop a comprehensive set of "use cases."

Use case documentation is a well-established methodology for analyzing and documenting controller interface requirements. It is through the development of each use case that all system interactions and sequences of events are detailed. Defining approaches to developing use cases is outside the scope of this document. Yet, it is critical to point out that it is necessary to consider all controller and/or leak detection analysts' processes requiring interactions with the LDS. This level of detail helps to ensure that the final system will meet these needs.

11.4 SYSTEM TESTING

As part of any implementation effort, testing is required. The intensity level, timing, and types of tests performed are functions of the leak detection system technology and operator's project, as well as engineering policies and procedures. Acceptance testing should be part of any implementation plan. Chapter 9, Leak Detection Performance, Testing, and Tuning discusses testing in more detail. However, the following are several general testing guidelines that one should consider for any LDS system testing:

- The range of tests should be an explicit decision during the implementation, planning, and documentation phases.
- It is imperative for a direct link to exist between the tests and one or more of the system specifications and performance metrics.
- Test and validation processes must exist for all of the system's requirements, specifications, and performance metrics.
- All tests must be documented with an explicit and sequential set of test steps.
- Each test must have a clearly defined pass/fail criteria established prior to execution.

With the testing requirements established, ICDs developed, and all functional and technical specifications documented, you are ready to develop a list of vendors that may be capable of supplying the LDS capabilities. (Note that you might have a vendor list already, but the list may not be complete.) The following section expands on one process for obtaining a more complete vendor listing.

11.5 VENDOR IDENTIFICATION AND ASSESSMENT

Regardless of the type of LDS selected, operators usually elect to obtain commercial products rather than proceed with a custom one-off solution. Generally, this makes financial sense because a vendor can leverage a broader installed system base to provide a higher level of LDS maintenance and system upgrades at a lower cost to each client.

From a long-term support perspective, obtaining a commercial product also makes sense. Although vendors have come and gone, typically, when a vendor goes out of business, another LDS vendor acquires their software licenses and client base and continues to provide support. More often than not, leak detection system support is not lost when a vendor ceases to operate. The same is not always the case for custom-developed systems.

It is also true that some vendors have obsoleted older systems and have ceased to provide enhancements and, eventually, support. When this occurs, the vendor often offers their existing clients the opportunity to upgrade to a new application. So, proceeding with the assumption that the owner desires to obtain a commercial system, how does one go about identifying the range of vendors who may be able to meet the system requirements, and how do you select the best-qualified firm?

To start with, you must identify what firms are available that provide the type of leak detection technology you are looking for. If you have a commercial or procurement group, you could request a market survey to identify a full list of potential vendors. Generally, it is rare that an operator changes or installs a new leak detection system frequently enough that the procurement group will have sufficient in-house knowledge to adequately support the request. Although this approach uses a cross-check by other knowledgeable personnel, it helps to ensure that all potential vendors are identified.

Another vendor identification method is to solicit the assistance of a specialized leak detection consulting firm. These firms focus on the industry and maintain databases of the range of vendors and what products they offer. The firms also tend to have a deep knowledge base regarding each vendor's system capabilities. Leveraging specialized leak detection consulting firms tend to have a high level of payback in reducing time requirements, ensuring a fuller coverage of vendors, and providing an independent view.

You can also leverage the ability to develop an in-house list by having personnel attend one or more of the pipeline conferences that routinely occur. Many leak detection vendors attend these conferences and are very willing to discuss their products.

You can augment the conference knowledge base by contacting your peers in other pipeline companies. The intent of contacting your peers is to identify a broader data set of potential vendors.

Once a list of potential vendors is available, the operator's procurement group often initiates a competitive bid process. This is especially true if the goal is to obtain a new leak detection system as opposed to upgrading an existing system. If a competitive bid process occurs, then several key activities must occur.

The first major activity is to parse the full vendor list and identify all vendors who will want to participate. Determining which vendors would want to participate typically requires formal communication. The procurement group handles this process.

Another, often parallel, task is to develop the bid response scoring tool. These tools should be directly linked to the requirement documents and tend to be very detailed. Table 11.1 provides a subset of a scoring tool example.

Table 11.1 combines the vendor response score with the operator's assigned importance level for each requirement. The vendor response score, in this case ranging from 0 to 5, is the evaluator's assessment of how well the vendor's response meets the specific requirement. A score of zero (0)

TABLE 11.1 Technical Scoring Example

Requirement	Vendor-Assigned Score (0–5)	Company-Assigned Importance Level (0–3)	Score
1.1 Perform leak detection in slack regions	4	3	12
Section 1 Subtotal	—	—	98
2.2 Vendor provides 24 × 7 technical support	2	3	6
Section 2 Subtotal	—	—	47
Total	—	—	212

24 × 7 indicates 24 h/day, 7 days/week.

indicates that either the vendor did not respond or the vendor is not capable of meeting the identified requirement. A score of five (5) indicates that the vendor's system not only meets the basic requirements but also exceeds the minimum needs.

The importance level assigned by the operator acknowledges that not every stated requirement is critical. Within the full range of system requirements, some are must-haves and others can be viewed as adding value, but the operator will consider other approaches if the vendor does not quite meet these capabilities. There are also items that would be nice to have, but the vendor's offering would be acceptable if these were not available.

The operator-assigned importance level value reflects the different needs. In our example, level three would equate to must meet (MM) requirements. Some evaluators highlight these with some notation such as MM or other indicator. This informs everyone that if the vendor cannot meet this requirement, then this eliminates further consideration. The requirements that are good to have and nice to have are set as level two and level one, respectively. An importance level of zero indicates that the requirement statement is descriptive or informational only and not a specific requirement.

The overall evaluation and scoring process typically includes the technical analysis and assignment of the values discussed, financial scoring, and occasionally areas of regulatory compliance or other internal criteria scoring. The review process considers each of these areas individually and as standalone categories. Once all reviews are complete, the total score is developed. Table 11.2 is an example of how to combine the categories.

TABLE 11.2 Summary Scoring Example

Category	Score (0–500)	Rank Percent	Score
Technical	375	0.5	187.5
Financial	225	0.3	67.5
Regulatory	110	0.2	22
Total			277

In Table 11.2 the score is the total score that the reviewers assigned each area. The rank percentage column is the level of consideration each specific category will contribute to the final score. Viewed in a slightly different way, this is the operator's assigned evaluation area level of importance. In this example, Technical has the highest level of importance, followed by Financial and Regulatory.

Although the project team is developing the evaluation tool, the procurement group could be conducting the next major activity, the competitive bid. The actual bid process is outside the scope of this book and tends to be operator policy–driven and procedure-driven. Yet, regardless of operator policy and procedures, this process is time-consuming. Historically, we have found that the process can easily take 6 weeks or more from transmittal of the competitive bids to receipt of vendor responses. Vendors also frequently request an extension in response time. Therefore, from a scheduling perspective, a 2-month schedule allocation needs to be included for this activity.

The next steps involve evaluating the responses received and selecting a preferred vendor. With the selection of a preferred vendor, negotiations begin.

During the negotiation phase, one should be aware of two points. First, during all negotiations, changes to the final requirements as well as terms and conditions will probably occur. These changes occur because, during the competitive bid and negotiations, each side learns more of what is required or possible from the other side. This learning drives changes.

The other point of consideration is that the vendor's competitive bid price is not necessarily final. As changes to the technical requirements occur, changes to the final pricing follow suit. The negotiation team should be aware of this and should be prepared for these changes.

Finally, one should be prepared for the fact that as the vendor and operator learn more about the details of the vendor's offering and the operator's requirements through the negotiation process, it may be necessary to revisit the selection of the preferred vendor.

Assuming that the negotiations have been successful, the project proceeds. The details of how the project proceeds are outside the scope of this book and are quite dependent on the specific project. At the end of the project, the formal commission of the system occurs.

11.6 COMMISSIONING

Ultimately, commissioning of a new or major upgraded leak detection system needs to occur. Depending on the operator and regulatory requirements, there are various approaches to system commissioning.

System commissioning is the process of verifying and documenting that the as-installed leak detection technology functions according to the design specifications, requirements, and the overall design intent. These include technical as well as operational requirements. As such, the commissioning scope of effort is dependent on the installed technology and the documented requirements and testing defined in Section 11.4 and in Chapter 9, Leak Detection Performance, Testing, and Tuning.

Regardless of the technology, there is a direct link between the project design requirements and specifications and the commissioning process. It is also essential for any commissioning effort to include written and agreed to details regarding:

- The specific test or tests that will be performed
- Participant roles and responsibilities
- What "system acceptance" specifically means
- The criteria that will be used to determine acceptance
- Procedures to be followed when items are identified that must be corrected and retested

Another commissioning function is the development and implementation of a detailed commissioning phase plan and schedule. The plan and schedule will define what tests occur in what sequence, anticipated duration of each test, and identification of who must be present for the tests.

Ultimately, the commission effort will answer the following question: does the leak detection system perform according to the operator's specifications and requirements?

Assuming that the commissioning process was successful, the project phase concludes and the system moves into the long-term support life cycle.

11.7 LONG-TERM SUPPORT ISSUES

Leak detection systems are long-term applications. These systems may continuously operate for 10 to 15 years or more. Therefore, the operator must understand the system's long-term support requirements. Long-term support includes personnel training, routine testing and calibration, emergency response, system upgrades, and other factors.

TABLE 11.3 Long-Term Support Training Requirements

Role	Training Requirement	Objective
Management	General operational and technical aspects Regulatory requirements	Ensure management has a core understanding of system operation and technical needs as well as associated regulatory requirements
Controller	Detailed operation and leak analysis procedures General technical Regulatory	Provide a detailed understanding of system operation, how to interact with the system and procedures to follow
Analysts	Detailed operation and leak analysis procedures Detailed technical Regulatory	Provide the most in-depth system knowledge that allows them to support the system and assist in system alarm and event analysis
Engineer	Detailed operation and leak analysis procedures Detailed technical Regulatory	Provide the most detailed level of system knowledge that allows them to support the system and assist in system alarm and event analysis
Field technical work force	General system understanding Detailed field instrument testing and calibration Regulatory	Ensure that these resources have the skill sets to trouble-shoot, upgrade, replace, and calibrate all leak detection associated field instrumentation
Field engineers	General system understanding Detailed field instrument leak detection support requirements Regulatory	Ensure that these resources understand the details associated with all supporting leak detection field devices

Everyone associated and involved with any part of the leak detection system must be aware of their specific role and associated responsibilities, and they should be trained in these. Table 11.3 provides a general view of the different roles and training requirements that may apply to any operator and staffing structure.

Training requirements, in general, are applicable regardless of the specific type of leak detection technology installed. However, specific details on maintenance and system upgrades are unique to the installed technology type. As an example, an external leak detection system generally requires less daily support than computational or computerized pipeline monitoring internal leak detection applications.

As noted, depending on the leak detection technology, different types of field equipment will be required. Additionally, the quantity and location of these devices will vary depending on the leak detection system requirements. A consistent requirement for any supporting field instrumentation is explicit written maintenance and upgrading policies and procedures as well as a routine maintenance schedule. These policies, procedures, and maintenance schedule assist in ensuring that the leak detection system infrastructure is properly designed, maintained, and upgraded as required.

Another requirement is a detailed system upgrade or change management policy and procedure. As previously noted, these systems have extensive life spans, and infrastructure upgrades and changes occur. Although the exact procedure is device- or system-dependent, the operator must ensure that a detailed change management policy and supporting procedures exist, that personnel are trained to implement them, and that they are followed.

When the installed leak detection system is one of the internal types, additional long-term support requirements exist. One critical long-term need is the presence of a vendor support agreement. With the exception of a leak detection system developed in-house, all vendor-supplied software is proprietary. As such, the operator's support personnel cannot access, make changes, or fix bugs found within the application. Having a vendor support contract provides the operator an avenue to request bug fixes, receive system upgrades, and request analysis support.

The complexity index of a internal leak detection system is also higher than that of most external systems. Therefore, specific skill sets and training are required to provide daily support and system event analysis. This specialized support tends to be the responsibility of the SCADA analysts, leak detection analysts, or leak detection engineers. With the addition of each leak detection system, the support personnel work load increases. Long-term support for these systems tends to require additional support personnel. Increased staffing requirements are common when installing a new internal leak detection system.

In summary, when installing a new leak detection system, long-term support needs will:

- Increase annual training requirements for all personnel who work with or support the system
- Require new leak detection performance and support policies
- Require new procedures to support the leak detection performance and support policies

- Require a higher level of maintenance support for leak detection field instruments
- Require specific management of change procedures

Furthermore, for new internal leak detection systems, it is a common long-term support outcome that an increase in internal analytical and/or engineering staff is required in addition to direct support personnel.

In summary, for an internal leak detection system, long-term support planning should always include a vendor support agreement. The vendor support agreement should include vendor technical personnel access and application upgrade services. Other possible vendor support activities could include leak event analysis support and annual training for controllers and analysts.

Implementing a leak detection system is a detailed process that encompasses many aspects of the operator's technical and personnel processes. These processes take time to implement and require attention for the life of the system. The operator who takes the time to define, detail, and plan the implementation effort will reduce the overall project cost and increase the probability of success. LDS project implementation failures often result from lack of definition and insufficient detailed planning as opposed to failures of the technology.

REFERENCES

[1] Henrie M, Carpenter P, Liddell P. Leak detection – 1: Alaska Lessons guide system selection, implementation. Oil Gas J July 18, 2010;108(26).

[2] American Petroleum Institute. Computational pipeline monitoring for liquids. Recommended Practice 1130 (API RP 1130):2007.

Chapter 12

Regulatory Requirements

Transportation of hazardous liquids and gases carries inherent risks to the population, pipeline facilities, third-party facilities and dwellings, and the environment. Due to these factors and others, various regulations have been developed and applied. These regulations may apply to all or a part of the hazardous material pipeline system. Areas of regulation may cover activities such as design, operation, maintenance, modification, and disposal.

This chapter presents a discussion of regulations in place around the world at the time of writing this book. The intent of this chapter is to provide the reader with an understanding of what regulations exist at this time.

Although the regulatory landscape is dynamic (as are all regulations), we can say that the sources of regulation vary according to the nation or nations in which the pipeline infrastructure exists. The regulatory bodies may include national, regional, and local government agencies, or even a combination of these regulatory agencies. The actual level of oversight may include a single entity, as may apply to an intrastate pipeline, or it may include different national regulations as well as local regulations for a pipeline system that crosses national borders, such as from Russia to Ukraine or Germany to Italy. Regulations for pipelines that transition from one nation to the next may also include international contracts and requirements, as well as right-of-way agreements.

As a cautionary note, regulations change, are modified, and new ones are written. The information provided here is not intended to be legal advice or used as a definitive reference list. Readers must perform their own due diligence within the context and environment they are working to verify or find any local, regional, and national regulatory requirements that would apply to their system of interest.

12.1 THE UNITED STATES OF AMERICA REGULATORY ENVIRONMENT

As noted previously, the United States relies extensively on hazardous liquid and natural gas pipeline transportation infrastructures. Some of these pipelines cross state borders, and some are fully contained within a state. Others

Pipeline Leak Detection Handbook. DOI: http://dx.doi.org/10.1016/B978-0-12-802240-5.00012-1

cross the United States and Canadian borders as well as the United States and Mexican borders.

As an example, there are pipelines that transport hydrocarbon commodities from Utah to Idaho and on to Oregon, across state boundaries. Within the United States, these are classified as "*interstate*" pipelines.

Conversely, as noted, there are pipelines that originate in, transverse, and terminate all within a single state. These are "intrastate" pipelines [1].

Within the United States, the distinction between interstate and intrastate is critical from a regulatory point of view. The distinction between interstate and intrastate pipelines clearly establishes the hierarchy level of who oversees safe operation of the pipeline infrastructure.

Specifically, for interstate pipelines, regulatory oversight resides with the Department of Transportation (DOT) Pipeline and Hazardous Material Safety Administration (PHMSA). Although PHMSA has many departments, the Office of Pipeline Safety (OPS) is PHMSA's administration arm for all hazardous materials and natural gas pipelines.

OPS is responsible for development and enforcement of interstate hazardous material pipeline regulations. They are also responsible for developing and promoting other risk management approaches targeted at assuring pipeline safety during the pipeline life cycle, which includes activities such as:

- Design
- Construction
- Testing
- Operation
- Maintenance
- Emergency response

OPS does not provide all interstate regulatory activities as a single entity. As part of their charter, they support the operation of, and coordinate with, the US Coast Guard in the National Response Center. OPS also serves as a DOT liaison with the Department of Homeland Security and the Federal Emergency Management Agency on matters involving pipeline safety.

OPS also develops and maintains partnerships with other federal, state, and local agencies, public interest groups, tribal governments, the regulated industry, and other underground utilities to address threats to pipeline integrity, service, and reliability and to share responsibility for the safety of communities.

A major OPS focus is on the administration of Pipeline Safety regulatory programs and establishment of the guiding regulatory agenda. They achieve this through development of regulatory policy options and initiatives, as well as through researching, analyzing, and documenting social, economic, technological, environmental, safety, and security impacts on existing/proposed regulatory, legislative, or program activities involving pipeline safety.

OPS also oversees pipeline operator implementation of risk management and risk-based programs. They also administer a national pipeline inspection and enforcement program.

Other OPS activities include providing technical and resource assistance for different state-specific pipeline safety programs to ensure oversight of intrastate pipeline systems and educational programs at the local level. OPS supports the development and conduct of various pipeline safety training programs for federal, state regulatory, and compliance staff, and the pipeline industry in general.

OPS serves as a focal point for pipeline safety studies and reports by the National Transportation Safety Board (NTSB), DOT Inspector General, and Government Accountability Office, and other oversight and/or stakeholder entities [2].

The following documents include applicable federal regulations for interstate hazardous materials and natural gas pipelines:

- 49 CFR 190—Pipeline Safety Programs and Rulemaking Procedures
- 49 CFR 192—Transportation of Natural and Other Gas by Pipeline
- 49 CFR 195—Transportation of Hazardous Liquids by Pipeline

12.1.1 US Interstate Federal Regulations

US interstate federal regulations are different for liquid pipelines and for natural gas pipelines. The following sections provide a brief look at the current interstate leak detection regulatory landscape as it applies to each type.

Interstate Hazardous Liquid Pipeline Regulations

Leak detection system regulatory requirements, as applicable to interstate hazardous liquid pipelines, are primarily located in 49 Code of Federal Regulations (CFR) 195.49 CFR 195.452 (3) (*Leak detection*). This regulatory section requires the operator of a hazardous liquid pipeline to have a means to detect pipeline leaks. Exactly how the pipeline operator is to perform or provide leak detection is not explicitly stated.

Although the hazardous liquid pipeline operator must provide leak detection, in general, the regulations expand this requirement to very specific locations classified as High Consequences Areas (HCA).

HCA pipeline definition originates in 49 CFR 195.450. HCA means any hazardous liquid pipeline whose location is one or more of the following:

- A commercially navigable waterway, which means a waterway where a substantial likelihood of commercial navigation exists
- A high-population area, which means an urbanized area, as defined and delineated by the US Census Bureau, that contains 50,000 or more people and has a population density of at least 1000 people/square mile

- Another populated area, which means a place as defined and delineated by the US Census Bureau, that contains a concentrated population, such as an incorporated or unincorporated city, town, village, or other designated residential or commercial area
- An unusually sensitive area, such as a drinking water or ecological resource area, that is unusually sensitive to environmental damage from a hazardous liquid pipeline release

Hazardous liquid pipeline HCAs, as part of Integrity Management, have additional regulations that extend the regulatory requirements of non-HCA pipeline areas. As stated in 49 CFR 195.452 [3] "An operator must evaluate the capability of its leak detection means and modify, as necessary, to protect the high consequence area." Leak detection systems are a specific element in the HCA mitigation measures. HCA leak regulations also include the following non-HCA regulatory requirements.

Non-HCA hazardous liquid pipeline leak detection regulatory regulations begin in 49 CFR 195.134. This portion of the regulation is specific to computational pipeline monitoring (CPM) leak detection systems and operators of all pipelines that transport single phase (without gas in the liquid). Specifically, an operator who installs a new CPM leak detection system or replaces an existing CPM leak detection system must do so in accordance with Section 4.2 of the American Petroleum Institute (API)-recommended practice number 1130 (API RP 1130) [3]. Compliance with API 1130 Section 4.2 is applicable to the leak detection system design. The regulations also require the owner/operator to apply any other design criteria addressed within API RP 1130.

49 CFR 195.444 also requires the owner/operator comply with API RP 1130 operation, maintenance, testing, record keeping, and dispatcher training requirements.

US federal regulations are silent regarding:

- Minimal detectable leak size
- Maximum time to detect
- Leak location performance capability
- Number of acceptable false alarms
- Number of leak detection systems that must be deployed
- Types of acceptable leak detection technologies

Interstate Gas Pipelines

Regulation of interstate gas pipelines resides with DOT and PHMSA as detailed in 49 CFR 192. Classification of gas pipelines consists of the following categories:

- Gathering line—a pipeline that transports gas from a current production facility to a transmission line or main

- Distribution line—a pipeline other than a gathering or transmission line
- Transmission line—a pipeline, other than a gathering line, that: (1) transports gas from a gathering line or storage facility to a gas distribution center, storage facility, or large-volume customer that is not downstream from a gas distribution center; (2) operates at a hoop stress of 20% or more of specified minimum yield strength (SMYS); or (3) transports gas within a storage field [4]

Interstate gas pipeline regulatory requirements are different from those of hazardous liquid pipelines. In this case, the federal leak detection regulations for gas pipelines are not as extensive as those for hazardous liquid pipelines. As an example, 49 CFR 192.903 identifies that if personnel response time to mainline valves on either side of an HCA exceeds 1 hr from the time of an event, then the system should provide additional measures such as a leak detection system.

Rather than extensive regulations for technology-based leak detection systems, 49 CFR 192 identifies that gas pipeline operators shall have a patrol program that looks for indications of leaks and for the need to perform leakage surveys on an annual basis not to exceed 15 months. These leak surveys will include the use of leak detection equipment. Although the regulations do not define what constitutes leak detection equipment, it is a requirement that the equipment must be able to detect the atmosphere of gas (wording is from the CFR). One assumes that gas sensors and infrared detectors would fit within the definition of leak detection equipment that can detect the atmosphere of gas.

US interstate gas pipeline federal regulations are silent regarding:

- Minimal detectable leak size
- Maximum time to detect
- Leak location capabilities
- Number of acceptable false alarms
- Number of leak detection systems that must be deployed
- Types of acceptable leak detection technologies

Natural gas interstate federal regulations do not explicitly require pipeline operators to install and operate continuous monitoring leak detection systems, such as a CPM system.

Intrastate Hazardous Liquid Pipelines

As noted previously, intrastate pipelines are those pipelines that are fully contained within one state. US federal pipeline safety statutes allow states to assume safety authority over intrastate hazardous liquid pipelines through PHMSA-signed Certifications and Agreements under 49 USC (United States Code) 60105 and 60106 [5,6].

These certifications and agreements require that the state will:

- At minimum, adopt federal safety standards applicable to their intrastate pipelines
- Enforce each standard
- Provide for penalty amounts of $100,000 per day up to a maximum of $1,000,000 for a related series of violations as set out under CFR 190.223 maximum penalties
- Have the same authority as that provided to the DOT
- Encourage and promote the establishment of a program designed to prevent damage
- Cooperate fully in the system of federal monitoring
- Provide annual progress reports as specified

Once the state and PHMSA have entered into the agreement, the state has regulatory authority over intrastate pipelines. For most states, and with respect to leak detection in hazardous material liquid pipelines, the regulatory requirement is that operators using computer-based leak detection systems must comply with API RP 1130. Operators who do not use computerized leak detection must perform the basic process of monitoring flow and pressure to detect large pipeline breaks [7].

Hazardous liquid pipeline operators are also obligated to have a prompt and effective means of detecting and responding to leaks. This is to include operating plans and procedures required by the pipeline safety regulations, which include an engineering analysis to determine if a computerized leak detection system is necessary and appropriate. If the engineering analysis does not support the deployment and use of a computerized leak detection system, then the operator shall perform a line balance calculation no less frequently than once per hour whenever product is flowing in the pipeline. This is the minimum regulatory intrastate hazardous liquid pipeline leak detection requirement that a state must implement. However, any state may elect to enact and enforce regulations that are more stringent. As an example, the Alaska Department of Environmental Conservation (DEC) and Department of Natural Resources (DNR) have regulatory jurisdiction over intrastate pipelines, and they have implemented additional regulatory requirements based on pipeline classifications.

The state of Alaska DOE has defined/classified two types of pipelines that have different leak detection regulatory requirements. The most stringent requirements are for transportation infrastructures classified as crude oil transmission pipelines. These requirements are located in the state of Alaska regulation 18 AAC 75.055 (*Leak Detection, Monitoring, and Operating Requirements for Crude Oil Transmission Pipelines*). This regulation requires that "a crude oil transmission pipeline must be equipped with a leak detection system capable of promptly detecting a leak, including:

1. if technically feasible, the continuous capability to detect a daily discharge equal to not more than 1% of daily throughput;
2. flow verification through an accounting method, at least once every 24 h; and
3. for a remote pipeline not otherwise directly accessible, weekly aerial surveillance, unless precluded by safety or weather conditions [8]."

The second state of Alaska pipeline classification defines "facility pipelines." Facility pipelines are defined as an onshore or offshore facility of any kind and related appurtenances including, but not limited to, a deep-water port, bulk storage facility, or marina, located in, on, or under the surface of the land or waters of the state, including tide and submerged land, that is used for the purpose of transferring, processing, refining, or storing oil [9]. A vessel, other than a nontank vessel, is considered an oil terminal facility only when it is used to make a ship-to-ship transfer of oil and when it is traveling between the place of the ship-to-ship transfer of oil and an oil terminal facility [9]. No specific state of Alaska DOE leak detection regulations exist for these facility pipelines.

Another state that has enacted regulations beyond the federal requirement is California. The state of California has expanded on DOT PHMSA leak detection regulations to require the utilization of leak detection on California intrastate pipelines. According to California regulation section 51013, "Any new pipeline constructed after January 1, 1984 ... shall include a means of leak detection ..." [10]. All leak detection requirements of 49 CFR 195 also apply.

The remaining states rely on the adoption of 49 CFR 195 as the foundation regulatory requirements.

12.2 CANADA

Oversight and regulation of Canada's interprovincial and international pipelines reside with the National Energy Board (NEB). Principal regulations reside within the National Energy Board (R.S.C., 1985, c. N-7) as amended in 2015-06-23 [11]. NEB's high-level objective is to ensure that any pipeline within its oversight is "...designed, constructed, and operated in a manner that is safe and security, protects the environment and the public, and is economically feasible and in the public interest" [12]. CSA Z662 annexes address leak detection requirements. The annexes are not mandatory; they are recommendations.

Additional pipeline regulations also exist within some provinces, such as Alberta and British Columbia, that comprise the CSA Z662 standard annexes, which are not mandatory at the national level but are mandatory and enforceable regulations at the province level.

Canada's regulations rely heavily on established industry standards. For pipelines, the primary standard is CSA Z662. Within Annex E of this standard resides recommended leak detection requirements. Although there are many similarities between Annex E and API RP 1130, some differences do exist. One difference is that leak detection systems that are defined in API RP 1130 as computational pipeline monitoring systems (CPM) are instead defined in Annex E as computerized leak detection systems (CLDS). Another key difference between the two standards is that Annex E focuses on requirements rather than the broader descriptive approach found in API RP 1130. It is important to note that Annex E, from the national standard perspective, is a nonmandatory annex. Nevertheless, as noted, several provinces have adopted the CAS Z662 annexes as regulatory-required.

Key features of Annex E include:

- The operator should consider all types of leak detection techniques for the pipeline of interest
- CLDS system implementation should be considered part of an overall leak detection strategy

Annex E also provides a prescriptive table of maximum calculation intervals, as shown in Table 12.1.

TABLE 12.1 CSA Z662 Annex E Data Retrieval and Calculation Windows

Segment Throughput Classifications, Data Retrieval, and Calculations Windows			
Class location/ segment type	All class locations		
	LVP and HVP transmission and gathering pipelines		
Nominal flow rate m^3/h	Flow >150 & all HVP	Flow <150 & >15	Flow <15
Fluid classification	LVP or HVP	LVP	LVP
Not to exceed data retrieval interval	5 min	1 h	24 h
Not to exceed calculation windows	5 min 1 h 24 h 1 week 1 month	1 h 24 h 1 week 1 month	24 h 1 week 1 month

Notes:
1. The data retrieval intervals and the calculation window are the maximum time suggested.
2. In general, shorter data retrieval intervals and shorter calculation windows will improve leak detection thresholds.
3. The calculation windows cannot be shorter than the data retrieval interval.
4. To maximize sensitivity, the calculation for the next longer window should accumulate the imbalances of the shorter window.
5. Adhere to intervals unless it can be technically demonstrated that overall leak detection effectiveness can be equal to or better than achieved when the interval is different.
6. It is not necessary to calculate all windows in the longer windows if they do not improve CLDS performance.

Annex E is also more prescriptive than API 1130 in allowed system uncertainties. The total system uncertainty values should not exceed 5% per 5 min, 2% per week, or 1% per month. The total uncertainty value is a cumulative uncertainty value, which includes the model calculation uncertainty, operational practices, instrumentation, and telecommunications transmission uncertainties.

In a parallel mode, the Canadian oil and gas industry has established a Canadian Pipeline Technology Collaborative (CPTC). "The CPTC functions as a collaborative science, technology, and research hub that supports the alignment of scientific research and technology development to sector priorities" [13]. One of the CPTC's programs is to develop and validate new technologies, which will improve the reliability and sensitivity of leak detection systems. This is CPTC's National Leak Detection Science, Technology, and Innovation Program. The CPTC is not a national or province regulatory oversight agency.

12.3 GERMANY

Germany has established pipeline safety rules in "Technische Regel für Rohrfernleitungen" (Technical Rules for Pipelines (TRFL)). TRFL applies to:

- Pipelines transporting flammable liquids
- Pipelines transporting liquids that are dangerous for water
- Many pipelines transporting gas

TRFL leak detection requirements are provided in chapter 11.5 of the document and include the following:

- Two autonomous, continuously operating systems that can detect leaks in steady-state conditions
- One of these systems, or a third one, able to detect leaks in transient conditions
- One system to detect leaks in shut-in conditions
- One system to detect gradual (creeping) leaks
- One system for fast leak localization

As noted, it is required that the leak detection systems must be redundant with at least one of the systems capable of detecting leaks in a transient state as well as during shut-in states.

Additional TRFL requirements include the need to provide leak location capabilities. TRFL also, in principle, requires that all leak detection supporting field instruments should be redundant as well. Yet, there are usually sufficient allowed variances that redundant field instrumentation is not a normal occurrence.

12.4 REGULATORY REQUIREMENTS IN OTHER JURISDICTIONS

Hazardous pipeline transportation systems exist around the world, and each of these systems has a corresponding risk associated with its utilization. In recognition of this, various other nations recognize that, for the most part, these systems require a level of regulation and oversight. The following identifies some of the nations and their associated hazardous material leak detection environments.

12.4.1 Brazil

Brazil's hazardous pipeline leak detection regulations are found in "Publication of the Technical Rules for Pipelines according to 0.5 of the Pipeline Decree." Within this decree:

- The pipeline operator must select and implement a leak-monitoring process, consistent with the level of operational complexity and the product carried. This process should be based on the leakage risk and response time to events for each passage through the use of equipment, systems, or procedures that have leak detection capability.
- The pipeline operator must ensure that the documentation of the monitoring process should be available for the operating activities, inspection, and maintenance of the duct.
- The pipeline operator must ensure that when this process is dependent on measurement equipment, it must be regularly calibrated.
- Regardless of the leak detection method, the carrier should periodically review their performance and make the necessary adjustments.
- The pipeline operator shall ensure that the procedures for the leak detection process are reviewed and updated when necessary, or at least every 3 years.

12.4.2 Great Britain

The Health and Safety Executive (HSE) "A guide to the Pipelines Safety Regulations 1996" defines Great Britain leak detection regulations. Within this guideline, leak detection requirements include:

- Safety systems also include leak detection systems where they are provided to secure the safe operation of the pipeline
- The method chosen for leak detection should be appropriate for the fluid conveyed and operating conditions
- Notification to HSE may not be required for minor adjustments to the pipeline leak detection system

Hazardous material and gas pipeline leak detection regulatory requirements comprise a dynamic area. Readers should familiarize themselves regarding the current state of regulatory requirements using this chapter as a guide.

REFERENCES

[1] Department of Transportation, Pipeline and Hazardous Materials Safety Administration, 49 U.S.C. 60101 US Code – Section 60101, Hazardous Liquid Pipeline Safety Act of 1979.

[2] Office of Pipeline Safety. http://www.phmsa.dot.gov/portal/site/PHMSA/menuitem.6f23687cf7b00b0f22e4c6962d9c8789/?vgnextoid=ca9fe4fca0380110VgnVCM100000762c7798RCRD.

[3] American Petroleum Institute. Computational pipeline monitoring for liquids. API Recommended Practice 1130 (API RP 1130).

[4] Department of Transportation, Pipeline and Hazardous Materials Safety Administration, 49 CFR 192. Transportation of natural and other gas by pipeline: minimum federal safety standards.

[5] 49 U.S. Code 60105, State pipeline safety program certificates.

[6] 49 U.S. Code 60106, State pipeline safety agreements.

[7] Advisory Bulleting, ADB-10-01. https://www.federalregister.gov/articles/2010/01/26/2010-1497/pipeline-safety-leak-detection-on-hazardous-liquid-pipelines#table_of_contents.

[8] State of Alaska, Selected Oil and Other Hazardous Substances Pollution Control Statues and Regulations.

[9] Alaska Statutes, AS 46.04.900. Definitions.

[10] California Codes, Section 51010-51019.1.

[11] National Energy Board Act (R.S.C., 1985, c. N-7) Amended 2015-06-23. http://laws-lois.justice.gc.ca/eng/acts/N-7/index.html.

[12] Energy and Mines Ministers' Conference. Safety and security of energy pipelines in canada: a report to ministers, Sudbury, Ontario; August 2014.

[13] Canadian Pipeline Technology Collaborative. http://www.thecptc.ca/.

Chapter 13

Leak Detection and Risk-Based Integrity Management

Leak detection is just one aspect of the complex problem of managing spill and rupture risk. Prudent pipeline owners and operators must strike a balance between their various needs, wants, desires, regulatory and legal constraints, and financial considerations when developing a program to manage and minimize the risks associated with unplanned commodity releases. This chapter digs into the complex problem of managing integrity breach risks by first discussing the magnitude of the pipeline spill or rupture probability based on published statistics, and the effectiveness of leak detection systems in detecting such breaches. It then discusses the place that a well-managed and multi-pronged leak and rupture detection system occupies in the larger context of integrity-based risk management issues.

One must keep in mind that a leak detection system does nothing to prevent leaks and ruptures. Instead, its purpose and usefulness target reducing the time required to detect and respond to a leak and to assist the operator in locating the leak, thereby reducing the consequence of a leak.

We acknowledge at the outset that this chapter analyzes risk from the perspective of costs of leaks reported in the United States to the US Department of Transportation (DOT) Pipeline and Hazardous Materials Safety Administration (PHMSA) from 2010 to 2015. The costs of leaks provided by that data are lower than the true costs because the costs reported to PHMSA exclude penalty and litigation costs. In addition, PHMSA costs do not include those that a pipeline operator may incur in pipeline and system upgrades to regain public trust following an event.

In addition, no attempt was made to quantify the cost of human life. Between 2010 and 2015, PHMSA reported 17 gas pipeline leak–related fatalities and seven liquid pipeline leak–related fatalities. Compared to the amount of product transported, many will argue, probably correctly, that the loss of life is less than would result from other possible transportation methods. However, each of these events is tragic. Moreover, given that some of these events are rightly categorized as disasters, there is public and regulatory pressure to operate a pipeline as safely as possible. We believe that

Pipeline Leak Detection Handbook. DOI: http://dx.doi.org/10.1016/B978-0-12-802240-5.00013-3

installing a leak detection system and making the long-term commitment required to fully realize the potential of the system have the associated benefit of demonstrating that the operators are responsibly taking steps to improve the safety of their operations.

However, such assessments are difficult to quantity and are outside the scope of this book. We turn now to an easier problem: quantifying risk based on the costs, frequencies, and sizes of pipeline leaks and spills contained within the PHMSA databases. Let us start by investigating the probability that an operator will experience a spill, as well as the impact that spills have on the economic well-being of the operation.

13.1 QUANTIFYING INTEGRITY BREACH RISK AND IMPACT

Quantifying the risk of a pipeline integrity breach is not trivial. In the United States, PHMSA is responsible for compiling pipeline safety and integrity-related accident statistics. These statistics are compiled from operator-supplied reports involving liquid pipeline spills and gas pipeline leaks in the United States that meet or exceed regulatory reporting limits. PHMSA makes the resulting spill data available to the public in the form of submitted reports compiled in a database on its website (http://www.phmsa.dot.gov/). These data provide a wealth of information that is invaluable for understanding the pipeline integrity breach risk as well as the cost of dealing with the consequences. In the next sections, we deal with some of the conclusions obtained by analyzing these data. However, note the following caveats: (1) the conclusions reached are applicable to pipeline operations in the United States and may have limited utility for other jurisdictions, and (2) the PHMSA data do not specifically address some aspects of integrity breaches that are uncommon in the United States (such as commodity theft).

13.1.1 Liquid Pipeline Spill Risk, Magnitude, and Cost

Table 13.1 summarizes an analysis of PHMSA data from 2010 through 2015. US hazardous liquid pipeline miles of operation grew slowly during this period, at a little less than 2% per year, with the total averaging approximately 190,000 miles. The data also show that the spill incident rate is relatively constant on a year-to-year basis. Spills for the entire US hazardous liquid pipeline system averaged 393 per year, with a relatively low 12% standard deviation, indicating high predictability for this incident rate on a year-by-year basis. On a per-mile basis, this implies a predictable incident rate of approximately 0.002 reportable spill incidents per mile of operating pipeline per year.

Spill volumes shown in this table only address unintentional losses and do not include intentional releases. Such additional releases may be required when spills occur in hilly or mountainous terrain, and additional drainage is required to effect repairs. The total unintentional volume of spilled commodity tends to

TABLE 13.1 PHMSA Hazardous Liquid Pipeline Spill Incident Statistics, 2010–16

Year	2010	2011	2012	2013	2014	2015	Averages	Standard Deviations
Pipeline miles in operation	181,986	183,500	186,221	192,417	199,243	199,243[a]	190,435	7698
Total spills	350	346	366	401	446	447	393	45.99
Spill incident rate (incidents per year/mile)	0.001923	0.001886	0.001965	0.002084	0.002238	0.002243	0.002057	0.000158
Total unintentional spilled BBL	100558	89111	45884	117467	47298	101571	83648	30094
Spilled BBL/ incident	287	258	125	293	106	227	216	81
Fatalities	1	1	3	1	0	1	1.2	1.0
Injuries	3	2	4	6	0	0	2.5	2.3
Total incident costs ($)	$1,075,193,990	$273,532,147	$144,910,768	$278,525,540	$129,809,687	$237,055,553	$356,504,614	$357,745,778
Spill cost/ incident	$3,071,983	$790,555	$395,931	$694,577	$291,053	$530,326	$962,404	$1,049,751
Spill cost/ operating mile ($/mile)	$5,908	$1,491	$778	$1,448	$652	$1,190	$1,911	$1,988

[a] *Used 2014 data because 2015 data were not available.*

be more variable than the incident rate would imply. The average number of spilled barrels on an annualized basis was 83,648 BBL per year (for an average release volume per spill of 216 BBL). The standard deviation of the average totalized annual barrels was 30,094 BBL (more than one-third of the average). The high standard deviation indicates that the variability of the spilled volumes is higher than the unpredictability of the spill incident rate on a fractional or percentage basis.

Total costs associated with these spills also tend to show a higher relative standard deviation than is typical for the incident or spilled volume rates. Total nationwide costs varied between a low of $130 million in 2014 and a high of a little more than $1 billion in 2010. This is a substantial range! The average annual total spill cost for all US liquid pipelines was $356,504,614 per year during 2010 to 2015. On a per-incident basis, this averages to nearly $1 million/spill incident. The corresponding estimated per-mile cost was $1911 per mile/year over this period. Note that given the low-inflation environment characteristic of the analysis period, there was no application of a spill cost inflationary adjustment.

It is worth noting that these costs are biased upward somewhat by one exceedingly costly incident in 2010 that cost nearly $1 billion in spill-related charges and damages. This is a dominating outlier when considered among all other spill costs. If we neglected this incident, then the data in 2010 would more closely match the data of other years covered by this analysis. This would reduce the per-incident and mile-based spill costs to approximately 60% of the values summarized here and in Table 13.1. However, automatic elimination of undesirable outliers can be misleading because they still represent a part of the probability distribution, albeit a low probability portion of the data. We will discuss this further, but need to point out here that outliers like this, sometimes referred to "black swans," are incidents that could bankrupt a company. Note that even if we eliminate this data point, the resulting operational cost would still be very high.

Can we assume that liquid commodity spill-related costs are generally proportional to the spill size? Fortunately, the PHMSA data include estimates provided by the pipeline operators of the spill size. The distribution of spill sizes for 2010 to early 2016 is shown in terms of the *complementary cumulative distribution function* (CCDF) in Fig. 13.1.

A variable's CCDF curve is also referred to as the *risk curve*, the *tail distribution*, the *exceedance*, or the *survival function*, and is defined as:

$$\mathrm{CCDF}(x) = 1 - \mathrm{CDF}(x) = 1 - \int_{-\infty}^{x} f_x(t)dt$$

EQUATION 13.1 Complementary Cumulative Distribution Function

where $f_x(t)$ is probability density function for random variable t and CDF(x) is the cumulative distribution function for $f_x(t)$ when evaluated at x. When

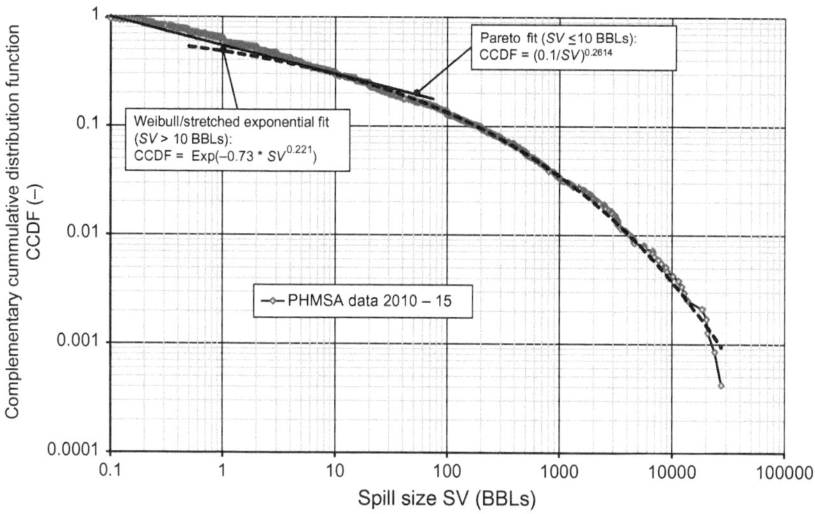

FIGURE 13.1 Spill volume CCDF curve (PHMSA data: 2010–15).

plotted as a function of the variable, the vertical coordinate of the CCDF risk curve defines the probability that a randomly generated instance of the variable t will be at least as big as any selected value x on the horizontal axis. Alternately, it is a probability that the random variable t will be greater than x. Consequently, the CCDF always starts at 100% on the left side of the chart and tails off to zero on the right. In this case, the independent variable is the spill volume SV or:

$$\text{CCDF}(SV) = 1 - \text{CDF}(SV) = 1 - \int_{-\infty}^{SV} f_{SV}(t)dt$$

EQUATION 13.2 CCDF as a Function of Spill Volume

Consequently, the CCDF curve for the spill volume (SV) specifies the probability that the volume will equal or exceed any particular value. Fig. 13.1 shows that the CCDF function (based on an integration of the ranked spill data) for the spilled barrels is well behaved and relatively smooth. It is often convenient when performing risk analysis to develop an analytic function describing the CCDF. In this case, a piecewise or composite curve fit involving a Pareto distribution for very small spills less than 10 barrels and a stretched exponential fit (the CCDF for a Weibull probability distribution) for larger spills work well. Pareto distributions [1] are heavy-tailed power curve fits of the form:

$$\text{CCDF}_{\text{Pareto}}(x) = \left(\frac{x_M}{x}\right)^{\alpha}$$

EQUATION 13.3 Pareto Probability Distribution CCDF

where x_M is the Pareto scale parameter (the Pareto distribution is not defined for $x < x_M$) and α is the Pareto shape parameter. Extensive use of Weibull distributions occur in reliability and survival analysis [1]; as noted previously, they have a CCDF defined as a stretched exponential function:

$$\mathrm{CCDF}_{\mathrm{Weibull}}(x) = e^{-(x/\lambda)^k}$$

EQUATION 13.4 Weibull Probability Distribution CCDF

In this equation, λ defines the Weibull scale parameter and k is the Weibull shape parameter. Note that a value of $k < 1$ indicates that the failure rate decreases with the size of the independent variable, which, in this case, is the spill size (*SV*).

Fig. 13.1 indicates that most reported spills are relatively small. Approximately 70% of all spills involve the release of 10 barrels or less, with a full 40% involving less than 1 barrel of spilled commodity. However, the right tail of the curve indicates that there is a 3% to 4% chance that the spill size will exceed 1000 BBL and an approximately 0.4% probability that the release will exceed 10,000 barrels. Very large spills more than 10,000 BBL are rare events constituting less than 0.4% of all events. However, based on the current incident rate in the United States (393 spills/year), statistically we can predict that one to two such occurrences may occur every year somewhere in the United States. For 100,000 barrels or more, truly enormous spills, the average period between incidents (based on an extrapolation of the chart trends) would be 25 years or more. Note that such extrapolation is beyond the range of the data, and is risky because it assumes that these historical statistical distributions will persist well into the future.

Fig. 13.2 shows the distribution of hazardous liquid spill remediation costs over the same period. The CCDF curve of the data is fit moderately well by a log-normal distribution [1] based on the mean and variance of the natural logs of the spill incident costs. However, the quality of the log-normal fit was determined to be of poorer quality in the extreme tail region, which is often the area of greatest interest for risk analysis. Consequently, we developed improved fit utilizing a log-t distribution [2,3]. The log-t distribution is the Student-t equivalent of a log-normal distribution and assumes that the t distribution describes the logarithms of the independent variable. The complementary cumulative distribution function for the log-t distribution is:

$$\mathrm{CCDF}_{\mathrm{Log}-t}(x|\mu_{\ln(t)}, \sigma_{\ln(t)}, \nu) = 1 - \mathrm{CDF}_{\mathrm{Log}-t}(x|\mu_{\ln(t)}, \sigma_{\ln(t)}, \nu) = 1 - \int_{-\infty}^{x} f_{\ln(t)}(t)dt$$

EQUATION 13.5 Log-t Probability Distribution CCDF

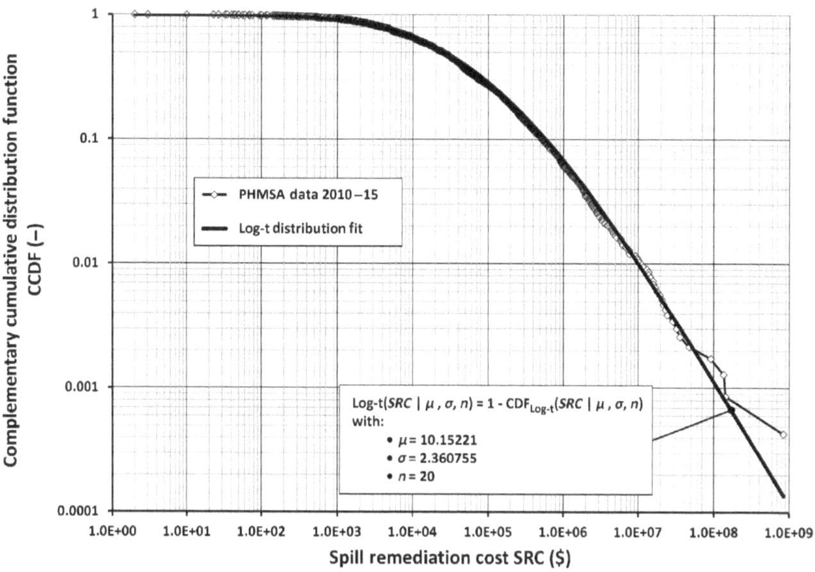

FIGURE 13.2 Spill cost CCDF curve (PHMSA data: 2010–15).

where:

$$f_{\ln(t)}(t) = \frac{\Gamma\left(\frac{\nu+1}{2}\right)}{\sqrt{\pi\nu}\,\Gamma\left(\frac{\nu}{2}\right)}\left(1+\frac{1}{\nu}\left(\frac{\ln(t)-\mu_{\ln(t)}}{\sigma_{(t)}}\right)^2\right)^{-\frac{\nu+1}{2}}$$

EQUATION 13.6 Support Probability Density for Log-t CCDF

In this equation, Γ is the Gamma function, $\mu_{\ln(t)}$ is the scale parameter (the average of all of the values of the log-transformed data $\ln(t)$), $\sigma_{\ln(t)}$ is the shape parameter (the standard deviation of all of the values of $\ln(t)$ relative to $\mu_{\ln(t)}$), and ν is the integer value of the degrees of freedom.

For $x = SRC$ (the spill remediation cost), Fig. 13.2 indicates that we obtain a good fit using the log-t distribution assuming $\nu = 20$. This figure confirms that spills are costly: 65% required more than $10,000 to handle damages and remediation work. However, it is in the tails where the costs become noticeable. The top 1% of spills consumed more than $10 million/event, and the top one-tenth of 1% of spills cost more than $100 million each. We note that, based on this distribution, the very expensive 2010 event (the last data point on the right) may indeed be an outlier because it deviates significantly from the trend curve. Based on this distribution, the average event period for spills costing $1 billion in 2015 US dollars is between 6 years (based on the data) and 25 years (based on the log-t tail fit). It is possible that a future analysis using more data collected over a longer period can resolve this issue.

It is reasonable to ask whether the spill volume and spill remediation costs are related. To answer this question, regression was performed using a logarithmic transformation of the PHMSA spill and cost data and then implementing a standard linear regression on the transformed data. Analysis indicates that an improved fit occurs when we add a small constant, SV_0, to the spill volume SV, so that the final form of the fit for the log-transformed data was:

$$\ln(SRC) = \ln(A) + B \cdot \ln(ASV)$$

EQUATION 13.7 Log-spill Cost vs. log-spill Volume Regression Fit

where A and B are regression constants, SRC is the spill remediation cost, and ASV is the adjusted spill volume:

$$ASV = SV + SV_0$$

EQUATION 13.8 Support Function for Spill Cost/Volume Regression Fit

Fig. 13.3 provides the results of this regression analysis. We see that there is a distinct slower-than-linear power relation (because $B = 0.6312 < 1.0$) that describes the cost as a function of the spill size. However, the weak r^2 value of 0.307 (where r is the correlation coefficient) implies that the regression reduces the standard deviation of the log-transformed data by only approximately 17%. This further implies that the spill size explains or accounts for only one small part of the total costs of the commodity release.

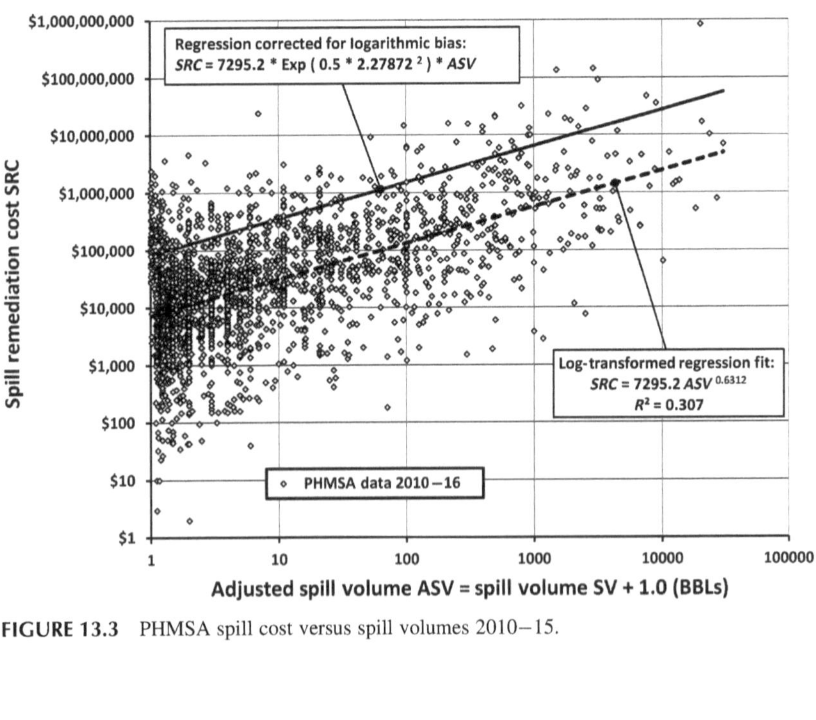

FIGURE 13.3 PHMSA spill cost versus spill volumes 2010–15.

Also, note that because this is a regression of transformed variables, the regression curve calculated for Eq. (13.7) is not a true mean cost curve. This is because the variation of the true data deviations around the regression line is log-normally distributed. This means that the transformed data regression line will not properly estimate the true average of the data, especially if there is a large residual variance around the regression line (which, according to Fig. 13.3, is clearly the case). Instead, it will tend to estimate the median of the data [4]. To specify the mean spill cost curve, we recognize that the residual standard deviation of the log-transformed data $\hat{\sigma}_{\ln(t)}$ around the regression line is:

$$\hat{\sigma}_{\ln(t)} = \sigma_{\ln(t)}(1-r^2)^{1/2}$$

EQUATION 13.9 Residual Standard Deviation for Log-transformed Data

where, again, $\sigma_{\ln(t)}$ is the standard deviation of all of the values of the log-transformed data relative to the average of the log-transformed values $\mu_{\ln(t)}$. Because the residual errors are log-normally distributed around the true mean, we apply a correction to Eq. (13.7) to obtain the estimate for the mean spill cost curve:

$$\mu_{SRC}(SV) = A \cdot e^{\frac{1}{2}(1-r^2)\sigma^2_{\ln(t)}} \cdot (SV + SV_0)^B$$

EQUATION 13.10 Mean Spill Cost vs. Spill Volume Equation

where $\mu_{SRC}(SV)$ is the regression estimate of the mean spill cost. The estimate of the true standard deviation of the data $\sigma_{SRC}(SV)$ around the estimated mean regression line can be likewise obtained through a similar correction:

$$\sigma_{SRC}(SV) = \left(e^{(1-r^2)\sigma^2_{\ln(t)}} - 1\right)^{1/2} \mu_{SRC}(SV)$$

EQUATION 13.11 Spill Cost Standard Deviation vs. Spill Volume Equation

Values for the constants A, B, SV_0, and σ_t can be obtained from Fig. 13.3. This figure clearly shows that while the median cost for small spills, approximately a few barrels, is approximately \$10,000, the *average* cost for these smaller spills is closer to approximately \$100,000, more than 10-times the median. This is because the costly spills, although relatively rare, are *very* expensive and significantly bias the mean results upward, but the outliers do not change the median cost. Across the range of 2- to 30,000-barrel spills (approximately 5 orders of magnitude), the mean and median costs increase nearly three orders of magnitude, so that the mean cost for the largest spills is approximately \$50 million (approximately 10-times the median cost for large spills, which is approximately \$5 million). As noted previously, however, the residual error of the regression mean curve is still substantial. Consequently, we again conclude that spill costs are strongly influenced by factors other than the released volume.

The following points provide a summary of pipeline liquid commodity spills in the United States for 2010 through 2015:

1. The spill incident rate is fairly constant from year to year, at approximately 0.002 incidents per mile/year. This works out to approximately 393 events/year.
2. Aggregate spill volumes are somewhat more variable on an annualized basis, ranging between approximately 45,000 and 100,000 BBL/year for an average rate of 84,000 BBL/year. Although the median spill is only approximately two barrels, the analysis indicates that due to the long tails in the spill distribution, the average spill is considerably larger, at approximately 213 barrels. This analysis indicates that the United States can expect to see approximately one to two spills in excess of 10,000 barrels annually.
3. Spill costs on an incident basis are highly variable, ranging from virtually nothing to as much as $840 million over the analysis period. Although the median spill cost is approximately $25,000 per event, the long tails in the distribution result in a much higher average spill event cost of approximately $960,000; we call this the Bill Gates effect. If you try to obtain the average annual income for the town where Mr. Gates lives, it will be higher than everyone else's salary. This is because his annual income dwarfs that of everyone else in the region. We see the same effect in spill costs. Analysis suggests that extremely costly events in excess of $100 million will occur approximately every 2 to 3 years.
4. Spill costs are partially dependent on the size of the spill. However, the residual variance of the regression function is indicative of significant dependence on factors other than spill volume.

With respect to the last item, factors influencing the apparent spill cost might include:

- Compilation errors in the PHMSA data because operators have not properly calculated or measured the spill volumes
- Inconsistencies in how various operators measure spill-related costs
- Variation in costs based on the level of damage incurred by third parties due to local property and environmental sensitivities that vary on a site or case basis
- Dispersion of damage to large numbers of third parties, leading to coalition building and its adverse consequences, such as class action lawsuits and higher settlement; such conditions will tend to occur for spills that propagate to open or navigable water or damage to commons such as groundwater or drinking water
- Damage to revered public properties or holdings, such as parkland, public monuments, or wildlife
- Injury or death

13.1.2 Liquid Commodity Spill Source Classification

PHMSA classifies spill sources as leaks, mechanical punctures, ruptures, overfills or overflows, and "other" spills. "Other" spills are those that fail to fit any identified PHMSA categories or those not classified by the pipeline operators who submitted the spill report. PHMSA leak classification includes connection failures, cracks, seal or packing leaks, or "pinholes." Cracks and seal leaks appear to be predominately located in pump station and terminal facilities. Pinholes are typically defined as very small holes usually caused by corrosion, but operators do not provide an orifice area or diameter. Mechanical punctures result from mechanical damage inflicted on the pipe due to site work, whereas ruptures occur by mechanical failure of the pipe under excess static or transient pressure. Overfills or overflows tend to be associated with commodity being transferred to some form of tankage past the full level of the tank. The excess commodity then overflows the tank.

Fig. 13.4 shows the result of a PHMSA data analysis regarding these spill source categories. We quickly note that, on an incident basis, leaks far outweigh all other primary categories, constituting a whopping 77% of all spill events. However, leaks are only responsible for 39% of the total spilled volume. This is primarily because the average spill volume is relatively low: only approximately 108 barrels. On an incident basis, pinholes and seal/packing leaks (generally confined to stations) constitute the largest categories of leaks, contributing a combined 45% of all spill events. However, these two categories contribute only 8% of all spilled volumes.

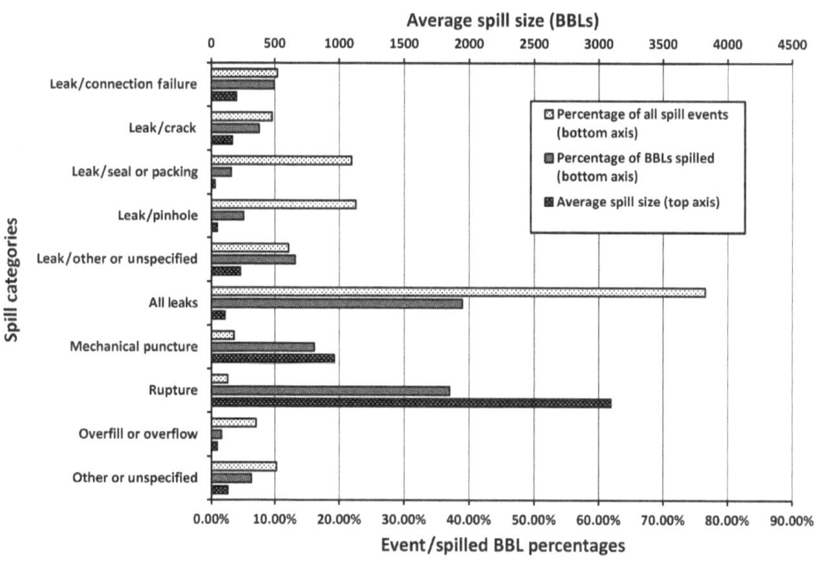

FIGURE 13.4 PHMSA spill source statistics 2010–15.

Mechanical punctures and ruptures, however, while accounting for only 6% of all spill events, are responsible for 53% of all barrels spilled. This is primarily because the average spill volumes per event are the highest of any spill category: 963 barrels for punctures and 3094 barrels for ruptures.

The analysis also determined that only 23% of the spills originated in the pipeline right-of-way (ROW). Most of the remainder (76%) originated in other operator property (stations, terminals). However, 69% of the total spill volume originated on the ROW, primarily because right-of-way spills are far larger than owner-controlled property spills (642 vs 87 barrels). This appears to be primarily because large-volume punctures and ruptures occur primarily on the ROW, whereas much more frequent but low-volume connection failures, cracks, and seal or packing leaks are primarily confined to stations.

13.1.3 Gas-Phase Commodity Integrity Breaches

PHMSA also tracks incidents for natural gas and related gas-phase commodities (synthetic gas, hydrogen, propane, etc.). Table 13.2 provides a summary of these analysis results. Gas releases in this table only address unintentional losses that meet PHMSA reporting requirements and do not include intentional releases such as blow downs, which are required for repairs. One thing we can see is that the estimated incident rate per mile is much lower than the equivalent metric for liquid spills: 0.0004 events per mile/year for gas pipelines versus 0.002 events per mile/year for liquid lines (see Table 13.1). However, these figures are not comparable because, although liquid spills are usually visible (particularly for liquids with low vapor pressure) and are always eventually detected, small leaks from gas pipelines are invisible, difficult to detect, and easy to ignore.

We also see that the annual cost per mile for natural gas integrity events is much smaller. Gas pipeline integrity breach cost per mile is approximately $400 per mile/year, whereas liquid line spill costs are closing in on $2000 per mile/year. Again, part of this is likely because: (1) there are generally no cleanup costs from a gas loss event unless there is property damage or loss of/injury to life and (2) there are simply fewer documentable events to address. Note that the high cost associated with events in 2010 is associated with one exceedingly high-cost event totaling $375 million. This is approximately 10-times the size of the next most costly event in the 2010 to 2015 period. The fact that 2010 was the most costly year during this period for both liquid and gas pipelines is assumed to be coincidental.

For these reasons, as well as the lack of regulatory requirements, leak detection systems have enjoyed far less penetration in the gas-phase transportation pipeline infrastructure than they have in the liquid pipeline business. PHMSA does not currently collect statistics on whether computational pipeline monitoring (CPM) systems are installed on pipelines experiencing incidents. Barring new regulation, we expect slow penetration of this technology

TABLE 13.2 PHMSA Natural Gas Pipeline Accidental Release Incident Statistics, 2010–16

Year	2010	2011	2012	2013	2014	2015	Averages	SD
Pipeline miles	317,925	317,570	319,347	320,186	319,401[a]	319,401	318,971	1005
Total release incidents	116	128	115	111	141	147	126	14.94
Total unintentional release (MCF)	2066487	1583609	1865160	1395615	2057205	1723546	1781937	266704
Fatalities	10	0	0	0	1	6	2.8	4.2
Injuries	61	1	7	2	1	14	14.3	23.4
Total cost	$413,151,925	$125,497,792	$57,969,638	$52,106,991	$57,423,969	$44,698,503	$125,141,470	$144,120,068
Incident rate (incidents per mile/year)	0.0003649	0.0004031	0.0003601	0.0003467	0.0004415	0.0004602	0.0003961	0.0000468
Released gas rate (MCF per mile/year)	6.4999	4.9867	5.8405	4.3588	6.4408	5.3962	5.5872	0.8405
Normalized cost ($ per mile/year)	$1,299.53	$395.18	$181.53	$162.74	$179.79	$139.94	$393.12	$453.66

[a] *Used 2014 data because 2015 data were not available.*

in the natural gas pipeline business in the near future. There are no federal regulations in the United States requiring gas pipeline leak detection, although there appears to be some movement in this direction.

13.1.4 Leak Detection Technology Versus Other Detection Mechanisms

So, where does leak detection technology fit into the current picture for detection of leaks, spills, and ruptures? For the natural gas industry, leak detection systems are not a significant part of the existing natural gas pipeline infrastructure, and how this might change in the future is uncertain. For liquid pipelines, we can evaluate at least part of this question by comparing CPM systems to other detection mechanisms. We can achieve this because PHMSA collects statistics on CPM systems (the largest implemented category of implemented LDS) via its "accident identifier" category, but does not specifically include all other types of leak detection systems.

Fig. 13.5 shows a summary of the means by which spills were identified in the PHMSA database. By event, the largest category of spill detections was by site personnel, who were responsible for more than 40% of the spill identifications.

The second largest category was either identified as "other" or left blank, with more than 35% of the events falling into this category. A brief, informal review of the notes accompanying the summarized reports indicates that many of these unclassified detections were effectively detected by operator field personnel, although a significant fraction fell into the other categories. Some were detected by field device alarms. Nearly all of the spills for which

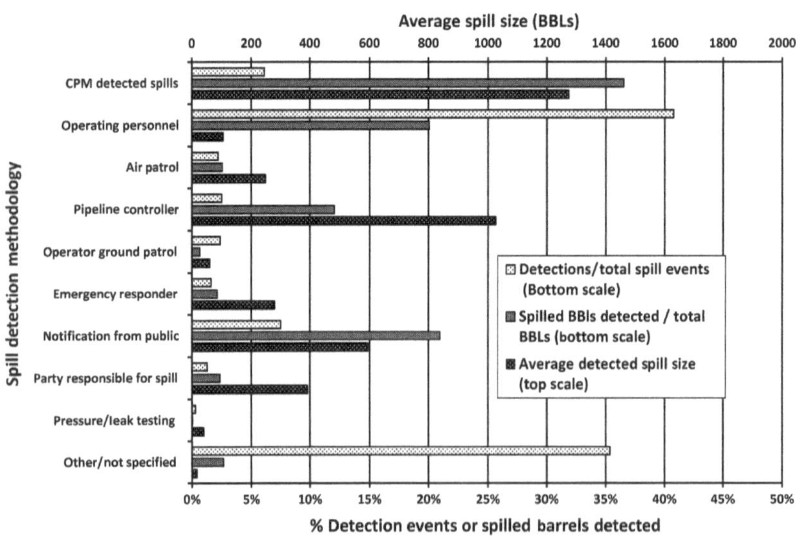

FIGURE 13.5 PHMSA spill detection and identification statistics 2010–15.

the accident identifier category was left blank were very small (approximately one BBL). Even when including incidents with an "accident identifier" of "other," the average spill size was only 16 barrels, primarily due to the influence of a limited number of larger spills.

The next largest category by event was notifications from the public, comprising 7.5% of all spills. CPM systems detected only 6% of all spill events.

We now consider the efficiency of all methodologies in terms of detecting spilled *volumes* because we already know from our previous analysis that spill volume is a factor (although clearly not the only one) in determining spill-related costs. CPM systems are now the best performers because they detect approximately 37% of all spilled volumes. Following this are notifications from the public and operating personnel, who directly observe 21% and 20% of all spilled volumes, respectively. Pipeline controllers are next, remotely detecting approximately 12% of the total via the pipeline SCADA system.

Right-of-way detections are an important sub-category for CPM spill detection because they occur outside of stations where operating personnel have a more limited presence. In fact, CPM system relative performance definitely improves on the ROW, as we see in Fig. 13.6, where CPM detections have climbed from 6% to 9% of the total ROW spill events. Field-operating personnel still have a healthy share of event detections, with 26% of all ROW spills. Interestingly, event detections by the public climb to a respectable 21%, which suggests that this is a resource that may be worth cultivating as part of a pipeline spill-risk control plan. From a spill *volume* aspect, the fraction of the total spilled commodity detected by a CPM system has gone from 37% of all volumes spilled to 47% of total releases on the ROW.

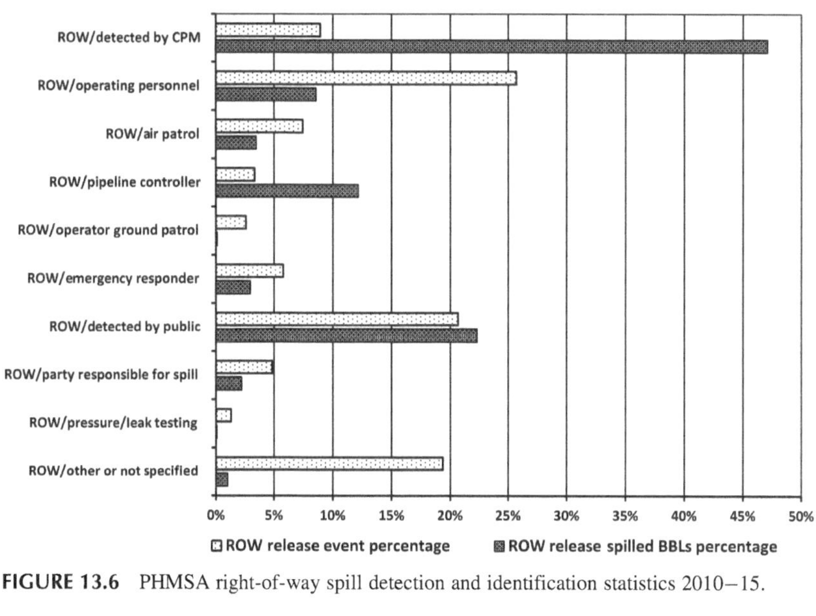

FIGURE 13.6 PHMSA right-of-way spill detection and identification statistics 2010–15.

Field-operating personnel detect only about 8% of all ROW spill volumes, while members of the public now detect over 22% of ROW spill volumes. In fact, third parties in general (members of the public, emergency responders, and parties responsible for the spill) catch roughly 28% of all escaped commodity volumes on the right-of-way.

It is important to note that the PHMSA accident event data indicate that only 31% (721 out of 2358 spills over approximately 6 years) of pipelines have a CPM system available. If we assume that this is an unbiased estimator of CPM system penetration (more on this in a bit) for all pipeline operations, then these CPM performance figures would need to be revised to account for the fact that CPM cannot catch a leak on a pipeline if it is not installed! One issue here is inconsistency in the data: of the spills apparently identified by PHMSA as "CPM leak detection system or supervisory control and data acquisition (SCADA)-based information (such as alarm(s), alert(s), event(s), and/or volume calculations)," 144 were detected by CPM systems. However, only 93 of these clearly indicate that a CPM system was present. We surmise that the discrepancy is associated with the "SCADA-based information" category, which includes other alarms and notifications, so we discard these potentially nonapplicable data. If we perform this correction and analyze only those pipelines that have a CPM system installed, then the event detection rate increases from 6% to 13%, and the volume detection efficiency increases from 37% to 50%. As noted, there is potential for bias in the CPM installation rate parameter (ie, there may be something special about pipelines that elect to install CPM, or about those that elect not to implement it). Consequently, we simply assume that the CPM performance metrics fall somewhere in these ranges. The average CPM-detected spill volume also increases to 1600 BBL, which is much higher than the average volume for any other detection category.

A brief analysis was performed to determine the relationship between spill volume and size. The distribution of leak flow rates is a particularly difficult function to understand, because this information is not required by PHMSA, probably because it is difficult to estimate. However, PHMSA does provide operating pressure and orifice size parameters for punctures and ruptures. Assuming a simple orifice relationship for these spills, the data indicate that there is a distinct relationship between the estimated leak rate and the spill size, albeit with a relatively low correlation coefficient and large residual variance on the regression line. Given the preponderance of small spills in the data, we very tentatively conclude that more than 50% to 70% of all liquid commodity spills occur at rates less than 20 to 50 BPH.

In summary:

1. When it comes to spill detection, there is no silver bullet. Spills are currently detected through a variety of methods (leak detection technology, by operator field personnel, pipeline controllers, ground and air patrols, and parties responsible for initiating the spill, as a result of pressure testing, and by the public and emergency responders).

2. A very large fraction of spill events (>60%) are identified by humans. Operating personnel are responsible for a significant fraction of the total (41%), but members of the public are responsible for a sizable fraction as well, indicating that the operator might want to come up with ways to enlist their assistance in detecting and reporting spills.
3. Although leak detection in the form of CPM systems may superficially appear to be on the unimpressive side on an event detection basis, they are particularly good at quickly detecting leaks associated with large spill volumes or high flow rates, and should definitely be a component of a hazardous liquid pipeline operator's portfolio of spill detection methods. Assuming actual installation of such a system, expected incident detection rates are between 6% and 13%, whereas volumetric detection rates are between 37% and 50% of total volumes lost.

13.2 UNDERSTANDING THE CONSEQUENCES OF A SPILL

No operator can do a good job of integrity-loss management if the consequences of a spill or rupture are not well understood. This section discusses approaches used to analyze the consequences of pipeline integrity breaches. We start by looking at pipelines with low vapor pressure, then move on to pipelines with high vapor liquid (HVL), and conclude with a look at gas-phase systems.

13.2.1 Low-Vapor-Pressure Liquid Pipeline Spills

Low-vapor-pressure (LVP) liquids such as crude oil and most hydrocarbon products consist of commodities with vapor pressure below atmospheric pressure and comprise the largest fraction of liquid commodities transported by pipeline. Once on the ground, these commodities have the advantage that their low vapor pressure reduces (but does not eliminate) the risk of explosion or out-of-control fire. That said, many LVP liquid pipeline commodities are still flammable, and all are considered to be dangerous in the sense that contact with them or their vapors can be hazardous to health and the environment, can contaminate groundwater and potable water supplies, are injurious to wildlife, and are expensive to clean up.

Key to understanding the impact of a low-vapor-pressure liquid commodity is being able to predict the outcome of a breach in integrity. Spill simulation tools fall into the following categories: (1) leak detection and spill volume evolution; (2) terrain surface flow spreading; and (3) subsurface infiltration and plume modeling. A tool for transient modeling of hazardous LVP pipeline spills has been described [5]. Note that the ability to handle draindown effects inside the pipe by performing reliable slack line calculations may be important. The described tool addresses transient spill volume calculation, with special focus on volumes lost during the leak detection, pipeline shutdown, pipeline isolation by closing of ESD (emergency shutdown) valves, and site draindown. It is particularly useful for pipelines in mountainous terrain, where pressure at the leak site during the draindown can drop very slowly, with resulting high intentionally spilled volumes and extended draindown times. Note that the described simulator is focused on land spills. Calculation of spill volumes from offshore lines can be challenging, particularly when stability and buoyancy effects must be addressed.

Basic gravity current models that address surface spreading and ground infiltration on horizontal solid surfaces are described by references [6–8]. More sophisticated and comprehensive models must address the transient evolution of surface spills on uneven terrain, including spreading, pooling, evaporation, weathering, ground infiltration, and other effects. A good summary of the modeling approaches to be taken in addressing these physical processes is provided by reference [9], and it discusses the state of the art provided at the time of its publication. A land spill simulator capable of addressing spreading, infiltration, and evaporation on an arbitrary terrain grid is also described in reference [10].

Special spreading models address oil slick spreading on bodies of water. A major part of such analysis involves addressing the fact that floating oil displaces the water to a great extent, thus modifying similar land-based spreading analysis [11] and convection due to river, lake, and ocean currents, as well as movement due to wind [12]. In addition, water spills are subject to weathering, evaporation, breakup due to wave action, and infiltration into the water itself. A sophisticated simulator may address the impact of dispersants as well as the placement of containment devices such as booms [13]. A recommended available simulator capable of handling trajectory and weathering analysis based on geographic information system (GIS) files is the National Oceanic and Atmospheric Administration (NOAA) General NOAA Operational Modeling Environment (GNOME) [14].

Note that although it is common to assume that hydrocarbons will float on water, asphalts and some crude oils can be dense enough to sink and/or move beneath the surface, complicating the spill analysis. In addition, many nonhydrocarbon hazardous liquids may be polar in nature and thus miscible with water, turning the spill modeling problem into a hazardous contaminant infiltration simulation, with molecular and turbulent diffusion as an essential part of the model [15].

The fundamental transport simulation of pipeline spill contaminants in groundwater and water reserves are handled through the same basic processes and physical models. A good case study involving contaminant infiltration and dispersion is provided elsewhere [16]. Supporting MODFLOW and MT3D software are also described elsewhere [17,18].

13.2.2 HVL Spills

High-vapor liquid spills are complicated by the higher than atmospheric pressure associated with the spills causing rapid evolution of gaseous plumes from the spills and by the fact that most such liquids are highly flammable, explosive, or poisonous, or all of these. Consequently, the risk of injury or damage due to leakage of HVL commodities is significantly elevated relative to low-vapor-pressure commodity spills, although cleanup costs may be lessened because most of the lost commodity either vaporizes or burns off, or both. Spreading of the spill is generally more limited in size due to the rapid evaporation rate. Means of calculation of volumetric losses from an HVL pipeline is discussed in reference [19], taking into account the evolution of liquid and mixed-phase commodity regions inside the pipe near the rupture site. A complete integration of pipeline loss, pooled spill size spreading with evaporation corrections, and plume development are discussed in detail by reference [20].

Useful software packages related to simulation of plume development and low/high-explosive limit definitions are HGSYSTEM [21,22] and ALOHA (Areal Locations of Hazardous Atmospheres) [23,24]. Note that although gas pipeline leak rate simulation is provided to some extent by these packages, spill rate evolution from HVL pipelines is not provided and must be separately determined by the user. Use of a constant escape rate based on normal operating pressure at the leak site will neglect the drop in pressure as the rarefaction wave develops inside the pipe and will tend to overestimate the extent of the plume.

13.2.3 Gas Pipeline Ruptures

Gas pipeline ruptures are easier to simulate than liquid pipeline ruptures because issues involving slack line, drain-down, spill pooling and spreading, evaporation, and other gravitational and two-phase effects can be neglected. Escaping flow at the rupture site in the compressible commodity is choked, and this must be addressed by the simulation [25]. Any good transient pipeline simulator should be capable of modeling such releases, although specialized simulators similar to the one described [5] for liquid pipelines still appear to be uncommon.

The HGSYTEM and ALOHA software packages previously described are good candidates to perform plume modeling for gas pipelines, and both packages do provide support for pressurized tanks and pipes. However, the source models are relatively simple and do not allow for the analysis of logistic effects such as the closurc of ESD valves.

13.2.4 Summary

It should be clear that due to physical modeling complexity and the fact that all pipeline implementations are unique, there is no one-size-fits-all tool available that will allow operators to handle all spill and rupture consequence modeling that an operator might find desirable. It is up to the operator to understand his/her own situation and then identify the required software requirements, obtain or develop the software resources, and perform the required calculations to determine the scope of its risk.

13.3 LEAK DETECTION AS A COMPONENT OF PIPELINE LOSS-OF-INTEGRITY RISK MANAGEMENT

A major problem with leak detection is that it is necessarily incomplete. The leak detection system or infrastructure contributes to reduction in potential injury, damage, and cost, but it is necessarily a horse-out-of-the-barn approach. Ideally, the commodity would never escape the pipeline pressure boundary and the leak detection system would never be triggered. Furthermore, detecting the leak, although it may reduce cost and damage by limiting the volume of the lost commodity, does nothing to address how cleanup costs can be further minimized by reducing additional volumes that can end up outside the pipe following the shutdown period.

A well thought-out spill risk management program will consist of the following critical components:

- An analytic base that supports the evaluation and analysis of the other components of the loss of integrity program
- A pipeline design that supports rapid detection of leaks and minimum impact from the commodity release
- A preventive maintenance program that minimizes the chances that a spill will occur due to a failure of one of the pipeline components
- The ability to rapidly and accurately detect and locate leaks
- A loss-of-containment response program that allows pipeline personnel to act rapidly by shutting down the pipeline and isolating the leak to minimize unintentional losses, mobilizes resources to the spill site, performs additional actions to minimize any additional intentional commodity losses, and implements cleanup and damage control

13.3.1 Analytical Basis

It nearly impossible to support an integrity loss risk-management program without a solid base of tools (typically institutional and physical knowledge and methods coupled with software) that can support all of the elements of the plan. These include the ability to perform both steady-state and transient modeling of the pipeline. In particular, the ability to simulate surge or "water hammer" events should be available to ensure that normal operation of the pipeline does not result in an over-pressure event with consequent rupture of the line.

Other tools may include the ability to simulate loss of containment, leak detection system response, operator response, logistical response to the leak site, and the external impacts of commodity releases, as described previously in Section 13.2. The operator should also be able to evaluate the efficiency of its leak detection technology and program using various tools, some of which are described in this book.

13.3.2 Pipeline Design to Minimize Loss of Containment Impact

The pipeline design should accommodate rapid detection and minimize volumes connected with a loss of containment and the external impacts of those volumes. In particular:

1. The pipeline design should accommodate both cleaning pigs and instrumented pigs designed to detect flaws, leaks, and corrosion defects.
2. Pipelines should be equipped with emergency shutdown (ESD) valves that permit critical sections of the pipeline to be isolated in the event of a leak. In particular, liquid pipelines in hilly or mountainous terrain are difficult to drain back into upstream or downstream tankage, and they can take very long to drain via the leak opening if the leak is located at a low point. In such cases, a higher density of valves should be considered. Remotely controlled valves are always preferred over manual valves at remote locations.
3. Corrosion protection in the form of coated pipe (to prevent corrosion) and cathodic protective systems should be included in the design.
4. The communications and instrumentation infrastructure should support a high-quality internal LDS, which optimally meets the pipeline's unique environmental, physical, and operation conditions.
5. It is easiest to install a comprehensive external fiber-based or cable-based LDS when the pipeline is initially designed and constructed, and this option should be actively considered at that time.
6. The external design of the pipeline should act to minimize commodity loss impacts to third parties. In particular, external drainage and containment should be considered at liquid pipeline river, lake, reservoir, and other water crossings to ensure that escaped volumes from upstream or downstream of the crossing do not enter the water.
7. River crossings should consider sleeved or double-walled pipe filled with an inert gas and an external monitoring system.
8. Pipe routing should avoid transits through inaccessible ravines, canyons, gorges, and other locations that make the ROW difficult to monitor and inspect.
9. Similarly, the design should assist in the detection of leaks and spills. Designs that guide escaped liquid volumes to containment areas should be considered. Design that enables oil to enter ravines, storm drains, and other locations that permit migration away from the spill site should be discouraged.
10. Cathodic protection should be included in the design.

13.3.3 Preventive Maintenance Program

Design of a preventive maintenance program should ensure that critical components do not fail and trigger a leak:

1. The operator should use cleaning pigs on a regular basis to remove wax from the pipe wall and prevent corrosion.
2. If the pipeline operates at a flow velocity much less than 3 feet/second, then the operator should increase the frequency of its cleaning pig runs to prevent settled water from contributing to corrosion.
3. Instrumented pigs should be run on an annual basis to monitor the state of any developing internal or external corrosion or high-stress deflections due to settlement.
4. Cathodic protection systems should be monitored to ensure that voltages and sacrificial anodes are properly maintained.
5. Remote ESD valves should be stroked regularly to ensure they will function in an emergency.
6. The right-of-way should always be freely accessible and should be kept clear of shrubs, bushes, trees, and other impediments to easy monitoring by pipeline personnel and even members of the public.
7. Signage to enable third parties to report a spill quickly to the control center or to proper authorities is encouraged.

13.3.4 Effective Leak Detection Program, Technology, and Procedures

When all else fails, it is important to detect the integrity breach as quickly as possible. The operator should not think of leak detection as just the technology, but as an optimized system comprising technology, procedural activities, the human element, and other resources. As we have seen previously in Section 13.1.4, leaks are identified by many methods, and many of them do not involve technology. There is no single, reliable, and rapid detection method that will detect a leak over the complete range of operating conditions. Some methods will detect large leaks quickly and reliably but will perform poorly for small leaks. Other methods may take more time for large leaks but will ultimately detect even small leaks. The natural question that should arise, as a result of the previous sections in this chapter, is: Can we obtain better leak detection if we combine more than one approach?

Let's try to get a conceptual or illustrative handle on this issue. Consider a simple hazardous LVP commodity pipeline with meters, pressure measurements, and temperature measurements at each end and intermediate pressure and temperature measurements at approximately 45-mile-long intervals. The pipeline is buried with the top of the pipe 3 feet below the surface, and the fill and soil surrounding the pipe are well-drained with an effective porosity of approximately 35%. Approximately 15% of the pipeline crosses an urban area with approximately 2500 people per square mile. Another 25% is in a more thinly populated outlying region at approximately 250 people per square mile. The remainder of the pipeline crosses a thinly populated rural region with only 25 people/square mile. Fast SCADA sampling at less than 5-s intervals potentially allows a rarefaction wave system to be installed. The bounding flow meters are noisy with outliers and have no ability to be proven. Beyond this, we assume no mechanical or instrumentation improvement, and we assume that we have sufficient pipeline design, operating data, and recorded SCADA and field data to develop rough performance maps based on the models and methods discussed in previous chapters.

We restrict our investigation to third-party/public spill detections, a rarefaction wave system, a few real-time transient model mass balance systems, and ground patrols conducted by the pipeline operator. An external LDS will not be evaluated due to the high cost of installation and risk associated with retrofitting an existing pipeline on a line-wide basis. In all cases, we will be looking for detections with 95% confidence. We also restrict our analysis to underground pipeline right-of-way spills (see Chapter 10: Human Factor Considerations in Leak Detection) on the assumption that station spills are usually restricted to the operator's property, and usually result in less property and other damage than right-of-way spills. It is also important to note that: (1) although many of the general conclusions discussed here can be extended to other low-vapor-pressure liquid pipelines, the specific details will vary from pipeline to pipeline, and (2) the results shown are likely not generalized to aqueous liquid, HVP commodity, or gas-phase pipelines.

Fig. 13.7 shows leak detection performance curves for this pipeline over a period of 5 days after leak onset. Let's consider what would happen in the absence of any leak detection capability on the part of the operator. In the absence of operator precautions, the public detection curves show how quickly spills of various sizes could be detected and notified to the operator by members of the public. The curves utilize the simple/conceptual underground release and the population based direct observation models discussed previously in Chapter 10, Human Factor Considerations in Leak Detection. The mobility factor for the simple direct detection model (ie, that portion of the population actually mobile and not confined to their homes) is set to 5%. We also assume that the average reportable spill capable of eliciting enough "surprise" to cause a member of the public to report is at least 5 feet across. We also assume a hard core of 25% of the public who will never report anything to the authorities.

The curves indicate that in the absence of any other detection methodology, even very large leaks in highly populated areas might take a few hours to detect. Keep in mind that the simple models used here automatically incorporate a large implied standard deviation that encompasses the period during the night when the mobility fraction approaches zero. Consequently, all such detections by the general public must be viewed as having a large degree of uncertainty.

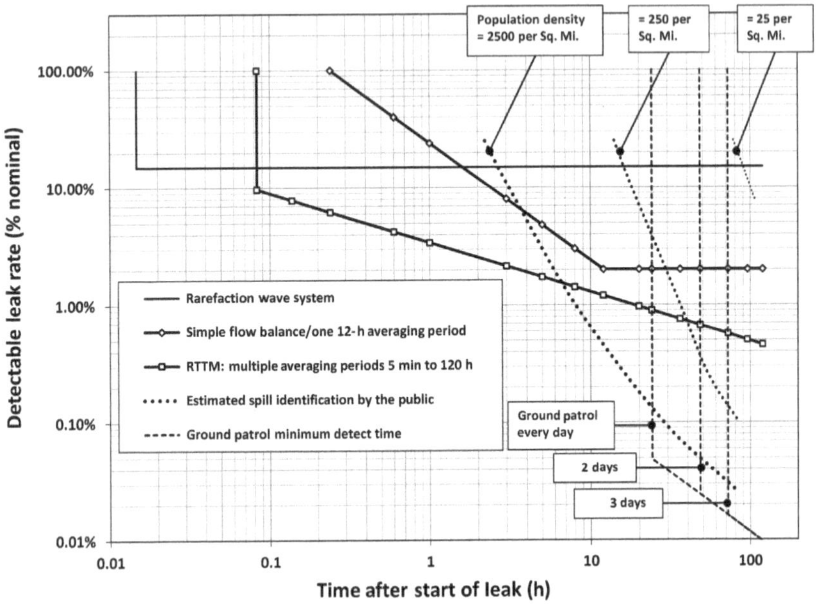

FIGURE 13.7 Example LVP liquid pipeline leak detection performance overlay.

For those portions of the pipeline where the population density is low, detection times by the public for even large spills could potentially be on the order of days. Although very small leaks of 1% of nominal flow or less might be detected by third parties in the highly populated sections within approximately 10 hours, in the mostly thinly populated sections of the pipeline such notifications might take a week or longer. Bear in mind that this conceptual analysis of direct detection by the general public utilizes an underground leakage model that is very simple, and that the population detection model is equally simple. That said, the models are not unreasonable and provide some basis against which leak detection via other approaches can be evaluated.

At this point, we should stop and note that even in the absence of any installed leak detection system, very large leaks could potentially be detected at the control console by an alert controller within a period that might be competitive with a CPM system or other leak detection technology. However, DOT studies [26] have revealed that although pipeline controllers may have an opportunity to infer the existence of a large leak via indirect indications provided by the SCADA system, such detections are highly dependent on individual controller expertise, and, in fact, controller action (or the lack thereof) in the absence of a CPM system has negatively contributed to the size and impact of spills in the past. In addition, the ability of a busy controller to quickly and reliably identify much smaller leaks less than

5% to 10% of flow is questionable at best in short time frames. We continue from this point by examining some internal leak detection system technologies while keeping in mind that an astute controller's leak detection performance curve is potentially close to but probably outside of (that is, to the right and above) some composite of these LDS technology curves.

Referring back to Figure 13.7, a rarefaction wave LDS is capable of providing ultra-fast leak detection (in less than 1 or 2 min depending on location of the detectors, the distance between the detectors, and the leak location) for larger leaks. However, attenuation of the negative pressure wave limits the 95% detection threshold to 15% of nominal flow. This does not imply that the system is useless below this threshold. However, the confidence will decline for smaller leaks until the system becomes essentially useless for leaks smaller than approximately 3% to 5% of nominal pipeline flow.

A relatively inexpensive option would be to install a simple flow balance system with no packing correction and one sampling or averaging period of 12 hours. This system is not as fast as the rarefaction wave system for large leak rates, but it provides improved performance to much less than the 15% of the nominal flow cutoff observed with the rarefaction wave system. For leaks smaller than 3–4% of nominal throughput, the flow balance system may be technically outperformed by direct detections obtained from the public in the highly populated portions of the pipeline, but not less densely populated larger geographic areas. The flow balance system bottoms out at the end of the 12-hour sampling period, for leaks of approximately 2% of pipeline nominal flow.

A potentially better solution is implementation of an RTTM-based mass balance system with multiple averaging or sampling periods. Let's assume that we have a large number of sampling periods ranging between 5 min and 120 h (5 days). This reduces much of the volume balance noise and allows for a shallower detection curve, but it has the potential disadvantage of offering more false alarms. We also assume some residual time correlation of the volume balance errors that result in an effective leak detection curve exponent of 0.35 (see Chapter 5: Statistical Processing and Leak Detection).

The new curve declines continuously to 120 h, after which it will level off at 0.4% of nominal flow. Note that in the areas of the pipeline with high and mid-range population density, it is more likely that tiny leaks below this leak flow rate will eventually be detected by the public. However, in the portions of the pipeline where there is very low population density, it is unlikely that leaks lower than 0.5% of nominal flow will be detected in less than 5 days.

At this point, we may have run out of technology options for our illustrative pipeline unless we consider external systems already ruled out on the basis of cost and risk. To address this, potential operator ground patrol performance is investigated based on the minimum time for an underground leak at a constant flow rate to break through to the surface (see Chapter 10: Human Factor Considerations in Leak Detection). The results in Fig. 13.7 indicate that total spill volume to the time of detection can potentially be limited based on the frequency of the ground patrol. The best improvement, not surprisingly, occurs with a patrol every day and is most beneficial in the portion of the pipeline where the population density is lowest. Given the cost of running daily patrols (1 person per day), the decision regarding how often ground patrols should be performed appears to be an issue best resolved based on considerations of the cost, the expected leak incident rate for the pipeline, the expected remediation cost per spilled barrel, the operator's appetite for risk, and any applicable regulations or legal requirements.

We see from this analysis why field personnel, members of the public, and other direct detection methods that do not involve technology catch so many leak events. The steep negative slopes shown in the log-transformed data in Fig. 13.7 for operator and public detections combined with the clear lags associated with these methods means that for low leak rates, human detection methods will excel, albeit at a potential significant cost in detection time. However, the human-mediated detection curves tend to exhibit a considerable lag of several hours or more and are far more variable in terms of expected leak detection performance. On the other hand internal LDS technology approaches are more likely to provide the fastest detection for larger leak rates, whereby extremely large spill volumes are also more likely.

Based on the illustrative analysis for this conceptual pipeline, the best combination of benefits appears to be provided by an RTTM system with multiple data sampling periods between 5 min and 5 days combined with ground patrols performed from once every day to once every week. However, additional ultra-fast response can be added by implementing a negative pressure wave system, and this should be considered as well for this particular pipeline.

Note that these conclusions apply to this particular example pipeline. Every pipeline is unique and the analysis described should be performed on an individual system basis. In addition, it would be a mistake to extrapolate the conclusions obtained here to pipelines transporting other much different commodities or in other environments. Finally, more confidence in the conclusions would be obtained by applying more rigor to the direct observation release and detection models.

We conclude that it is difficult for an operator to choose the best leak detection solution if the performances of the various options are not understood, and that the best solution may involve a multifaceted implementation of more than one approach. German regulations recognize this fact and require two distinct leak detection systems for hazardous material pipelines. See Chapter 12, Regulatory Requirements for more discussion of regulations around the world.

13.3.5 Effective Response Plan

The operator must be able to respond quickly and effectively in the event of a leak indication from technology, company personnel, or third parties. This implies the following:

1. The operator must be well versed in the operation of any LDS technology, personnel must be trained to diagnose leak alarms quickly, and there should be established procedures to shut down and isolate the pipeline as well as to alert the pipeline's emergency response team.
2. Members of the response team should be clearly identified and trained on company response procedures.
3. The individual responsible for making decisions and taking control should be identified as well and given the appropriate authority to implement the response plan.
4. A company spokesperson should be identified as part of the response plan.
5. It is advisable for the operator to have a set of procedures for various anticipated commodity release scenarios (prepared in advance). These should include field isolations, usage of booms and dams, methods and identified circumstances to implement controlled releases, and other field actions.
6. A list of public emergency provider contacts (fire department, police, etc.) as well as appropriate government and regulatory officials should be available at all times. These officials should be immediately informed when it is clear that a loss-of-integrity event has occurred.

13.4 CONCLUSION

Much of this chapter has been dedicated to the economic impact of unplanned commodity losses. For operation in the United States, we have seen that although most spills are small, extremely large spills more than 10,000 BBL can occur. Spill costs are proportionate to some degree to spill volume sizes, but there is a significant remaining variance in the costs that indicates that other factors play an even larger role in determining costs. Similarly, although median spill costs are only approximately $25,000 per event, the average spill costs approximately $960,000 primarily because outliers in the right tail are extremely large. The largest incident costs can be expected to top $1 billion. We need to acknowledge, of course, that these patterns may differ in other parts of the world.

Furthermore, most spill events, especially those involving small commodity volumes, are directly or remotely detected by people, and for good reason. In most places, there are always people around. As individuals, humans may be unreliable detectors, but in aggregate, and in large numbers, the likelihood that even a small spill will go unreported for a very long time is low.

Consequently, utilization of structured procedural detection methods, such as ground patrols, and/or carefully thought-out consideration of third-party detections by the general public and others, should be part of the leak detection arsenal for most pipelines.

However, direct detections of spilled or excaped commodity by people, particularly in the form of the public, are also probabilistic and slow, especially of the commodity is occluded immediately following its initial release. Consequently, CPM/internal leak detection systems will generally tend to outperform detections by operator personnel and third parties whenever leak rates are larger (on the order of a percent of nominal pipeline flow or more, depending on the quality of the instrumentation and LDS technology). We note that targeted external systems, particularly fiber-based or cable-based systems, can outperform internal leak detection technology and people for even low leak rates, especially if installed as part of original pipeline construction. However, such systems are very expensive to retrofit, present high risk during installation to an existing pipeline, and will not work in every application.

We conclude that the leak detection function is one essential part of every pipeline's risk management program. The operator should be aware that leak detection is not a panacea, and other components of its risk management program are just as important. In addition, the perceptive operator will recognize that pipeline leak detection is best implemented as a system consisting of technology, formal procedures implemented by the operator's staff through various mechanisms (pipeline controller monitoring, ground patrols, aerial patrols, station and terminal sweeps), and as a result of chance detections by third parties, including the general public. All of these methods have their place. It is up to the operator to develop strategies using all of these methods to efficiently identify pipeline leaks.

REFERENCES

[1] Dougherty ER. Probability and statistics for the engineering, computing and physical sciences. Prentice-Hall; 1990.

[2] Mitzenmacher M, Tworetsky B. New models and methods for file size distributions. In: Allerton 2003.

[3] Platen E, Sidorowicz R. Empirical evidence on student-t log returns of diversified world stock indexes. In: Quantitative finance research center research paper 194, ISSN 1441-8010, March 2007.

[4] Baskerville GL. Use of logarithmic regression in the estimation of plant biomass [J]. Can J For Res 1972;2:49–53.

[5] Carpenter P, Chrisman K, Salmatanis N. Comprehensive leak and spill quantification toolset. In: Pipeline simulation interest group annual conference, Napa Valley, California, 24 May–27 May 2011.

[6] Huppert HE. Gravity currents: a personal perspective. J Fluid Mech 2006;554:299–322.

[7] Grimaz S, Allen S, Stewart J, Dolcetti G. Predictive evaluation of surface spreading extent for the case of accidental spillage of oil on the ground. AIDIC Chem Eng Trans 2007; vol. 11:389–94.

[8] Hussein M, Jin M, Weaver JW. Development and verification of a screening model for surface spreading of petroleum. J Contam Hydrol 2002;57(3–4):281–302.
[9] Simmons CS, Keller JM, Simmons CS, Keller JM. Status of models for land surface spills of nonaqueous liquids. Pacific Northwest Regional Laboratory/Operated by Battelle for the U.S. Department of Energy, PNNL-1435, August 2003.
[10] Havlíček J, Balatka, M. Stochastical approach to spread modeling of the leaked contaminant spill. Advanced Research in Scientific Areas, December 2012.
[11] Chebbi R. Viscous-gravity spreading of oil on water: modeling and challenges. Int J Chem Eng Appl, April 2014;vol. 5(2).
[12] Mazurek J, Smolarek L. Oil spill models: a state of the art of the grid map as a function of wind, current and oil parameters. TransNav, March 2013;7(1).
[13] Reeda M, Johansena Ø, Brandvika PJ, Dalinga P, Lewisb A, Fioccob R, et al. Oil spill modeling towards the close of the 20th century: overview of the state of the art. Spill Sci Tech Bull, April 1999;5(1):3–16.
[14] GNOME software, available at: http://response.restoration.noaa.gov/gnome.
[15] Thibodeaux LJ. Environmental chemodynamics: movement of chemicals in air, water and soil. 2nd ed. John Wiley & Sons, Inc.; 1996.
[16] Abu-El-Sha'r WY, Hatamleh RI. Using modflow and MT3D groundwater flow and transport models as a management tool for the Azraq Groundwater System. Jordan J Civil Eng 2007;1(2).
[17] MODFLOW software, available at: http://water.usgs.gov/ogw/modflow/.
[18] MT3D software, available at: http://hydro.geo.ua.edu/mt3d/.
[19] Cleaver RP, Cumber PS, Halford A. Modelling outflow from a ruptured pipeline transporting compressed volatile liquids. J Loss Prevent Proc Ind 2003;16:533–43.
[20] Morrow TB. Development of an improved pipeline breakflow model: final report. SwRI-6917/3 1982.
[21] Witlox HWM, McFarlane K. Interfacing dispersion models in the HGSYSTEM hazard-assessment package. Atmos Environ, 28, 18, 2947–2962
[22] HGSYSTEM plume modeling and dispersion software, available at: www.hgsystem.com.
[23] ALOHA (Areal Locations of Hazardous Atmospheres) 5.4.4: Technical Documentation. NOAA Technical Memorandum NOS OR&R 43; November 2013.
[24] ALOHA plume modeling and dispersion software, available at: https://www.epa.gov/cameo/aloha-software.
[25] Nouri-Borujerdi A. Transient modeling of gas flow in pipelines following catastrophic failure. Math Comput Model 2011;54:3037–45.
[26] Supervisory Control and Data Acquisition (SCADA) in Liquid Pipelines, National Transportation Safety Board Safety Study NTSB/SS-05/02, PB2005-917005, 29 November 2005.

Index

Note: Page numbers followed by "*f*" and "*t*" refer to figures and tables, respectively.

D

E

G

H

I

O

P

S

T

U

V

W

FSC
www.fsc.org
MIX
Papier | Fördert
gute Waldnutzung
FSC® C083411

Kolibri 360 GmbH
Ferdinand-Jühlke-Straße 7
99095 Erfurt, Deutschland
produktsicherheit@kolibri360.de